Subcellular Biochemistry

Volume 19

Endocytic Components: Identification and Characterization

Edited by

J. J. M. Bergeron
Department of Anatomy
McGill University
Montreal, Canada

and

J. R. Harris
Institute of Zoology
University of Mainz
Mainz, Germany

SPRINGER SCIENCE+BUSINESS MEDIA, LLC

The Library of Congress cataloged the first volume of this title as follows:

Sub-cellular biochemistry.

London, New York, Plenum Press.
v. illus. 23 cm. quarterly.
Began with Sept. 1971 issue. Cf. New serial titles.
1. Cytochemistry—Periodicals. 2. Cell organelles—Periodicals.
QH611.S84 574.8'76 73-643479

ISBN 978-1-4613-6319-4 ISBN 978-1-4615-3026-8 (eBook)
DOI 10.1007/978-1-4615-3026-8

This series is a continuation of the journal *Sub-Cellular Biochemistry*, Volumes 1 to 4 of which were published quarterly from 1972 to 1975

Originally published by Plenum Press, New York in 1993
Softcover reprint of the hardcover 1st edition 1993

Subcellular Biochemistry

Volume 19

Endocytic Components: Identification and Characterization

SUBCELLULAR BIOCHEMISTRY

Recent Volumes in This Series:

Volumes 5–11 Edited by Donald B. Roodyn

Volume 12 **Immunological Aspects**
Edited by J. R. Harris

Volume 13 **Fluorescence Studies on Biological Membranes**
Edited by H. J. Hilderson

Volume 14 **Artificial and Reconstituted Membrane Systems**
Edited by J. R. Harris and A.-H. Etémadi

Volume 15 **Virally Infected Cells**
Edited by J. R. Harris

Volume 16 **Intracellular Transfer of Lipid Molecules**
Edited by H. J. Hilderson

Volume 17 **Plant Genetic Engineering**
Edited by B. B. Biswas and J. R. Harris

Volume 18 **Intracellular Parasites**
Edited by J. L. Avila and J. R. Harris

Volume 19 **Endocytic Components: Identification and Characterization**
Edited by J. J. M. Bergeron and J. R. Harris

Contributors

J. J. M. Bergeron Department of Anatomy, McGill University, Montreal, Canada H3A 2B2

J. S. Blum Department of Cell Biology and Physiology, Washington University School of Medicine, St. Louis, Missouri 63110. *Current address*: Virginia Mason Research Center, Seattle, Washington 98101

J. W. Burgess Department of Anatomy, McGill University, Montreal, Canada H3A 2B2

Pierre J. Courtoy Cell Biology Unit, International Institute of Cellular and Molecular Pathology and University of Louvain Medical School, 1200 Brussels, Belgium

R. Diaz Department of Cell Biology and Physiology, Washington University School of Medicine, St. Louis, Missouri 63110

Carlos Enrich Departmento de Biologia Celular, Facultad Medicina, Universitat de Barcelona, Spain

W. Howard Evans National Institute for Medical Research, Mill Hill, London NW7 1AA, United Kingdom

Michel Jadot Laboratoire de Chimie Physiologique, Facultés Universitaires Notre-Dame de la Paix, B-5000 Namur, Belgium

M. N. Khan Department of Anatomy, McGill University, Montreal, Canada H3A 2B2. *Current address:* Phoenix International Sciences, Montreal, Canada H4R 2N6

W. H. Lai Department of Anatomy, McGill University, Montreal, Canada H3A 2B2

Frederick R. Maxfield Department of Pathology, College of Physicians and Surgeons, Columbia University, New York, New York 10032

L. S. Mayorga Facultad de Cincias Medican, Instituto de Histologia y Embriologia, 550 Mendoze, Argentina

Sandra Misquith Laboratoire de Chimie Physiologique, Facultés Universitaires Notre-Dame de la Paix, B-5000 Namur, Belgium

B. I. Posner Department of Medicine, McGill University, Montreal, Canada H3A 2B2

Nita H. Salzman Department of Physiology, College of Physicians and Surgeons, Columbia University, New York, New York 10032, and Department of Pharmacology, New York University School of Medicine, New York, New York 10016

Sandra L. Schmid Departments of Cell and Molecular Biology, The Scripps Research Institute, La Jolla, California 92037

P. D. Stahl Department of Cell Biology and Physiology, Washington University School of Medicine, St. Louis, Missouri 63110

Robert Wattiaux Laboratoire de Chimie Physiologique, Facultés Universitaires Notre-Dame de la Paix, B-5000 Namur, Belgium

Simone Wattiaux-de Coninck Laboratoire de Chimie Physiologique, Facultés Universitaires Notre-Dame de la Paix, B-5000 Namur, Belgium

Paul H. Weigel Department of Human Biological Chemistry and Genetics, University of Texas Medical Branch, Galveston, Texas 77555-0647

Preface

The rapid advance of knowledge relating to the endosome, its cellular localization, separation, and functions, has led to the appropriate inclusion of this exciting topic within the *Subcellular Biochemistry* series. Furthermore, the continuity of emphasis on "membrane-related" topics within the series is thereby maintained. A list of internationally respected authors has been compiled, with the aim of covering within a single book the principal areas currently being explored within the field of endosome research.

The book commences with a chapter by S. L. Schmid, who describes in detail the use of free flow electrophoresis for the physical separation and biochemical definition of the endosomal compartment. P. J. Courtoy then deals specifically with the endosomal compartment in rat hepatocytes; he covers a range of analytical approaches in an attempt to derive an overall structural model of the rat liver endosome. This is followed by a chapter by J. S. Blum and colleagues in which they deal with proteolysis in endosomes and discuss its significance for vesicle fusion and vesicle-mediated transport during endocytosis in reconstituted systems. The application of a range of different quantitative fluorescence techniques for the characterization of endocytosis is presented by N. H. Salzman and F. R. Maxfield, who thoroughly survey this important technical approach. This is followed by a chapter by P. H. Weigel on endocytosis and the function of the hepatic asialoglycoprotein (galactosyl/N-acetylgalactosaminyl) receptor. The endocytic components of liver nonparenchymal cells are characterized in detail by R. Wattiaux and colleagues, who present a comprehensive survey of the endocytic behavior of this population of cells. Complementing this chapter is one by C. Enrich and W. H. Evans in which the endocytic compartment of normal and regenerating liver is discussed. In the final chapter, M. N. Khan and colleagues discuss the potential role of endosomes in transmembrane signaling. Having defined the organelles involved in receptor-mediated endocytosis, these authors are concerned primarily with the functions of endosomes in cellular signaling.

It is hoped that the broad coverage presented by the material included in the eight chapters of this book will enable it to be of interest and use to cell biologists and biochemists concerned with this rapidly advancing field, as well as other scientists wishing to keep abreast of progress in this area of membrane biochemistry. In addition, the book should be of use to honors degree and graduate students in cell biology and biochemistry.

J. J. M. Bergeron
J. R. Harris

Montreal, Canada, and
Mainz, Germany

Contents

Chapter 1

Toward a Biochemical Definition of the Endosomal Compartment: Studies Using Free Flow Electrophoresis

Sandra L. Schmid

Chapter 2

Analytical Subcellular Fractionation of Endosomal Compartments in Rat Hepatocytes

Pierre J. Courtoy

Chapter 3

Reconstitution of Endosomal Transport and Proteolysis

J. S. Blum, R. Diaz, L. S. Mayorga, and P. D. Stahl

Chapter 6

Characterization of Endocytic Components of Liver Nonparenchymal Cells

Robert Wattiaux, Michel Jadot, Sandra Misquith, and Simone Wattiaux-de Coninck

Chapter 7

The Endocytic Compartments of Normal and Regenerating Liver

Carlos Enrich and W. Howard Evans

Chapter 8

Potential Role of Endosomes in Transmembrane Signaling

M. N. Khan, W. H. Lai, J. W. Burgess, B. I. Posner, and J. J. M. Bergeron

Chapter 1

Toward a Biochemical Definition of the Endosomal Compartment

Studies Using Free Flow Electrophoresis

Sandra L. Schmid

1. OVERVIEW

The eucaryotic cell is highly compartmentalized, containing many functionally and structurally distinct membrane-bound organelles. Subcellular fractionation is a powerful technique being applied to the purification of individual organelles so that they might be studied in biochemical detail. A complete biochemical definition of organelles of the vacuolar system involved in membrane transport along either the endocytic or exocytic pathways requires their definition in both static and dynamic terms. An organelle is defined in static terms by its own specific set of resident proteins, which give it its structural and functional identity. An organelle is defined in dynamic terms by its temporal location along an intracellular membrane traffic route based on the kinetics of protein transit into and through each organelle along either the endocytic or exocytic pathway. To completely define an organelle in static terms would require its purification to homogeneity and subsequent biochemical characterization. To adequately define an

Sandra L. Schmid Departments of Cell and Molecular Biology, The Scripps Research Institute, La Jolla, California 92037.
Subcellular Biochemistry, Volume 19: Endocytic Components: Identification and Characterization, edited by J. J. M. Bergeron and J. R. Harris. Plenum Press, New York, 1993.

organelle in dynamic terms does not necessarily require that an organelle be purified to homogeneity but instead requires a rapid, high-yield procedure that clearly resolves the organelle from other organelles along a given pathway. This review discusses the novel method of free flow electrophoresis that we have applied to both the preparation of highly purified endosomal fractions for their biochemical characterization and to the analytical characterization of the kinetics of transit of endocytic tracers through the organelles of the endocytic pathway.

2. TOWARD A BIOCHEMICAL DEFINITION OF THE ORGANELLES OF THE ENDOCYTIC PATHWAY

2.1. Organelles of the Biosynthetic Pathway Are Reasonably Well Defined

To fully and accurately define the specific and distinct biochemical composition of any organelle would eventually require its purification to homogeneity. Although few of the organelles of the vacuolar system including the endoplasmic reticulum, the Golgi apparatus, endosomes, and lysosomes have been purified to homogeneity, the organelles along the exocytic/biosynthetic pathway are nonetheless reasonably well defined in both dynamic and static terms. A number of factors have aided researchers in deriving these definitions. First, the organelles of the exocytic pathway are morphologically distinct, which facilitated the detailed mapping of this pathway using the technique of electron microscopic (EM) autoradiography to follow the transit of newly synthesized proteins from the endoplasmic reticulum through the Golgi cisternae and on to secretory granules and the cell surface in pancreatic acinar cells (Jamieson and Palade, 1967a,b). Second, sequential posttranslational modifications to newly synthesized glycoproteins as they transit through the various organelles along the biosynthetic pathway have enabled derivation of a more detailed biochemical map based on the kinetics of occurrence of these sequential modifications (reviewed by Kornfeld and Kornfeld, 1985). Third, the identification of enzymatic functions specific to one organelle, termed *marker enzymes* (for examples, see Fliescher and Kervina, 1974) has aided the biochemical definition of these organelles in static terms. Fourth, the development of simple and sensitive assays for detecting these marker enzymes has facilitated the biochemical dissection of these organelles using cell fractionation procedures. Finally, the purification of these marker enzymes themselves has led to the availability of organelle-specific antibodies for use in immunoelectron microscopy (reviewed by Farquhar, 1985; Dunphy and Rothman, 1985). Thus, coming full circle, the biochemical analysis has provided reagents that lend even greater detail to the morphological mapping described above.

2.2. Organelles of the Endocytic Pathway Are Poorly Defined

In contrast, the organelles of the endocytic pathway, or *endosomal compartment,* lack biochemical definition in both static and dynamic terms. These organelles do not have clear morphological distinctions (reviewed by Steinman *et al.*, 1983; Helenius *et al.*, 1983). The internalized proteins that pass through them do not, in general, undergo any biochemical modifications. Marker enzymes specific to the organelles of the endosomal compartment have not yet been identified. Finally, endosomes are difficult to isolate using standard cell fractionation techniques because their general properties of density and size are similar to many other organelles (for example, the Golgi, plasma membrane, endoplasmic reticulum). Thus more than a decade since their characterization as a prelysosomal sorting compartment (Steinman *et al.*, 1976; Goldstein *et al.*, 1979), the controversy still exists as to whether endosomes are in fact bona fide organelles, defined by their own specific set of resident proteins, or are instead transient structures derived from internalized plasma membrane, transported through the cell and destined to fuse with lysosomes. Resolution of this controversy could clearly be accomplished through the detailed biochemical definition, in both static and dynamic terms, of the organelles along the endocytic pathway. This, in turn, will require the development of methodologies for endosome purification.

This review focuses on one approach recently developed (Marsh *et al.*, 1986; Schmid *et al.*, 1988) for the purification of endosomes from tissue culture cells based on the novel criteria of surface charge using free flow electrophoresis (FFE). The following sections first review the technique as we have used it for both the large-scale purification of endosomes from tissue culture cells and analytical studies on transport through endosomal subpopulations. The advantages, limitations, and potentials for the use of FFE are then discussed and compared to those of other techniques being developed for endosome purification. Finally, a model is presented for membrane traffic along the endocytic pathway, based on new insights gained through the use of subcellular fractionation of the endosomal compartment by FFE.

2.3. Operational Definitions for the Organelles along the Endocytic Pathway

In lieu of any biochemical definitions, the endosomal compartment instead has been defined operationally as a heterogeneous class of membrane-bound vesicles, associated tubules, and multivesicular bodies that contain endocytic tracers en route to lysosomes. Further operational distinctions have also been made that identify functionally distinct subpopulations of the endosomal compartment. The terminology used in designating these subcompartments has yet to be generally accepted, and here I make an attempt to list them in terms most commonly found in the literature. These terms are variously used by a number of investigators and, with apologies, no attempt has been made to reference their originators. *Primary endocytic vesicles* (0.1–0.2 μm in size) including coated vesicles and stripped coated vesicles are by definition derived from coated pits and from the plasma membrane. They are the first endocytic compartment labeled by internalized tracers and as such have also been termed *early endosomes* (I believe this nomenclature to be inappropriate for two reasons; first, it leaves no room for the possibility that endosomes exist as independently derived organelles; and second, it doesn't allow for distinction of endosomes and coated vesicles with regard to their sorting functions). Endocytic tracers are next seen in larger vesicles (0.4–1 μm in size) and associated tubules. Although it is clear that intracellular membrane fusion events are required for the delivery of endocytic tracers into these larger vesicular structures, it is unknown whether these fusion events occur exclusively between incoming primary endocytic vesicles to form these larger structures *de novo* or whether they represent the fusion product of primary endocytic vesicles with a preexisting membrane-bound organelle. Although these larger endosomal structures have been referred to by a variety of names, the most common, albeit the least descriptive, is the term *early endosomes,* since they are the first compartment encountered by endocytic tracers that can be morphologically and structurally distinguished from coated vesicles. Upon arrival in early endosomes, incoming receptors and ligands first encounter an acidic environment sufficient to facilitate the pH-dependent dissociation of many receptor-ligand complexes. Thus the first step in sorting of receptors and ligands is accomplished in the early endosome. Free receptors are in general recycled back to the cell surface for reutilization, in an as yet poorly defined class of vesicles termed *recycling endosomes.* Most released ligands are targeted through the remaining endocytic pathway toward lysosomes. The role of early endosomes in membrane sorting has been clearly defined two ways: morphologically using double-label immunogold electron microscopy on frozen ultrathin sections (Geuze *et al.*, 1983, 1984) and immunofluorescence (Yamashiro *et al.*, 1984), and biochemically using either density-shift procedures (Stoorvogel *et al.*, 1987) or immunoadsorption (Mueller and Hubbard, 1986) to demonstrate the sorting of two cointernalized endocytic tracers into distinct vesicular compartments. For these reasons, early endosomes are also commonly referred to as the compartment of uncoupling of receptor and ligand (*CURL*), or *sorting endosomes.* The next operationally distinct endosomal subcompartment is the *late endosome* or as the preceding discussion would have it, the *postsorting endosome.* This endosomal compartment is thought to interact with microtubules and be carried into the perinuclear region of the cell where it eventually fuses with lysosomes (Matteoni and Kreis, 1987; Gruenberg *et al.*, 1989). Late endosomes can be distinguished from lysosomes by histochemical techniques using arylsulfatase or acid phosphatase to specifically label lysosomes. Even these compositional differences, however, are more quantitative than qualitative since recent results suggest that both newly synthesized lysosomal enzymes and membrane proteins are delivered to the late endosome compartment en route to lysosomes (Griffiths *et al.*, 1988; Geuze *et al.*, 1988; see also below). The exact crossover point between the endocytic pathway and the biosynthetic pathway leading to lysosomes has not been delineated and may represent yet another subclass of late endosomes (reviewed by Kornfeld and Mellman, 1989). This compartment, which contains endocytic tracers, lysosomal membrane glycoproteins, and lysosomal enzymes still associated with the mannose-6-phosphate receptor, all en route to lysosomes, has been termed the *intermediate compartment* or *prelysosomal endosome*. The ontogenic relationship between early and late endosomes or between potential subclasses of late endosomes remains poorly defined.

The above operational distinctions between endosomal subclasses clearly indicate functional heterogeneity among the organelles of the endosomal compartment. This in turn implies the existence of compositional distinctions between the various subcompartments that would be required to carry out such subclass-specific functions as membrane sorting, receptor-recycling, delivery of contents to lysosomes, fusion with Golgi-derived vesicles carrying lysosomal precursor enzymes, or fusion with plasma membrane–derived vesicles carrying newly internalized receptors and ligands. Despite the implied existence of discrete, functionally-distinct organelles along the endocytic pathway, not a single endosome-specific marker protein or enzyme has yet been identified. Two very different and complementary approaches are now being taken to identify endosome-specific proteins to define biochemically the endosomal compartment. One approach (recently reviewed by Gruenberg and Howell, 1989) is to establish cell-free assays that measure endosome-specific functions such as coated vesicle–early endosome fusion (Mayorga *et al.*, 1988; Redelmeier and Schmid, unpublished results), early endosome–early endosome fusion (Davey *et al.*, 1985; Gruenberg and Howell, 1986; Braell, 1987; Diaz *et al.*, 1988; Woodman and Warren, 1988), endosome-lysosome fusion (Mullock *et al.*, 1989), and so forth so that these may be used to identify the endosome-specific proteins that mediate these reactions. The second approach, which is the subject of this review, involves the development of novel techniques for the purification of endosomal subcompartments so that they might be fully biochemically characterized and their endosome-specific marker proteins identified. The validity of the first approach has been borne out by its success in the identification of cis-, medial-, and trans-Golgi-specific marker enzymes involved in the sequential modification of glycoproteins during transit along the biosynthetic pathway (reviewed by Kornfeld and Kornfeld, 1985; Dunphy and Rothman, 1986). The validity of the second approach has been borne out by the successful identification of organelle-specific proteins through the production of organelle-specific antibodies raised against highly purified membrane fractions. Examples include endoplasmic reticulum and Golgi-specific markers (Louvard *et al.*, 1982) and lysosomal membrane-specific proteins and lysosomal glycoproteins (Lewis *et al.*, 1984).

3. LARGE-SCALE PURIFICATION OF ENDOSOMES FROM TISSUE CULTURE CELLS

Most procedures for subcellular fractionation (see for example Fleischer and Kervina, 1974) separate organelles on the basis of size and bouyant density by means of differential centrifugation and velocity or equilibrium density centrifugation in a variety of media. These two criteria for separation have been most successfully applied to the purification of organelles that are either large and light, for example plasma membrane sheets from rat liver (Hubbard *et al.*, 1983); small and heavy, for example dense secretory granules from exocrine cells (Bartlett and Smith, 1974; Cameron and Castle, 1984) or coated vesicles from a variety of tissues (Pearse, 1983; reviewed by Daiss and Roth, 1983); very small and very light, for example synaptic vesicles from neuronal cells (Carlson *et al.*, 1978); or very large and very heavy, for example nuclei (Tata, 1974) and mitochondria (Guerra, 1974). Unfortunately, following homogenization of tissue culture cells, most of the organelles of the biosynthetic and endocytic pathways including the endoplasmic reticulum, lysosomes, endosomes, the Golgi apparatus, and the plasma membrane are of very similar size and density. Purification of endosomes in particular, using the standard procedures derived for subcellular fractionation of rat liver, has proven difficult. For example, in a procedure requiring three density-centrifugation steps, one gel filtration step, and three days, endosomes were purified 37-fold from human KB cells and estimated to be only approximately 50% pure (Dickson *et al.*, 1983). We therefore sought to apply an alternative criteria to distinguish and separate the organelles of the endocytic pathway; that is, to use the novel technique of free flow electrophoresis to separate organelles based on their net surface charge. The feasibility of this approach had been suggested by earlier results of Harms *et al.* (1980) who successfully applied FFE to the purification of lysosomes.

3.1. Purification of Endosome Subpopulations by Free Flow Electrophoresis

Free flow electrophoresis is a method by which particles (proteins, nucleic acids, membrane-bound organelles) are separated based on their net surface charge density (Hannig and Heidrich, 1974). Operationally, a thin continuously flowing curtain of buffer is run through a chamber, which for the VAP21 electro-

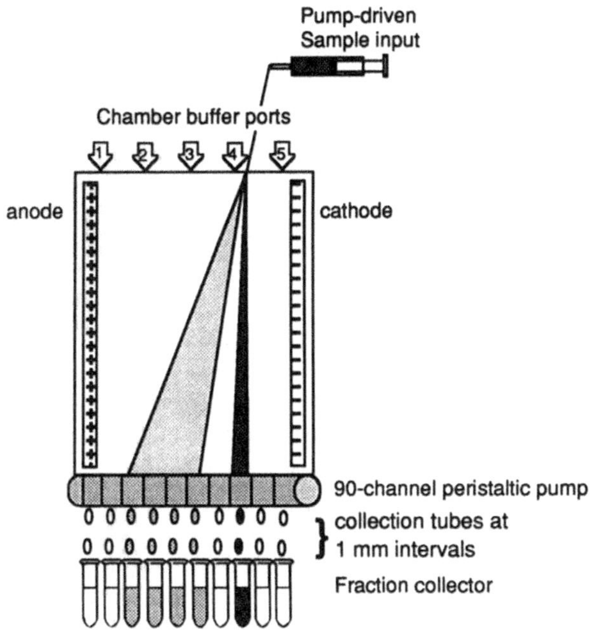

FIGURE 1. Schematic diagram of Elpho Vap 21 Free Flow Electrophoresis apparatus (Bender and Hobein, Munich, distributed by Protein Technologies, Inc., Tucson, Arizona).

phoresis unit (Bender and Hobein, Munich; distributed by Protein Technologies, Tucson, Arizona) used in our studies has dimensions of 100 × 250 × 0.5 mm (Figure 1). An electric field (for our studies 130 mA, 1500–1800 V) is applied perpendicular to the flow of buffer. The sample is injected from a pump-driven syringe in a thin stream at the cathodal side of the top of the chamber and is deflected anodally relative to its surface charge density as it encounters the electric field. There are several reviews that provide a more detailed discussion of the principles governing resolution using free flow zonal electrophoresis. (See for example Hannig and Heidrich, 1974, 1977; Crawford, 1985.)

3.1.1. Selective Labeling of Endosomal Subpopulations with Endocytic Tracers

As indicated, there are no known marker enzymes for endosomes nor any markers that might distinguish potential endosomal subpopulations. Therefore, before their purification can be attempted, the endocytic organelles of interest must be selectively labeled with endocytic tracers, either by receptor-mediated endocytosis or by fluid-phase uptake. The operational definitions of endosomal subpopulations described above are used as criteria for specifically labeling endosomes or endosomal subpopulations. For example, early endosomes by definition are preferentially labeled after very brief exposure of cells to endocytic tracers. Both early endosomes and recycling endosomes can be labeled by incubating cells with radiolabeled transferrin that is internalized and recycled along with its receptor through early endosomes (Klausner *et al.*, 1983; Ciechanover *et al.*, 1983). By definition, the endosomal compartment is uniformly labeled after internalizing endocytic tracers for incubation times no longer than the minimum transit time to lysosomes. The minimum transit time to lysosomes is determined for internalized ligands by measuring the kinetics of their appearance in characteristically dense lysosomal fractions using Percoll gradient centrifugation (see for example Storrie *et al.*, 1984; Ukkonen *et al.*, 1986). (It should be noted that this too is only an operational definition of the lysosome.) Endosomes can also be selectively labeled by internalization at reduced temperatures (<20°C) since delivery to lysosomes is blocked under these conditions (Dunn *et al.*, 1980). However, the exact location of the temperature block relative to access of ligands to late endosomal subpopulations is uncertain. A list of endocytic tracers used in our studies, their advantages, and our criteria for selective labeling of endosomal compartments in Chinese hamster ovary (CHO) cells is given in Table I. It is important to note that the kinetics of internalization and delivery to lysosomes

Table I
Endocytic Tracters Used to Label Endosomes in CHO Cells

Endocytic tracer	Advantages	Criteria for specific labeling
Fluid-phase markers		
FITC-dextran	pH sensitive probe	Early endosomes ≤5 min; late endosomes 10–15 min; all endosomes 30–60 min at 20°C
Horseradish peroxidase (HRP)	Sensitive enzymatic marker, morphological marker	Early endosomes ≤5 min; late endosomes 10–15 min; all endosomes 30–60 min at 20°C
Membrane-bound markers		
^{125}I-transferrin (^{125}I-Tfn)	Specific to recycling pathway	Labeled to equilibrium surface-bound removed.
FITC-transferrin (F-Tfn)	pH sensitive probe	Labeled to equilibrium, surface-bound removed
^{35}S-Semliki Forest virus (35S-SFV)	Targeted to lysosomes, pH sensitive probe, morphological marker	Early endosomes ≤5 min, surface-bound removed. Late endosomes 10–15 min, surface-bound removed

varies greatly among cell types. Therefore these kinetic parameters must first be firmly established for each cell type in order to establish operational definitions for selective labeling of the various endosomal subpopulations.

3.1.2. Protocol for the Two-Step Purification of Endosomes and Lysosomes from CHO Cells

We first describe the protocol we have developed for the large-scale purification of early and late endosomes and lysosomes from CHO cells. The procedure has been successfully extended to the purification of these organelles from rat liver (Fuchs *et al.*, 1989). Later in this section we discuss two steps in this purification scheme in more detail; namely, the homogenization conditions and the requirement for mild proteolysis. Finally, we discuss how this procedure can be readily adapted for the rapid analytical purification of early and late endosomes enabling their definition in dynamic terms based on the kinetics of migration of endocytic tracers through them.

The purification scheme, the time required for each step and the recoveries of membrane are listed in Table II. CHO cells are readily grown in suspension and 10^9 cells can be harvested from approximately 2 liters of growth media. The entire procedure can be accomplished in one (very long) day's work. However, since the fractions are stable after the final purification step on Percoll gradients, these fractions can be assayed and harvested the next day saving about 2 hrs time in the first day. Preparation of the microsomal fraction does not contribute to the purification but instead serves to concentrate the postnuclear supernatant (PNS) before fractionation on the sucrose gradient. The sucrose flotation gradient affords an approximate 10-fold purification of both endosomes and Golgi with a comparable yield of endosomes and lysosomes allowing the purification of both these organelles in high yield following FFE. This is in contrast to our previous procedure (Marsh *et al.*, 1987) using a sucrose gradient consisting of 1.2M/1.0M/0.6M sucrose layers which gave optimal purification of endosomes (about 40-fold) and considerably reduced yield of lysosomes (9%). This procedure results in a reasonably high yield (10–15%, 200–500 μg) of highly enriched early endosomes, late endosomes, and lysosomes.

The elution profiles of endosome and lysosomal markers following the FFE and Percoll gradient centrifugation steps of this procedure are shown in Figure 2. The distribution of contaminating membranes after FFE is indicated by the square brackets and is as previously documented (Marsh *et al.*, 1987). Not shown in this experiment is the fact that the broad distribution of endosomes, labeled after continuous internalization of HRP, reflects the existence of two distinct endosomal subpopulations designated early and late based on the kinetics of labeling with endocytic tracers (Schmid *et al.*, 1988, see Figure 4). Early endosomes are less anodally shifted and are enriched in the fractions designated

Table II
Procedure for Endosome Purification Using Free Flow Electrophoresis and Percoll Gradient Centrifugation

Time/step: (total time) (hr)	Procedure	Membrane fraction/quantity
1: (1)	~10^9 CHO cells (~4 ml cell pellet) labeled by internationalization of HRP for 15 min at 37°C	
0.5: (1.5)	Homogenize in 5 volumes (20 ml) TEA/sucrose buffer Prepare 3000g × 10 min PNS	Homogenate/400 mg, PNS/260 mg
1: (2.5)	Collect 100,000g × 45 min microsomal pellet onto 1M sucrose cushion	
2: (4.5)	Resuspend microsomal pellet in 1.25M sucrose and overlay with 1.1M sucrose and 0.6M sucrose. Centrifuge flotation gradient at 40K rpm × 1.5 hr in Beckman SW41 rotor	Microsomes/62 mg
.5: (5.0)	Harvest 1.1M/0.6M interface, and dilute to 1mg/ml protein and 0.25M sucrose Digest 5 min at 37°C with 2.5–4.0 μg/ml TPCK-trypsin, stop with 2X STI	1.1/0.6 interface/12 mg
1.5: (6.5)	Inject at 2–4 ml/hr into FFE chamber at 130 mA/~1800 volts	
2: (8.5)	Assay and pool desired fractions, collect by centrifugation at 100,000g × 1 hr onto 1M sucrose cushion. Resuspend in TEA/sucrose containing 20% Percoll	Pool 1/8.5 mg, Pool 2/1.5 mg
2: (10.5)	Centrifuge at 18K rpm × 1.5 hr in Beckman Type 50 ti rotor	
1.5: (12.0)	Fractionate gradient, assay and pool desired fractions *n.b.* These highly enriched fractions are stable and can be assayed and pooled the next day	
1: (13.0)	Dilute and pellet 100,000g × 1 hr. Harvest membranes from top of Percoll pellet	Early endosomes/0.42 mg late endosomes/0.23 mg lysosomes/0.20 mg

Pool 2, whereas late endosomes are more anodally shifted and elute in the position corresponding to Pool 1. As can be seen, late endosomes and lysosomes (Pool 1) are well resolved from contaminating endoplasmic reticulum, plasma membrane, Golgi, and mitochondrial fractions. The major contaminant of early endosomes remains the Golgi apparatus, and this problem is only partially re-

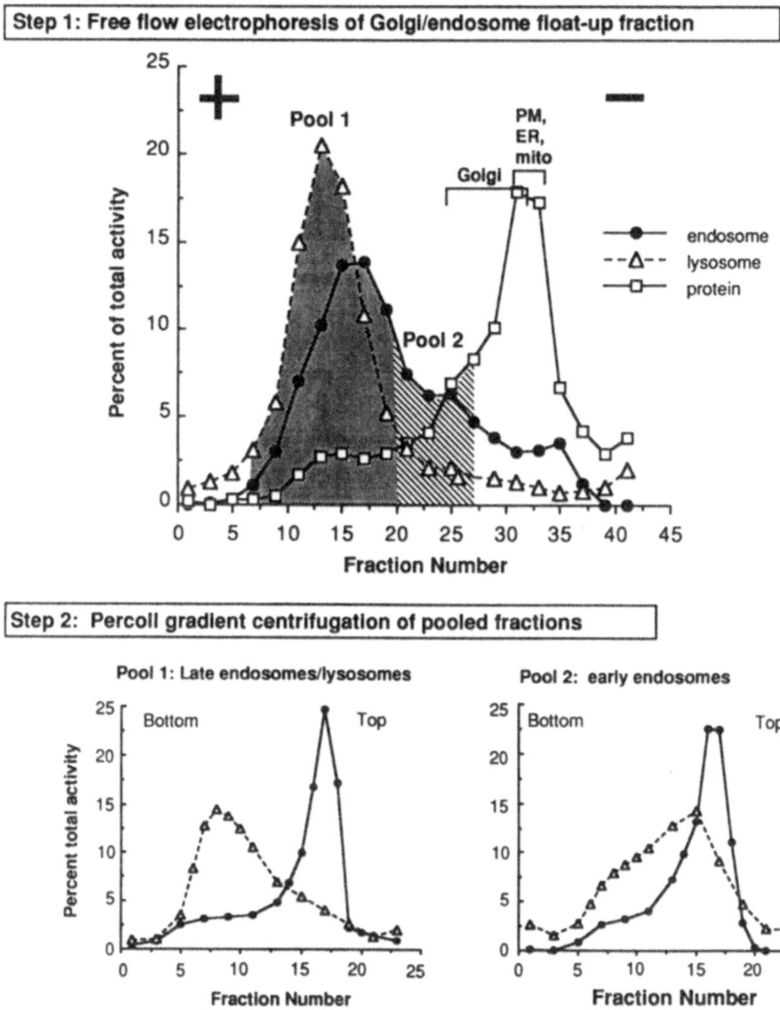

FIGURE 2. Two-step purification of endosomes and lysosomes from enriched light membrane fractions using FFE. *Top:* elution profile of lysosomes (△ = β-hexosaminidase activity); endosomes (● = 15 min internalized FITC-dextran); and protein (□ = Bradford Assay). *Bottom left:* Percoll gradient profile of Pool 1 fractions (symbols as above). *Bottom right:* Percoll gradient profile of Pool 2 fractions (symbols as above).

solved. Percoll gradient centrifugation of Pool 1 clearly resolves dense β-hexosaminidase–containing lysosomes from late endosomes. Interestingly, the small amount of total β-hexosaminidase activity that coelutes in Pool 2 with early endosomes following FFE appears to reside in a membrane fraction that is distinct from dense lysosomes. This result demonstrates that the electrophoretic heterogeneity observed following FFE is a true reflection of organelle heterogeneity and not simply incomplete resolution of like fractions. In addition, as alluded to above, it demonstrates the potential pitfalls associated with the use of marker enzymes and the use of operational criteria such as organelle density on Percoll gradients to define organelles such as the lysosome. Are these less dense, less electrophoretically mobile β-hexosaminidase–containing organelles newly formed lysosomes, or do they represent prelysosomal endosomes, transport vesicles, or a combination of these?

Endosomes purified by FFE are approximately 70-fold enriched over total homogenate and 80–90% pure as judged by electron microscopy using DAB cytochemistry on HRP-containing endosomal fractions (Marsh *et al.*, 1987). These fractions have been used to gain an improved biochemical definition of endosomal subpopulations using lactoperoxidase (LPO)-catalyzed iodination of endosome-specific proteins (Schmid *et al.*, 1988). For these experiments, LPO was taken up by CHO cells under conditions that uniformly label the endosomal compartment (15 min at 37°C). Various LPO-containing endosome subpopulations were resolved by FFE prior to radiolabeling with ^{125}I. The resulting protein patterns displayed by 2D-SDS-polyacrylamide gel electrophoresis demonstrated that both early and late endosomes are clearly distinct in protein composition from the plasma membrane and from each other (Schmid *et al.*, 1988).

3.1.3. Importance of Homogenization

Many organelles are subject to fragmentation during cell homogenization, which can greatly affect their properties during subsequent fractionation procedures. The most obvious examples of organelle fragmentation are the vesiculation of the endoplasmic reticulum and the plasma membrane. Although poorly documented, fragmentation may also affect organelles of the endocytic pathway. For example, tubular extensions of both endosomes and lysosomes may be subject to shearing and separation from the vacuolar portions of endosomes. These two subfractions might then be resolved during subcellular fractionation and lead to the erroneous conclusion that they represent distinct organelles. This possibility is supported by the results of Belcher *et al.* (1987), who isolated three endosomal fractions from estradiol-treated rats. Internalized ligand first appeared in low and then intermediate density fractions consisting of vacuoles and multivesicular bodies with associated membrane appendages. The highest density fraction consisted of membranes resembling the appendages seen in each of the

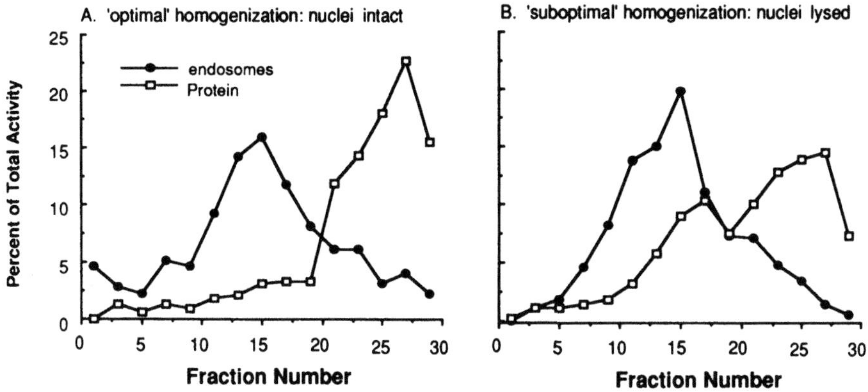

FIGURE 3. The effect of homogenization conditions on the elution of total protein and endosomes following FFE. CHO cells were labeled for 15 min with FITC-dextran prior to homogenization using the ball bearing homogenizer for either three passes (A) to obtain approximate 80% cell lysis while leaving nuclei intact or six passes (B) to obtain 100% lysis and disruption of about 20% of the nuclei.

other two fractions. Fragmentation of the nucleii and the consequent release of DNA can generally disrupt further purification because of DNA-mediated vesicle aggregation. When free flow electrophoresis is to be employed, not only is aggregation a problem, but released DNA can coat membranes, thereby altering their surface charge properties. In practice, this results in a higher proportion of contaminating membranes migrating anodally in the electric field (Figure 3). For each of these reasons it is important to optimize cell breakage conditions for any subcellular fractionation procedure and to ensure that these conditions can be reproduced at the onset of each preparation. Homogenization is affected by the method of cell breakage, the buffer in which cell breakage is performed, and the ratio of cell volume to homogenization buffer. In our studies we have used the ball-bearing homogenizer developed by Balch and Rothman (1985) and ratio of 5 volumes of TEA/sucrose (10 mM triethanolamine-acetic acid, 1 mM EDTA, 250 mM sucrose) to cell pellet. For consistency, cell and nuclear lysis are monitored by phase-contrast microscopy during the course of each homogenization. Under these conditions we routinely obtain 75–90% cell lysis without disrupting nucleii.

3.1.4. Requirement for Trypsinization

The least-satisfying requirement for this purification scheme is that cell lysates must be briefly treated with trypsin in order to obtain separation during FFE. Without this step, CHO microsomal membranes elute together in a single sharp, undeflected peak (Schmid and Mellman, 1987). Similar results have been obtained by Evans and co-workers who used FFE to analyze sucrose gradient-enriched endosomal fractions derived from rat liver. In the absence of any proteolysis step, they were unable to resolve endosomes from the bulk of the protein, which eluted as a broad band following FFE (Evans and Flint, 1984). A similar proteolysis step is required for the procedure developed by Beardmore *et al.* (1987) in which endosomes are loaded with colloidal gold-conjugated antibody to the EGF receptor and purified based on their increased mass by pelleting through dense sucrose. These authors showed electron microscopic evidence that the proteolysis was required to disrupt filamentous (cytoskeletal?) networks in which endosomes were trapped.

We have performed a variety of control experiments (see Schmid and Mellman, 1987) which give us confidence that these very gentle conditions of proteolysis are not severely damaging to endosome structure or function because (1) This mild proteolysis has little effect on the protein composition of endosomes and even proteins with known trypsin-sensitive cytoplasmic domains are unaffected. (2) ATP-dependent acidification, a biochemically measurable endosomal function highly dependent on endosome integrity, is not affected by these digestion conditions (Schmid and Mellman, 1987; Fuchs *et al.*, 1989). In fact, control experiments were performed under conditions used for isolating endosomes from a microsomal pellet, which requires incubation with at least 20 μg/ml trypsin. Five- to tenfold lower concentrations of trypsin are required when enriched endosomes are used as the starting material for FFE. In addition, other investigators have demonstrated the presence of trypsin-sensitive factors on endosomal membranes required for fusion activity *in vitro*. Demonstration of this trypsin sensitivity, however, requires incubation of endosomal fractions under considerably harsher conditions than those we employ. For example Woodman and Warren (1988) used 100 μg/ml trypsin for 1 hr at 4°C to inhibit fusion activity by 50% while Diaz *et al.* (1988) used 400 μg/ml for 2 hr at 4°C to inhibit fusion activity by 77%. Taken together, these results strongly suggest that we might expect minimal disruption of endosomal function and structure as a result of the mild proteolysis required for their purification. Nevertheless, the use of proteases for preparing biological samples remains unsatisfactory. How can the proteolysis step be eliminated? Preliminary attempts at disrupting the filamentous network by treatment of cells or lysates with microtubule- and microfilament-disrupting drugs have been unsuccessful (Fuchs and Mellman, unpublished results). Harms and colleagues (1980) have successfully used FFE to resolve lysosomes from contaminating membranes without proteolysis, enabling their purification from primary human fibroblasts. Their procedure involves extensive washing (five repeats) of lysosomal fractions by low-speed centrifugation (20,000 g × 10 min). Attempts to wash endosomal fractions using the higher centrifugal forces required to pellet them resulted in loss of acidification activity and latency (Fuchs and Schmid, unpublished results). Clearly more work is required toward both of these approaches and the development of other approaches aimed at avoiding the necessity for proteolysis in this purification.

3.2. A Rapid Analytical Method for Analysis of Protein Transport Through Endosomal Subpopulations

The procedure indicated in Table II is readily modified to provide a rapid and efficient method for the analysis of the kinetics of transport through the endosomal compartment. As few as 3×10^7 cells can be analyzed when sensitive endocytic tracers such as HRP or radiolabelled ligands such as ^{125}I-Tfn or ^{35}S-SFV are used to label endosomes. A microsomal fraction is prepared as described in Table II and adjusted to 0.25M sucrose (in TEA buffer) and 1mg/ml protein prior to a brief incubation (5 min at 37°C) in the presence of 20–25 μg/ml trypsin. The microsomal fraction is then loaded directly onto the FFE apparatus at a flow rate of 1–1.5 ml/hr. In practice, cells are labeled for various time points or under various conditions and the different cell homogenates are processed to prepare microsomal fractions. These microsomal fractions are maintained on ice until just prior to FFE when they are treated with trypsin and applied directly to the FFE apparatus. Since each analytical run requires only about 30 min, up to ten individual samples can readily be analyzed in a single day using this procedure. On a given day there is little or no variation in the elution profile of total protein or of the endogenous markers β-hexosaminidase and alkaline phospodiesterase which are routinely assayed allowing for direct comparison between samples.

The sequential appearance of internalized SFV into early and then late endosomes as detected by FFE (as first described in Schmid *et al.*, 1988) is shown in Figure 4. Later studies using the pH-dependent change in conformation of the SFV spike glycoprotein, E1, as a more specific probe for the arrival of SFV into acidic endosomes demonstrated that the early peak seen after 2 min of internalization would be further resolved into neutral early endosomes (probably primary endocytic vesicles) and acidic early endosomes (Schmid *et al.*, 1989). The kinetics obtained using this approach have aided in refining, in dynamic terms, the definition of early and late endosomes and their relative positions in the endocytic pathway.

4. ADVANTAGES AND DISADVANTAGES OF FREE FLOW ELECTROPHORESIS

The advantages of free flow electrophoresis for the purification of early and late endosomes for tissue culture cells are many. Endosomes prepared by FFE are obtained in high yield (10–15%, 200–500 μg from 10^9 cells) and high purity

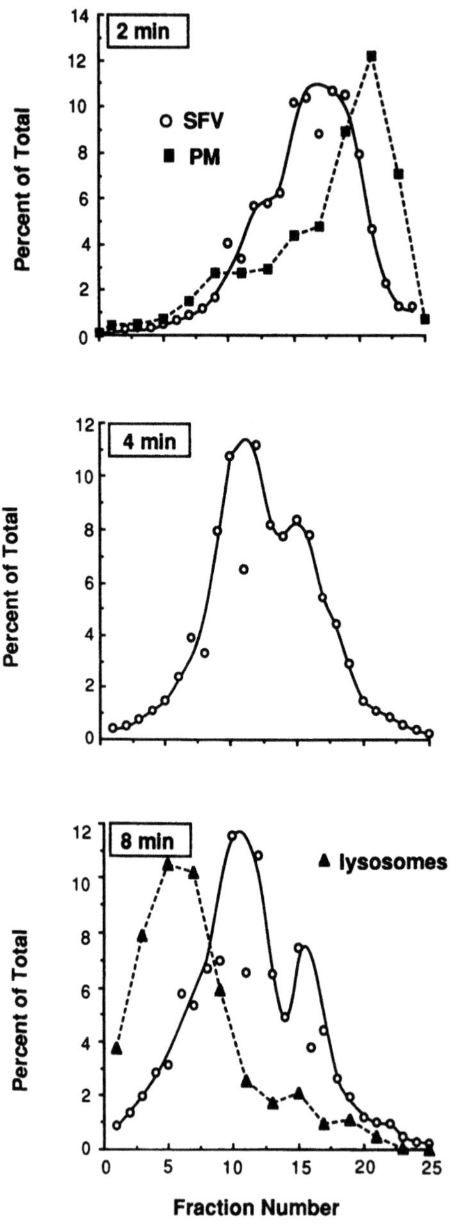

FIGURE 4. Analytical FFE to determine the kinetics of transit of SFV through the endosomal compartment. SFV was bound to monolayer CHO cells at 4°C, unbound virus was removed, and the cells incubated for the indicated times at 37°C, before rapid cooling to 4°C and removal of surface-bound virus with subtilisin and preparation of the sample for analysis by FFE. Plasma membrane distribution was determined by alkaline phosphodiesterase activity (■) and lysosomal distribution by β-hexosaminidase activity (▲). These data are simplified from Schmid *et al.* (1988). Adapted with permission.

(70-fold, 80–90% pure) with little preparation time as compared to conventional methods of cell fractionation. By comparison, purification of endosomes using gel-filtration chromatography and sucrose-gradient centrifugation required three days, and yielded approximately 250μg from 10^9 human KB cells which were 37-fold enriched and judged to be 50% pure (Dickson *et al.*, 1983). A second advantage of FFE is that the procedure can readily be scaled up or down. For smaller scale preparations, time-consuming steps such as preparation of the microsomal pellet and even sucrose-gradient centrifugation can be deleted resulting in enhanced yields without affecting the purity of the final product. Endosomes purified by FFE retain their biological activity as assessed by ATP-dependent acidification (Marsh *et al.*, 1987; Fuchs *et al.*, 1989). Finally, endosome purification by FFE relies on the intrinsic properties of the endosomal membranes and not, as will be seen below, in the nature of the ligands or receptors passing through them. Thus, we gain information about the static or steady-state properties of the endosomal subpopulations (i.e., that early and late endosomes are structurally distinct) from our ability to resolve them.

A general advantage of FFE is that whereas most current methods for subcellular fractionation rely on differences in size and density, FFE separates organelles on the basis of surface charge. As indicated above, many different subcellular organelles cannot be distinguished on the basis of size and density. The surface charge of an organelle will vary with the distinct chemical composition of each organelle membrane: It is unlikely that any two organelles of distinct protein composition will have the same surface charge, yet similar organelles should have the same charge. This expectation is consistent with the results we obtain on the electrophoretic mobility of β-hexosaminidase-containing vesicles. Although the vast majority of β-hexosaminidase activity elutes in a sharp, anodally deflected peak following FFE, β-hexosaminidase activity is also detected trailing in the less anodally shifted fractions. Subsequent Percoll-gradient analysis demonstrates that this subpopulation of β-hexosaminidase-containing vesicles does not result from a heterogeneous charge distribution of dense lysosomes but that they are in fact a distinct, low-density vesicle population.

If each organelle is compositionally distinct, FFE has the potential to resolve all subcellular organelles. By analogy, one might compare the resolution of cellular proteins obtained using SDS-polyacrylamide gel electrophoresis, which separates only on the basis of size, to that obtained using isoelectric focusing and two-dimensional gel electrophoresis, separating both on the basis of size and charge. Unfortunately, the potential resolving power of FFE has yet to be realized. Constraints such as ionic conditions for electrophoresis, the requirements of maintaining vesicle integrity, vesicle aggregation, and so forth limit its resolving power. As will be discussed below, other modes of FFE, distinct from the continuous preparative FFE procedure described here, may enhance its resolving power. These methods such as isotachophoresis and isoelectric focusing have yet to be fully explored for the purification of subcellular organelles.

The major disadvantage of FFE is the cost of the apparatus (about $70,000). This severely limits its availability. A second disadvantage is that optimal conditions for organelle separation must be worked out empirically. Variables including ionic strength, pH, and composition of the separation media and sample preparation must be worked out in pilot experiments made tedious by the need to assay all fractions in order to determine the distribution of marker enzymes. The more rapid and sensitive these assays are, the more efficient the optimization procedure will be. On-line scanners for detecting absorbance at 280 nm or light scattering have been developed that assist somewhat in these experiments (e.g., the Elphorscan, Bender and Hobein, Munich, distributed by Protein Technologies, Tucson, Arizona). However, the information is restricted to total protein distribution and these too are expensive (about $35,000). Optimal conditions once established for one purification problem are unlikely to be optimal for another purification problem.

5. COMPARISONS WITH OTHER METHODS FOR ENDOSOME PURIFICATION

Endosomal subpopulations have been successfully purified using two other approaches: density-shift protocols and immunoadsorption. Density-shift methods involve loading endosomal subpopulations with endocytic markers such as HRP, used to catalyze the deposition of a dense DAB-reaction product in the endosomal lumen, (Courtoy *et al.*, 1984; Quintart *et al.*, 1984; Ajioka and Kaplan, 1987), colloidal gold (Beardmore *et al.*, 1987), or very fine ferrite particles (Sato *et al.*, 1986) to alter their densities or magnetic properties so that they might readily be resolved from other cellular components. Immunoadsorption involves isolation of endosomal membranes using antibodies directed toward the cytoplasmic domain of either endogenous receptors (Mueller and Hubbard, 1986) or proteins specifically inserted into the endosomal membrane (Gruenberg and Howell, 1986).

One of the disadvantages of both these approaches is that they rely on the operational definitions of the transit of receptors and ligands through endosome subpopulations as the basis for their purification. This, in turn, relies on the synchrony of internalization and labeling with a given ligand. For example, by definition early endosomes would be specifically enriched by density-shift procedures following brief internalization of HRP. However, results from FFE show that even after very brief internalization times (2–5 min), some HRP is present in electrophoretically distinct late endosomes and so both subpopulations would, in fact, be represented. Thus, this approach is limited by the difficulty in precisely localizing an endocytic tracer to one endosomal compartment. Similarly, immunoadsorption procedures have been used to distinguish receptor-positive from receptor-negative endosomes (Mueller and Hubbard, 1986). Both early endosomes and recycling endosomes are receptor-positive, however, and both subpopulations would be purified using this criteria. In addition, a substantial amount of internalized ASOR is recycled along with its receptor (Townsend *et al.*, 1984) thus accounting for the fact that only 60% of internalized ASOR could be resolved into receptor-negative endosomes (Mueller and Hubbard, 1986). It should be noted that the elution profile of recycling endosomes has not been directly examined using FFE. The distribution of [^{125}I]-Tfn-containing endosomes is slightly different from that of early endosomes labeled by a 4-min pulse of HRP (Schmid *et al.*, 1988, Figure 1) in that the elution profile of [^{125}I]-Tfn-containing endosomes displayed a prominent cathodal shoulder. There is as yet no evidence that this technique has resolved these two endosomal subpopulations.

Density-shift and immunoadsorption procedures have been most successfully applied to addressing the issue of "who's with whom, when." That is, the kinetics of sorting of two distinct endocytic tracers (for example a receptor and its ligand, or a ligand destined for transcytosis from one destined for the lysosome) have been determined by labeling one ligand and using the second ligand/receptor for purification. Examples of this include the sorting of [^{125}I]-ASOR from HRP-Tfn (Stoorvogel *et al.*, 1987), [^{125}I]-EGF from Tfn-HRP (Ajioka and Kaplan, 1987) and the sorting of [^{125}I]-ASOR from the ASGP-receptor (Mueller and Hubbard, 1986). At early time points after internalization both tracers are present in the same compartment and the radiolabeled tracer can be coshifted or coimmunoadsorbed. At later time points, after sorting has occurred, the two ligands are in separate compartments and the radiolabeled ligand is no longer copurified. Thus, valuable information for the establishment of a dynamic definition for endosomal subpopulations can be obtained from these studies.

Both these approaches, however, have limitations in their usefulness for providing a biochemical definition for endosome subpopulations in static terms. Although density-shift procedures could be scaled up to produce large amounts of endosomal membranes, the DAB-reaction product has been shown to cross-link endosomal membrane components (Ajioka and Kaplan, 1987). As a result, protein constituents are rendered insoluble and cannot be analyzed by SDS-polyacrylaminide gel electrophoresis or immuneprecipitation, and enzyme activities are therefore lost (Ajioka and Kaplan, 1987). Density-shift procedures using colloidal gold require that large amounts of gold-conjugated ligand be internalized in order to effect a density shift. The usefulness of this procedure is therefore restricted to cell types with high concentrations of specific surface receptors. Furthermore, it would not be readily amenable to large-scale preparations. Similar limitations apply to endosome purification by immunoadsorption.

Gruenberg and Howell (1985) have demonstrated that a relatively high concentration of antigens must be present on the endosomal membrane for efficient immunoadsorption. Large-scale purification by this method is limited by the quantities of antibodies and support materials (e.g., Staph-A, magnetic beads, and so forth) required.

6. WHAT HAVE WE LEARNED? A MODEL FOR MEMBRANE TRANSPORT ALONG THE ENDOCYTIC PATHWAY IN CHO CELLS

One model based on the kinetics of transit of the membrane-bound endocytic tracer Semliki Forest virus ([^{35}S]-SFV) along the endocytic pathway in CHO cells is shown in Figure 5. This model is consistent with the new biochemical and kinetic information obtained through the purification of endosomal subpopulations using FFE. Receptor-bound SFV is rapidly internalized via coated pits and coated vesicles ($t_{1/2}$ < 3 min). Intracellular SFV is first detected in a neutral endocytic compartment electrophoretically distinct from early endosomes. As depicted in the model, these most likely represent primary endocytic vesicles (Schmid *et al.*, 1989). The half-time for passage of SFV through this compartment is approximately 3 min. Although some SFV can be detected in acidic (pH < 6.2) early endosomes even at very early time points after internalization (< 2 min), the acid form of SFV was clearly resolved from neutral SFV. These results suggest that in CHO cells the rate of transport to electrophoretically distinct acidic early endosomes is asynchronous and can be very rapid and that the less anodally shifted primary endocytic vesicles detected by FFE are not acidic. Transit of SFV through early endosomes occurs with the half-time of approximately 5–8 min. Since early endosomes are compositionally distinct from the cell surface (Schmid *et al.*, 1988) the model indicates that this involves fusion of primary endocytic vesicles with preexisting early endosomes. The existence of a bona fide early endosome organelle in turn suggests the existence of distinct biosynthetic pathways that deliver newly synthesized early endosome–specific proteins to that compartment. As indicated, these novel biosynthetic pathways could involve direct delivery from the Golgi apparatus to early endosomes or they could involve delivery and/or recycling of proteins via the late endosomal compartment. After an approximate 2 min lag, SFV is next delivered to a third electrophoretically distinct subpopulation termed *late endosomes*. Late endosomes are generally more acidic than early endosomes, however our data suggest that both early and late endosomes are heterogeneous with respect to their luminal acidity. While late endosomal pH is generally <5.4, early endosomal pH ranges from >6.2 (relatively neutral) to <5.4 (fairly acidic). This rather large range in early endosomal pH could be caused by a number of

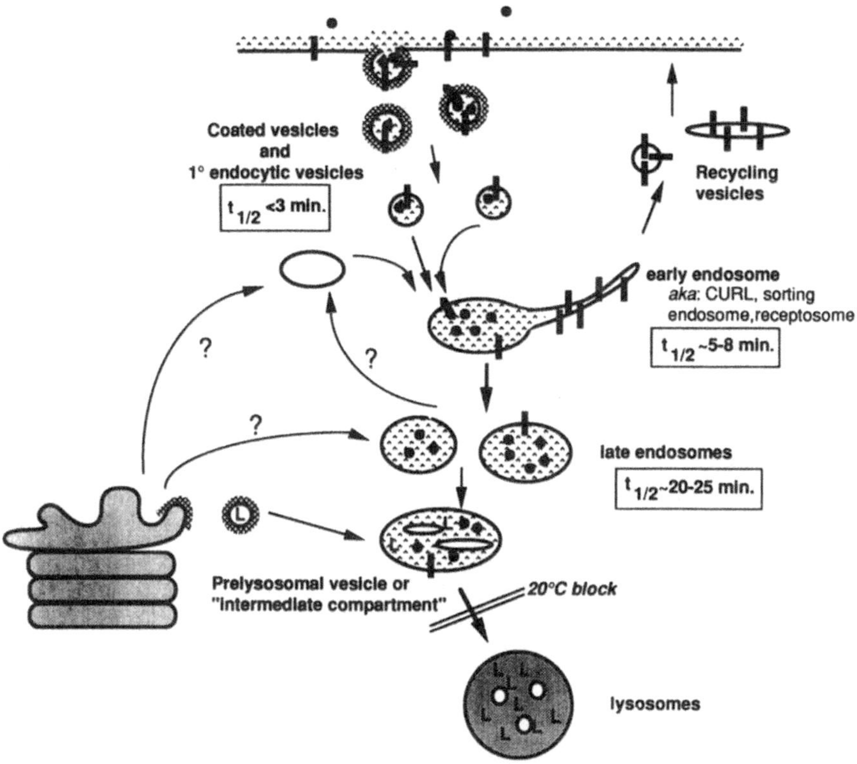

FIGURE 5. A model for membrane traffic into and out of the organelles of the endocytic pathway in CHO cells.

factors including differences in the numbers or activities of H^+-pumps in individual endosomes, differences in the size (i.e., luminal volumes) of individual endosomes, or differences in the number of regulatory pumps such as the Na^+K^+-ATPase (Fuchs *et al.*, 1989) relative to H^+-pumps present in individual endosomes. Late endosomes are again compositionally distinct from early endosomes and so a pathway from the Golgi to late endosomes is indicated. Whether this is the same pathway that carries lysosomal enzymes and lysosomal membrane glycoproteins to lysosomes via the intermediate compartment, as indicated in Figure 5 or a distinct biosynthetic pathway is not known.

Biochemical evidence for the delivery of lysosomal proteins to late endosomes in CHO cells comes from a result in which endosomal subpopulations were labeled by LPO-catalyzed iodination as described above and various proteins immunoprecipitated from them (Schmid and Mellman, unpublished data). Plasma membrane, early endosomes, late endosomes, or lysosomes were labeled following internalization of LPO for 0 min (iodination at 4°C), 4 min, 4 min followed by an 11 min chase, or 10 min followed by a 30 min chase, respectively. Endosomes were also labeled by internalization of LPO for 30 min at 20°C. Enriched endosome and lysosomal fractions were prepared and iodinated as described (Schmid *et al.*, 1988) and detergent lysates were prepared for sequential immunoprecipitation using the following antibodies: rabbit anti-M6P-R, and mouse monoclonals eh1C9 and eh3E9. The rabbit anti-M6P-R antibody recognizes the cation-independent, 215 kD, mannose-6-phosphate receptor and was prepared by C. Gabel, Columbia University. eh1C9 and eh3E9 are CHO-specific monoclonal antibodies prepared from mice immunized with endosome/lysosome fractions prepared by FFE (Schmid, Fuchs, and Mellman, unpublished results). eh1C9 recognizes an abundant 58 kD glycoprotein of the lep class of proteins (Lippincott-Schwartz and Fambrough, 1987) localized to the cell surface, early and late endosomes, and lysosomes by colloidal gold-immunolocalization on frozen thin sections (Schmid, K-P. Zimmer, R. Fuchs and Mellman, unpublished results). eh3E9 recognizes a less abundant approximately 96 kD glycoprotein of the "lgp" class based on its colocalization with lgp120 by double immunogold-labeling of frozen thin sections (Schmid, K-P. Zimmer, R. Fuchs and Mellman, unpublished results). Table III summarizes these results. As expected from the immunolocalization, the eh1C9 antigen could be precipitated from each of the various labeled vesicle populations; the plasma membrane, early and late endosomes, 20°C compartment, and lysosomes. Also as expected, the M6P-R could be precipitated from plasma membrane, early and late endo-

Table III
Localization of Antigens Along the Endocytic Pathway

	(Antibody) Rabbit serum (Antigen) CI M6P-R	eh1C9 lep 58	eh3E9 lgp 96
Labeled membrane fraction			
Plasma membrane LPO-catalyzed iodination at 4°C	+	+	–
Early endosome LPO internalized or 4 min at 37°C	+	+	–
Late endosomes LPO internalized for 4 min, then 11 min chase at 37°C	+	+	+
20°C compartment LPO internalized for 30 min at 20°C	+	+	+
Lysosomes LPO internalized for 10 min then 30 min chase at 37°C	–	+	+

somes, the 20°C compartment, but not lysosomes. These two expected results serve to validate the experimental approach. Interestingly, the eh3E9 antigen could not be detected on either the plasma membrane or in early endosomes, but could be detected in late endosomes, lysosomes, and the 20°C compartment. These data suggest that lysosomal proteins are delivered to late endosomes encountered between 4 and 15 min after internalization of ligand. Furthermore, as indicated in the model, they suggest that the 20°C temperature block in CHO cells can be placed downstream from the intersection of the biosynthetic and endocytic pathways in CHO cells. Further studies employing additional probes for endosomal subcompartments and more refined kinetics of labeling are clearly required to improve the resolution of this technique.

The relationship between early endosomes and late endosomes remains unclear. Our studies indicate that although early and late endosomes have some compositional overlap, both early and late endosome-specific proteins could be identified (Schmid *et al.*, 1988). Compositional overlap could arise if, as suggested by Gruenberg *et al.* (1989), late endosomes are derived from the vacuolar portion of early endosomes forming large carrier vesicles that subsequently attach to microtubules in order to transport endosomal content into the perinuclear region of the cell. Early endosome-specific proteins might remain in the tubular portions while late endosome-specific proteins could be added at any stage by delivery of proteins from the Golgi. In this way, the endosomal content would remain intact while the membrane is remodeled around it, thus providing a biochemical basis for the functional differences between late endosome subpopulations.

Although the subcellular fractionation approach reviewed here has given some insight into the *mapping* of the endocytic pathway and the organelles involved, it provides little or no insight into the *mechanisms* that govern the specific membrane interactions involved in membrane transport and sorting along the endocytic pathway. An understanding of these mechanisms will only be gained through the reconstitution and dissection of these processes in cell-free systems. Although beyond the scope of this review, as indicated above, such studies are well underway and insights gained from them have been recently reviewed elsewhere (Gruenberg and Howell, 1989).

7. FUTURE DIRECTIONS

7.1. Other Approaches for the Use of FFE

Alternative methods of free flow electrophoresis can provide greater resolution of electrophoretically distinct species and can potentially be applied to the purification of endosome subpopulations. Although isoelectric focusing is possible with the apparatus currently available, the high cost of ampholines would seem to render this mode of electrophoresis impractical. In addition, vesicles might be expected to aggregate and precipitate as they reach their isoelectric point. An alternative method more readily applicable to the isolation of organelle subpopulations is isotachophoresis or ITP (reviewed by Bocek *et al.*, 1985). For ITP the separation chamber contains two buffers of differing composition and pH. The principle is similar to that employed in the stacking gel during discontinuous SDS-PAGE (reviewed in Holloway and Battersby, 1984). A buffer containing the leading ion is fed through the three buffer ports on the anodal side of the separation chamber, while a buffer containing the terminating ion is fed through port five (closest to the cathode) (see Figure 1). The sample (diluted in the leading ion buffer) is injected into the chamber through port four. The leading and terminating ions are chosen on the basis of their electrophoretic activities (determined by their pKa's): the leading ion being more mobile and the terminating ion being less mobile than the particles to be resolved. Once steady state is achieved, the particles (in this case, endosomal subpopulations) are then concentrated and separated at charge boundaries based on their electrophoretic mobilities relative to the leading and terminating ions. Spacer ions of intermediate mobilities can be mixed with the sample prior to injection into the chamber to obtain even greater resolution (Bocek *et al.*, 1985). The increased resolution obtained by this technique offers great potential for the purification of endosomal subpopulations. Although ITP has been successfully applied to the purification of serum proteins (Kopwillem *et al.*, 1976) and virus particles (Heydt *et al.*, 1988) it has yet to be tested for its applicability to organelle separation. Again, the limitation that separation conditions must be empirically derived has perhaps delayed attempts at using ITP for the purification of endosomes.

7.2. The Search for an Endosome-Specific Marker

The biochemical definition of endosomal subpopulations is far from complete. How many endosome subpopulations are there? How are they ontogenically related to one another? If a biosynthetic pathway to endosomes exists, what is its route? If endosome-specific proteins exist, how do they avoid delivery to and/or degradation in lysosomes? Many questions remain unanswered, not the least of which is the major question of whether endosomes are in fact bona fide organelles or simply specialized transport vesicles. Unambiguous answers to many of these questions will require the identification of an endosome-specific protein constituent and subsequent analysis of its intracellular itinerary, both during biosynthesis and once it has reached the endosomal compartment. The existence of endosome-specific proteins has been indicated by analysis of endosomal fractions (Schmid *et al.*, 1988). None, however, has yet been identified and our attempts to isolate endosome-specific monoclonal antibodies from mice immunized with highly enriched endosomal fractions have yielded antigens of the lep and lgp class, as indicated above. Lysosomal proteins are an expected constituent of late endosomes as they pass through these organelles en route to lysosomes. If these proteins are more antigenic, they will be more readily detected in the hybridoma screen. Although this approach of immunizing mice with enriched subcellular fractions and screening hybridomas for organelle-specific antibodies has been successful in the identification of lysosome-specific antigens (Lewis *et al.*, 1984), it relies on the relative immunogenicity of the various antigens, and the experimenter is at the mercy of the mouse immune system. The use of more highly enriched endosome subpopulations as antigens, or better yet, individual endosome-specific candidate proteins extracted from 2D-gels should help to solve these problems. Such endosome-specific markers could then be used in biochemical and morphological studies to better define biosynthetically and functionally the various organelles along the endocytic pathway.

ACKNOWLEDGMENTS. All of the author's work reviewed in this chapter was performed with the help of my skilled collaborators, Drs. Renate Fuchs, Klaus-Peter Zimmer, and Mr. Philippe Mâle, during my postdoctoral tenure in Ira Mellman's laboratory in the Department of Cell Biology, Yale University Medical School. I am indebted to each of them and to Dr. Ari Helenius for many helpful discussions which have helped me to formulate the ideas presented in this review. Work was supported by NIH grants (to I. M.) and the Lucille P. Markey Charitable Trust (to S. L. S.). I am also indebted to the initial support of a Helen Hay Whitney Postdoctoral Fellowship and am currently a Lucille P. Markey Scholar.

8. REFERENCES

Ajioka, R. S., and Kaplan, J., 1987, Characterization of endocytic compartments using the horseradish peroxidase–diaminobenzidine density shift technique, *J. Cell Biol.* **104**:77–85.

Balch, W. E., and Rothman, J. E., 1985, Characterization of protein transport between successive compartments of the Golgi apparatus: asymmetric properties of donor and acceptor activities in a cell-free system, *Arch. Biochem. Biophys.* **240**:413–425.

Bartlett, S. F., and Smith, A. D., 1974, Adrenal chromattin granules: Isolation and disassembly, *Meth. Enzym.* **31**:379–389.

Beardmore, J., Howell, K. E., Miller, K., and Hopkins, C. R., 1987, Isolation of an endocytic compartment from A431 cells using a density modification procedure employing a receptor-specific monoclonal antibody complexed with colloidal gold, *J. Cell Sci.* **87**:495–506.

Belcher, J. D., Hamilton, R. L., Brady, S. E., Hornick, C. A., Jaeckle, S., Schneider, W. J., and Havel, R. J., 1987, Isolation and characterization of three endosomal fractions from the liver of estradiol-treated rats, *Proc. Natl. Acad. Sci. USA*, **84**:6785–6789.

Bocek, P., Gebauer, P., Dolnik, V., and Foret, F., 1985, Recent developments in isotachophoresis, *J. Chrom.* **334**:157–195.

Braell, W. A., 1987, Fusion between endocytic vesicles in a cell-free system, *Proc. Natl. Acad. Sci. USA,* **84:**1137–1141.

Cameron, R. S. and Castle, J. D., 1984, Isolation and compositional analysis of secretion granules and their membrane subfractionation from the rat parotid gland, *J. Membr. Biol.* **79:**127–144.

Carlson, S. S., Wagner, J. A., and Kelly, R. B., 1978, Purification of synaptic vesicles from elasmobranch electric organ and the use of biophysical criteria to demonstrate purity, *Biochem.* **17:**1188–1199.

Ciechanover, A., Schwartz, A. L., Dautry-Varsat, A., and Lodish, H. F., 1983, Kinetics of internalization and recycling of transferrin and the transferrin receptor in a human hepatoma cell line, *J. Biol. Chem.* **258:**9681–9689.

Courtoy, P. J., Quintart, J., and Baudhuin, P., 1984, Shift of equilibrium density induced by 3,3′-diaminobenzidine cytochemistry: a new procedure for the analysis and purification of peroxidase-containing organelles, *J. Cell Biol.* **98:**870–876.

Crawford, N., 1985, Use of free flow electrophoresis (FFE) in studies of surface and intracellular membranes and specific surface functional domains in blood platelets and leucocytes, in *Cell Electrophoresis* (W. Schutt and H. Klinkmann, eds.) pp. 225–246, Walter deGuyter and Co., Berlin, New York.

Daiss, J. L., and Roth, T. F., 1983, Isolation of coated vesicles: comparative studies, *Meth. Enzym.* **98:**337–349.

Davey, J., Hurtley, S. M., and Warren, G., 1985, Reconstitution of an endocytic fusion event in a cell-free system, *Cell,* **43:**643–652.

Diaz, R., Mayorga, L., and Stahl, P., 1988, *In vitro* fusion of endosomes following receptor-mediated endocytosis, *J. Biol. Chem.* **263:**6093–6100.

Dickson, R. B., Beguinot, L., Hanover, J. A., Richert, N. D., Willingham, M. C., and Pastan, I., 1983, Isolation and characterization of a highly enriched preparation of receptosomes (endosomes) from a human cell line, *Proc. Natl. Acad. Sci., USA,* **80:**5335–5339.

Dunn, W. A., Hubbard, A. L., and Aronson, N. N., Jr., 1980, Low temperature selectively inhibits fusion between pinocytic vesicles and lysosomes during heterophagy of ^{125}I-asialofetuin by the perfused rat liver, *J. Biol. Chem.* **255:**5971–5978.

Dunphy, W. G., and Rothman, J. E., 1985, Compartmental organization of the Golgi stack, *Cell,* **42:**13–21.

Evans, W. H., and Flint, N., 1985, Subfractionation of hepatic endosomes in Nycodenz gradients and by free flow electrophoresis, *Biochem. J.* **232:**25–32.

Farquhar, M. G., 1985, Progress in unraveling pathways of Golgi traffic, *Ann. Rev. Cell Biol.* **1:**447–488.

Fleischer, S., and Kervina, M., 1974, Subcellular fractionation of rat liver. *Meth. Enzym.* **31:**6–40.

Fuchs, R., Male, P., and Mellman, I., 1989, Acidification and ion permeabilities of highly purified rat liver endosomes, *J. Biol. Chem.* **264:**2212–2220.

Fuchs, R., Schmid, S., and Mellman, I., 1989, A possible role for Na^+, K^+-ATPase in regulating ATP-dependent endosome acidification, *Proc. Natl. Acad. Sci. USA,* **86:**539–543.

Geuze, H. J., Slot, J. W., and Strous, G. J. A. M., 1983, Intracellular site of asialoglycoprotein receptor-ligand uncoupling: double-label immunoelectron microscopy during receptor-mediated endocytosis, *Cell,* **32:**277–287.

Geuze, H. J., Slot, J. W., Strous, G. J. A. M., Peppard, J., von Figura, K., Hasilik, A., and Schwartz, A. L., 1984, Intracellular receptor sorting during endocytosis: comparative immunoelectron microscopy of multiple receptors in rat liver, *Cell,* **37:**195–204.

Geuze, H. J., Stoorvogel, W., Strous, G. J. A. M., Slot, J. W., Bleekemolen, J. E., and Mellman, L., 1988, Sorting of mannose 6-phosphate receptors and lysosomal membrane proteins in endocytic vesicles, *J. Cell Biol.* **107:**2491–2501.

Goldstein, J. L., Anderson, R. G. W., and Brown, M. S., 1979, Coated pits, coated vesicles and receptor-mediated endocytosis. *Nature.* **279:**679–685.

Griffiths, G., Hoflack, B., Simons, K., Mellman, I., and Kornfeld, S., 1988, The mannose 6-phosphate receptor and the biogenesis of lysosomes, *Cell,* **52:**329–341.

Gruenberg, J. E., and Howell, K. E., 1985, Immuno-isolation of vesicles using antigenic sites either located on the cytoplasmic or the exoplasmic domain of an implanted viral protein. A quantitative analysis. *Eur. J. Cell Biol.* **38:**312–321.

Gruenberg, J. E., and Howell, K. E., 1986, Reconstitution of vesicle fusions occurring in endocytosis with a cell-free system, *EMBO J.* **5:**3091–3101.

Gruenberg, J. E., and Howell, K. E., 1989, Membrane traffic in endocytosis: insights from cell-free assays, *Ann. Rev. Cell Biol.* **5:**453–482.

Gruenberg, J. E., Griffiths, G., and Howell, K. E., 1989, Characterization of the early endosome and putative endocytic carrier vesicles *in vivo* and with an assay of vesicle fusion *in vitro, J. Cell Biol.,* **108:**1301–1316.

Guerra, F. C., 1974, Rapid isolation techniques for mitochondria: Technique for rat liver mitochondria, *Meth. Enzym.* **31:**299–305.

Hannig, K., and Heidrich, H.-G., 1974, The use of continuous preparative free flow electrophoresis for dissociating cell fractions and isolation of membranous components, *Meth. Enzym.* **31:**746–761.

Hannig, K., and Heidrich, H.-G., 1977, Continuous free flow electrophoresis and its application in biology, in *Cell Separation Methods Part IV. Electrophoretic Methods* (H. Bloemendal, ed.), pp. 93–116, Elsevier/North Holland, Biomedical Press, Amsterdam.

Harms, E., Kern, H., and Schneider, J. A., 1980, Human lysosomes can be purified from diploid skin fibroblasts by free flow electrophoresis, *Proc. Natl. Acad. Sci. USA,* **77:**6139–6143.

Helenius, A., Mellman, I., Wall, D., and Hubbard, A., 1983, Endosomes, Trends in *Biochem. Sci.* **8:**245–250.

Heydt, A., Wagner, H., and Paul, H. L., 1988, Concentration and purification of plant pathogenic viruses by field step electrophoresis, *J. Virol. Meth.,* **9:**13–22.

Holloway, C. J., and Battersby, R. V., 1984, Preparative isotachophoresis, *Meth. Enzym.* **104:**281–301.

Hubbard, A. L., Wall, D. A., and Mah, A., 1983, Isolation of rat hepatocyte plasma membranes. I. Presence of the three major domains, *J. Cell Biol.* **96:**217–229.

Jamieson, J. D., and Palade, G. E., 1967a, Intracellular transport of secretory proteins in the pancreatic exocrine cell. I. Role of the peripheral elements of the Golgi complex. *J. Cell Biol.* **34:**577–596.

Jamieson, J. D., and Palade, G. E., 1967b, Intracellular transport of secretory proteins in the pancreatic exocrine cell. II. Transport to condensing vacuoles and zymogen granules, *J. Cell Biol.* **34:**597–616.

Klausner, R. D., van Renswoude, J., Ashwell, G., Kempf, C., Schechter, A. N., Dean, A., and Bridges, K. R., 1983, Receptor-mediated endocytosis of transferrin in K562 cells., *J. Biol Chem.* **258:**4715–4724.

Kopwillem, A., Merriman, W. G., Cuddeback, R. M., Smolka, A. J. K., and Bier, M., 1976, Serum protein fractionation by isotachophoresis using amino acid spacers, *J. Chrom.* **118:**35–46.

Kornfeld, R., and Kornfeld, S., 1985, Assembly of asparagine-linked oligosaccharides, *Ann. Rev. Biochem.* **54:**631–664.

Kornfeld, S., and Mellman, I., 1989, The biogenesis of lysosomes, *Ann. Rev. Cell Biol.* **5:**483–525.

Lewis, V., Green, S. A., Marsh, M., Vihko, P., Helenius, A., and Mellman, I., 1985, Glycoproteins of the lysosomal membrane, *J. Cell Biol.* **100:**1839–1847.

Lippincott-Schwartz, J., and Fambrough, D. M., 1987, Cycling of the integral membrane glycoprotein, LEP100, between plasma membrane and lysosomes: kinetic and morphological analysis, *Cell* **49:**669–677.

Louvard, D., Reggio, H., and Warren, G., 1982, Antibodies to the Golgi complex and the rough endoplasmic reticulum, *J. Cell Biol.* **92:**92–107.

Marsh, M., Schmid, S. L., Kern, H., Harms, E., Male, P., Mellman, I., and Helenius, A., 1987, Rapid analytical and preparative isolation of functional endosomes by free flow electrophoresis, *J. Cell Biol.* **104:**875–886.

Matteoni, R., and Kreis, T. E., 1987, Translocation and clustering of endosomes and lysosomes depends on microtubules, *J. Cell Biol.*, **105:**1253–1265.

Mayorga, L. S., Diaz, R., and Stahl, P. D., 1988, Plasma membrane-derived vesicles containing receptor-ligand complexes are fusogenic with early endosomes in a cell-free system, *J. Biol. Chem.* **263:**17213–17216.

Mueller, S. C., and Hubbard, A. L., 1986, Receptor-mediated endocytosis of asialoglycoproteins by rat hepatocytes: receptor-positive and receptor-negative endosomes, *J. Cell Biol.* **102:**932–942.

Mullock, B. M., Branch, W. J., van Schaid, M., Gilbert, L. K., and Luzio, J. P., 1989, Reconstitution of an endosome-lysosome interaction in a cell-free system, *J. Cell Biol.* **108:**2093–2099.

Pearse, B. M. F., 1983, Isolation of coated vesicles, *Meth. Enzym.* **98:**320–326.

Quintart, J., Courtoy, P. J., and Baudhuin, P., 1984, Receptor-mediated endocytosis in rat liver: purification and enzymic characterization of low density organelles involved in uptake of galactose-exposing proteins, *J. Cell Biol.* **98:**877–884.

Sato, S. B., Sako, Y., Yamashina, S., and Ohnishi, S.-I., 1986, A novel method of isolating specific endocytic vesicles using very fine ferrite particles coated with biological ligands and the high-gradient magnetic separation technique, *J. Biochem.* **100:**1481–1492.

Schmid, S. L., and Mellman, I., 1987, Isolation of functionally distinct endosome subpopulations by free flow electrophoresis, in *Cell Free Analysis of Membrane Traffic* (D. J. Morre, K. E. Howell, and G. M. W. Cook, eds.), pp. 35–49, Alan R. Liss, New York.

Schmid, S. L., Fuchs, R., Male, P., and Mellman, I., 1988, Two distinct subpopulations of endosomes involved in membrane recycling and transport to lysosomes, *Cell,* **52:**73–83.

Schmid, S. L., Fuchs, R., Kielian, M., Helenius, A., and Mellman, I., 1989, Acidification of endosome subpopulations in wild-type Chinese hamster ovary cells and temperature-sensitive acidification-defective mutants, *J. Cell Biol.* **108:**1291–1300.

Steinman, R. M., Brodie, S. E., and Cohn, Z. A., 1976, Membrane flow during pinocytosis. A stereological analysis. *J. Cell Bio.* **68:**665–687.

Steinman, R. M., Mellman, I. S., Muller, W. A., and Cohn, Z. A., 1983, Endocytosis and the recycling of plasma membrane, *J. Cell Biol.* **96:**1–27.

Stoorvogel, W., Geuze, H. J., and Strous, G. J. A. M., 1987, Sorting of endocytosed transferrin and asialoglycoprotein occurs immediately after internalization in HepG2 cells, *J. Cell Biol.*, **104:**1261–1268.

Storrie, B., Pool, R. R., Sachdeva, M., Maurey, K. M., and Oliver, C., 1984, Evidence for both prelysosomal and lysosomal intermediates in endocytic pathways, *J. Cell Biol.* **98:**108–115.

Tata, J. R., 1974, Isolation of nuclei from liver and other tissues, *Meth. Enzym.* **31:**253–262.

Townsend, R. R., Wall, D. A., Hubbard, A. L., and Lee, Y. C., 1984, Rapid release of galactose-terminated ligands after endocytosis by hepatic parenchymal cells: Evidence for a role of carbohydrate structure in the release of internalized ligand from receptor, *Proc. Natl. Acad. Sci. USA* **81:**466–470.

Ukkonen, P., Lewis, V., Marsh, M., Helenius, A., and Mellman, I., 1986, Transport of macrophage Fc Receptors and Fc receptor-bound ligands to lysosomes, *J. Exp. Med.* **163:**952–971.

Woodman, P. G., and Warren, G., 1988, Fusion between vesicles from the pathway of receptor-mediated endocytosis in a cell-free system, *Eur. J. Biochem.* **173:**101–108.

Yamashiro, D. J., Tycko, B., Fluss, S. R., and Maxfield, F. R., 1984, Segregation of transferrin to a mildly acidic (pH 6.5) para-Golgi compartment in the recycling pathway, *Cell,* **37:**789–800.

Chapter 2

Analytical Subcellular Fractionation of Endosomal Compartments in Rat Hepatocytes

Pierre J. Courtoy

1. DEFINITION OF THE ENDOSOMAL COMPARTMENTS

This chapter focuses on the contribution of complementary analytical fractionation procedures to the dissection of the endosomal compartments in rat hepatocytes. It also describes their composition and properties, and concludes with the proposal of a structural model. Recent reviews on endosomes and opinions on some unsolved issues can be found elsewhere (e.g., Gruenberg and Howell, 1989; Hubbard, 1989; Courtoy, 1991; Griffiths and Gruenberg, 1991; Murphy, 1991; Stoorvogel *et al.*, 1991).

1.1. Heterogeneity of Endosomal Compartments in Rat Hepatocytes

In the past ten years, endosomes have been somehow rediscovered and moved to the front place in research, with the realization of their crucial role in a large number of key events of receptor-mediated endocytosis. Established func-

---

Pierre J. Courtoy Cell Biology Unit, International Institute of Cellular and Molecular Pathology and University of Louvain Medical School, 1200 Brussels, Belgium.
Subcellular Biochemistry, Volume 19: Endocytic Components: Identification and Characterization, edited by J. J. M. Bergeron and J. R. Harris. Plenum Press, New York, 1993.

tions of endosomes involve: (1) ligand-receptor uncoupling, sorting and segregation, receptor recycling, and receptor-mediated transcytosis (Harford *et al.*, 1983; Geuze *et al.*, 1983, 1984; Courtoy *et al.*, 1985); (2) intracellular addressing of newly synthesized lysosomal matrix enzymes and lysosomal membrane proteins (Brown *et al.*, 1986; Geuze *et al.*, 1988; Griffiths *et al.*, 1988); (3) convergence of auto- and heterophagic pathways (Gordon and Seglen, 1988); (4) iron uptake (Dautry-Varsat *et al.*, 1983; Klausner *et al.*, 1983; Sibille *et al.*, 1986); (5) membrane crossing into the cytosol of several toxins and viral genomes (Olsnes and Sandvig, 1985; Marsh and Helenius, 1989); and (6) transient or sustained sequestration of internalized constituents such as receptors (Grant *et al.*, 1990), ligands (Opresko and Karpf, 1987), or transmembrane transporters (Slot *et al.*, 1991). Insofar as constituents are sorted by some endosomes and dispatched into other endosomes committed to various destinations, the endosomal apparatus necessarily comprises distinct functional compartments.

In rat hepatocytes, ligands and receptors internalized together at the basolateral (sinusoidal) membrane are sorted in tubulospherical endosomes and addressed to at least three distinct destinations, namely (1) degradation in lysosomes, (2) recycling to the basolateral domain, and (3) transcellular transfer to bile canaliculi—strictly speaking, diacytosis (Jacques, 1968)—but hereafter referred to as transcytosis, according to the common use. These processes already require at least five functionally distinct intermediate endocytic compartments: primary endocytic vesicles (defined here as primary endosomes of basolateral origin), the secondary tubulospherical endosomes generated by the coalescence of primary endosomes and acting as a sorting station or CURL (Geuze *et al.*, 1983) and finally, the various epigones, i.e., the distinct tertiary endocytic vesicles committed either to degradation, recycling/regurgitation, or transcytosis (Courtoy *et al.*, 1985). In addition, endocytosis must also take place at the apical (canalicular) membrane, and internalized constituents should be either recycled back to the canalicular membrane or discharged into lysosomes. This implies the occurrence of additional sets of distinct pericanalicular endosomes (Quintart *et al.*, 1989).

Various studies have demonstrated the heterogeneity of rat liver endosomes with respect to physical properties (Debanne *et al.*, 1984; Kindberg *et al.*, 1984; Evans and Flint, 1985; Limet *et al.*, 1985; Wall and Hubbard, 1985; Bergeron *et al.*, 1986; Branch *et al.*, 1987; Kennedy and Cooper, 1988), fusion competence (Mullock *et al.*, 1989), and occurrence of plasma membrane markers (Quintart *et al.*, 1989). It remains, however, to define structurally the limits and/or overlaps of those distinct compartments, to correlate structural features with functional properties, and to identify (a set of) possible endogenous markers.*

*We define here *tracers* as exogenous probes, incorporated into one or several subcellular compartments, for example by endocytosis. We reserve the use of *markers* to endogenous constituents, selected for their restricted distribution to essentially one subcellular compartment (see postulate of single location, de Duve, 1964), e.g., marker enzymes.

1.2. Probes

In the current analyses, endosomes are still generally defined by their ligand content, in effect, according to the pathway followed and the stage of processing (whether by maturation, vesicular transfer, or a combination of the two processes: see Hubbard, 1989). An important limitation to this approach is the asynchronous uptake and processing of the exogenous tracers by the liver *in situ*. This problem is partially circumvented by more complex procedures such as single-pass liver perfusion (Perez *et al.*, 1988), but is less important along the degradation pathway because of the large physical differences of the sequential compartments.

We usually use ligands conjugated with horseradish peroxidase that are radiolabeled, cleared of aggregates as well as of unconjugated moieties by gel filtration, and injected intravenously at 0.5–2 μg/g body mass. At various times after injection the liver is flushed for about 1 min by the portal vein with PBS containing 1 mM EGTA (to remove tracer either free in the extracellular space, or bound in a Ca^{2+}-dependent manner to the plasma membrane exposed to the chelator during the brief flush). Flushing considerably reduces the amount of tracer recovered after differential sedimentation in the final supernatant (soluble fraction). Parts of the liver are then fixed by perfusion and processed for peroxidase cytochemistry or homogenized and analyzed by subcellular fractionation.

The degradation pathway is commonly traced by the injection of radiolabeled asialofetuin or asialoorosomucoid, which are cleared from the blood by the galactose-specific receptor of the hepatocytes. We often prefer galactosylated bovine serum albumine (GalBSA), which is conveniently prepared in relatively large amounts and exhibits a similar, if not higher, affinity to the galactose-specific lectin (Courtoy, 1984). Efficient labeling of the transcellular pathway of rat hepatocytes is possible by selecting preparations of rat or human polymeric IgA (pIgA) that do not appreciably bind to the galactose-specific receptor of rat hepatocytes (Courtoy, 1984; Kennedy and Cooper, 1988), and show a high fractional excretion into bile (typically 40–60% of injection). Some misaddressing to the lysosomes is nevertheless unavoidable (Schiff *et al.*, 1986; Mullock *et al.*, 1989). We have recently reported, however, that $F(ab')_2$ fragments of affinity-purified antibodies against the rat-free secretory component are better tracers of transcytosis, because they are even more efficiently taken up by the liver and show negligible misaddressing to lysosomes (Courtoy *et al.*, 1992).

When using peroxidase conjugates, we add to the injection 2 μg dialyzed yeast mannan/g body mass, to inhibit the unwanted uptake of the conjugate by liver sinusoidal cells. Indeed, both endothelial and Kupffer cells can recognize mannosyl residues on the peroxidase moieties by their polyspecific receptor (Shepherd *et al.*, 1981; Quintart *et al.*, 1983). The almost exclusive uptake of peroxidase conjugates by hepatocytes in the presence of mannan and their distinct intracellular addressing are illustrated at Figure 1.

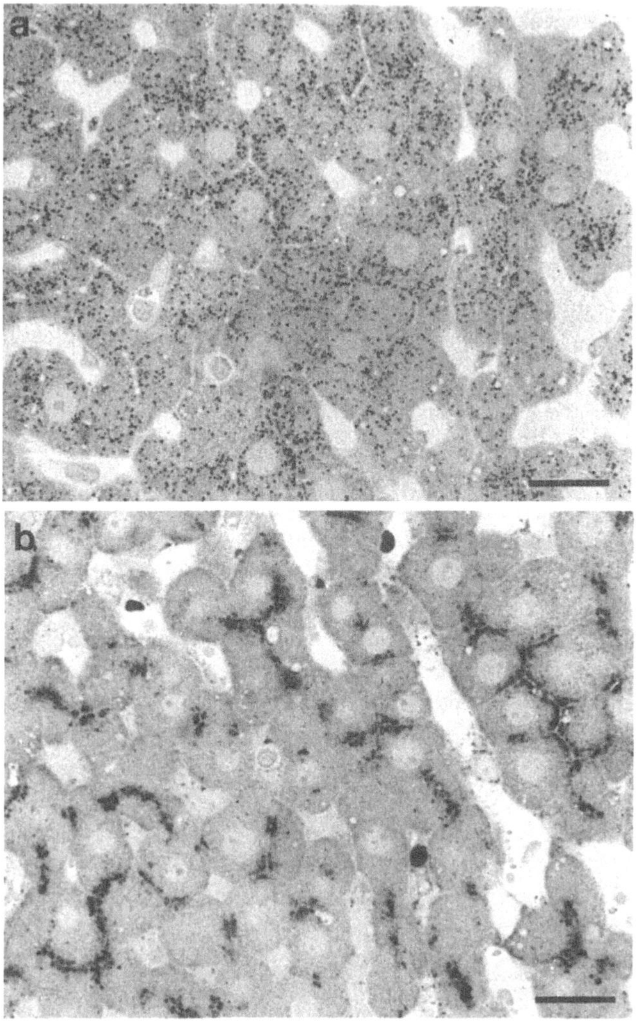

FIGURE 1. Tissue distribution of ligand-peroxidase conjugates. The liver was perfusion-fixed at 20 min after the intravenous injection of 2 μg/g body mass of either GalBSA-HRP (a) or pIgA-HRP (b). Due to the addition of an equal amount of yeast mannan, there is virtually no uptake of GalBSA-HRP by sinusoidal cells, and little uptake of pIgA-HRP in these cells. Labeling by GalBSA-HRP is concentrated in the hepatocytes, in discrete vesicles at some distance from bile canaliculi, whereas the reaction product generated by pIgA-HRP closely delineates bile canaliculi, which appear empty. Reproduced from Quintart *et al.* (1989), by copyright permission of the European Journal of Biochemistry. Bars indicate 20 μm.

1.3. Electron Microscopy

Several reports have described in detail the ultrastructural features of hepatocyte endosomes labeled by peroxidase conjugates in the intact liver and in subcellular fractions, purified by conventional procedures or by the diaminobenzidine-induced density shift. Peroxidase has been conjugated to galactose-exposing proteins (Wall *et al.*, 1980; Stockert *et al.*, 1980; Quintart *et al.*, 1983; Quintart *et al.*, 1984; Courtoy *et al.*, 1985; Kennedy and Cooper, 1988), to polymeric IgA, (Courtoy *et al.*, 1985; Hoppe *et al.*, 1985; Limet *et al.*, 1985; Quintart *et al.*, 1989), to EGF (Dunn and Hubbard, 1984) and to transferrin (Ajioka and Kaplan, 1987; Stoorvogel *et al.*, 1991). The latter probe has not yet been applied successfully to rat hepatocytes.

Studies with tracers of the degradation pathway like galactose-exposing proteins and EGF generally show an early increase in size in the subsinusoidal cortex, from the labeling of small vesicles and tubules to that of larger tubulospherical structures, followed by a migration to the pericanalicular region where tracers are found in multivesicular bodies and eventually lysosomes. With IgA conjugates, most labeled vesicles remain of smaller size throughout transcytosis. Whatever the label, endosomes are frequently associated with microtubules, and indeed their saltatory motion has been shown in other cell types (de Brabander *et al.*, 1988). Disruption of the microtubular network in rat hepatocytes delays the degradation of galactose-exposing proteins (Kolset *et al.*, 1979) and the transcellular transfer of pIgA (Perez *et al.*, 1988).

2. ANALYTICAL SUBCELLULAR FRACTIONATION

2.1. Methodological Remarks

Analytical subcellular fractionation is the only approach that provides quantitative data on the subcellular distribution of several constituents in a large biological sample. Before reviewing the contribution of the various fractionation procedures, it is important to stress the possible experimental bias and misleading interpretations inherent in the use of tracers in every fractionation study of endocytic organelles, as well as to evaluate the advantages and limitations of the liver as a starting material.

2.1.1. Application of Analytical Cell Fractionation to Endocytosis

Labeling of organelles by noncovalently bound tracers exposes several difficulties in analytical subcellular fractionation, irrespective of the cells studied.

First, leakage of soluble tracers from endocytic particles* can occur during tissue homogenization and the resuspension of pellets. Leakage would lead to underestimating purification and might produce artificial distributions in case of readsorption.

Second, comparison of the purification of various endocytic compartments in the same cell based on the relative specific activity of tracer over protein becomes misleading if the tracer is either diluted, as shown for bulk-membrane endocytosis (de Chastellier *et al.*, 1987) and to a larger extent for fluid-phase endocytosis (Griffiths *et al.*, 1989), or if it becomes concentrated with respect to the mass of protein of the organelle (Steinman *et al.*, 1974; Besterman *et al.*, 1981). It is obvious that the specificity of uptake and intracellular sorting implies several possible stages of concentration by selective inclusion, or dilution by selective exclusion.

Third, especially at intermediate intervals of uptake, the asynchrony of intracellular movement (Kielian *et al.*, 1986) causes the association of tracers with more than one compartment. This can be in marked deviation from single location, a central postulate in analytical subcellular fractionation (de Duve, 1964). Whereas a bimodal distribution in a linear gradient is sufficient evidence for a dual location, a well-defined unimodal peak is no proof of single location (see the comparison of early and late transcytotic endosomes, Quintart *et al.*, 1989). Conversely, each endosomal compartment probably shares endogenous markers with the other compartments with which it interacts (see Figure 9). Those difficulties must be kept in mind for a critical evaluation of results.

2.1.2. Advantages of the Rat Liver

The choice of the liver for analytical studies of endosomes offers several advantages. Some apply to the fractionation study of any type of subcellular particle: (1) tissue is readily available in relatively large quantities; (2) after perfusion with buffered saline, hepatocytes contribute close to 95% of tissue protein (see Blouin *et al.*, 1977); (3) liver of young rats can be gently homogenized to release at least 80–85% of particles at the cost of minimal particle damage, as indicated by less than 5% of lysosomal enzymes being recovered in the high-speed supernatant (Amar-Costesec *et al.*, 1974a)—such an excellent compromise is more difficult to achieve with cultured cells; (4) results can be compared with one of the largest collections of data on markers accumulated in

*To avoid confusion in terminology, we shall follow here the suggestion of de Duve (1964) and restrict the denomination *constituent* to biochemical entities (i.e. molecules) and *component* to supramolecular structures (organelles, particles). Furthermore, the denomination *organelles* will be reserved to the subcellular components in the intact cells, and *particles* to their counterpart after isolation in subcellular fractions. For further safety, we shall try, whenever possible, to define subcellular fractions by their physical properties, not by the expected content in endocytic particles.

the field of subcellular fractionation over the last 35 years; (5) distribution in the rat liver can be compared with fractionation data on cultured rat hepatocytes (see Limet *et al.*, 1982; Kindberg *et al.*, 1984).

Other features are of special interest to the study of endocytosis: (1) uptake of soluble ligands is very fast and efficient because of the combination of the abundant blood supply to the liver, the free exchange of proteins through the fenestrated endothelium and the permanent mixing of Disse's space by the "massage" effect described by Wisse *et al.* (1983), as well as the abundance of receptors at the sinusoidal surface of hepatocytes: This makes it easy to obtain a high specific labeling of tissue endosomes; (2) shortening of the endocytic pulse can be achieved using the perfused liver system with a single pass of medium (Perez *et al.*, 1988); (3) the rat liver is a convenient system to study transcytosis, because hepatocytes are well polarized and equipped with tight junctions impermeable to proteins, and because transcytozed ligands can be easily recovered using bile duct canulation.

2.1.3. Limitations of the Rat Liver

Obviously, analytical studies on endosomes from the total rat liver also have their limitations. First, the fractional volume of endosomes and secondary lysosomes may be severalfold lower in the hepatocytes *in situ* (approximately 1% for each compartment) than in most cultured cells (up to 4%): A higher enrichment is thus required to bring preparations of liver endosomes to a level of purity comparable to that of cultured cells.

Second, sinusoidal cells exhibit quite a high rate of fluid endocytosis (Ose *et al.*, 1980; Munniksma *et al.*, 1980) and selectively take up protein aggregates as well as several glycoproteins (Shepherd *et al.*, 1981): As already mentioned, this difficulty can be circumvented by targeting ligands to hepatocytes through receptor-mediated endocytosis, by eliminating aggregates using gel filtration of the probes, and by the use of appropriate competitors. In addition, the kinetics of uptake and transfer to lysosomes is much faster in liver sinusoidal cells than in hepatocytes and involves endosomes of distinct equilibrium densities, thus avoiding confusion between liver-cell populations at intermediate times of uptake (5–20 min; Quintart *et al.*, 1983; Magnusson and Berg, 1989). Lysosomes of sinusoidal cells and hepatocytes can also be resolved from one another by selectively increasing the equilibrium density of the former upon accumulation of invertase (Jadot *et al.*, 1986) or colloidal gold (Henning and Plattner, 1974), taking advantage of mannose-specific endocytosis or the intense non-specific capture of colloids (colloidoplexy) by sinusoidal cells, respectively.

Third, the favored access to the sinusoidal fluid and the relatively higher membrane area of sinusoidal cells makes it difficult to identify, in the liver homogenate, covalently labeled hepatocyte surface proteins. On the other hand,

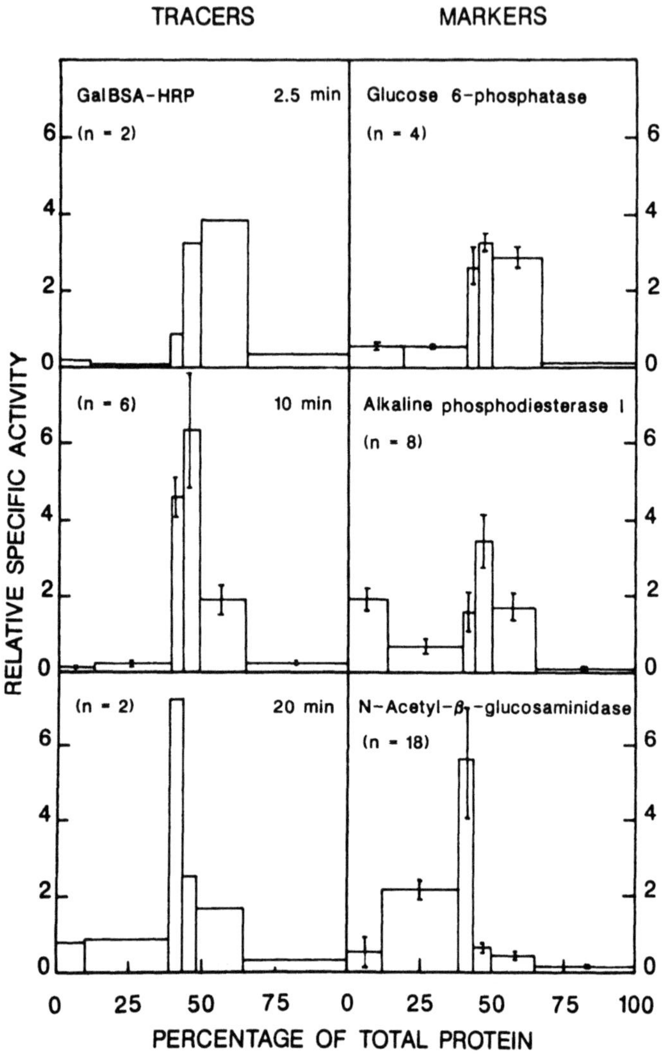

FIGURE 2. Differential sedimentation of endocytic organelles of rat hepatocytes. At 2.5, 10 or 20 min after the intravenous injection of GalBSA-HRP (0.5 μg/g body mass) and mannan (2 μg/g body mass), the liver was flushed with PBS-EGTA and about 2 g were homogenized. Each g of liver contained 207 ± 37 mg protein, 21 U glucose 6-phosphatase, 7.7 U alkaline phosphodiesterase I, and 3.1 U N-Acetyl-β-glucosaminidase. Liver uptake corresponded to 11% of injected tracer at 2.5 min, 24% at 10 min, and 33% at 20 min. The homogenate was resolved by differential sedimentation, into an N pellet (6000 *g* × min, washed twice), an M pellet (33,000 *g* × min, washed once), an L pellet (275,000 *g* × min, washed once), an F fraction (fluffy layer corresponding to particles loosely sedimenting over the packed L pellet after its second sedimentation), a P′ pellet ("restricted" microsomes, i.e., without particles of the fluffy layer, 3 × 10^6 *g* × min) and a final supernatant.

a convenient endogenous marker enzyme restricted to the hepatocyte basolateral membrane is not really available, and 5′-nucleotidase, an excellent marker of the canalicular membrane of hepatocytes, also occurs in appreciable amounts on sinusoidal cells (Matsuura *et al.*, 1984; Chen *et al.*, 1987).

Fourth, except when using the perfused organ system, the rat liver is less amenable to controlled perturbations such as quick temperature shifts and some drugs. One should also remember that moderate differences in receptor content and/or access of hepatocytes along the liver lobule (Van der Sluis *et al.*, 1988), combined with the unequal sensitivity of the three acinar zones to various perturbations, could lead to a marked tissue heterogeneity.

2.1.4. Complementarity of Approaches

Every possible difference in the natural or artificially altered physical properties of the subcellular particles can be exploited to resolve endosomal compartments of rat hepatocytes. The most popular are differences in size and equilibrium density. The rapid increase in size of endosomes and/or increasing deviation from round shape along the degradation pathway can be most conveniently demonstrated by differential cell sedimentation. Differences in the density of endosomes are more subtle. Fortunately, major alterations in the equilibrium density of endocytic organelles or particles can be provoked by *in vivo* or *in vitro* perturbations, respectively. One can also hope that the development of free flow electrophoresis and other techniques based on differences in surface properties will increasingly provide independent analyses of rat liver endosomes. No doubt a comprehensive picture will emerge from the convergence of information gathered from the combination of all procedures.

2.2. Differential Sedimentation

The temporal distribution of injected ligands depends on the pathway they follow. Along the degradation pathway, major differences can be demonstrated by differential sedimentation (Figure 2). Galactose-exposing proteins are successively recovered in the microsomal pellet ("restricted" P′, peak at 2 min after injection), the fluffy layer (F, 5–10 min), the light mitochondrial pellet (L, 10–15 min), and finally the heavy mitochondrial pellet (M, after 20 min). Because

Recoveries ranged between 90 and 103% of the homogenate. Distributions of tracers and markers are presented according to de Duve *et al* (1955), and are average of 2–16 experiments (indicated in parentheses). The patterns of marker enzymes are compatible with those reported by Amar-Costesec *et al.* (1974a). Compared to the usual L fraction, the fluffy layer is markedly depleted in lysosomal enzyme markers. It is also more enriched in plasma membrane markers than conventional microsomes. Reproduced in part from Courtoy (1990), by copyright permission of Cambridge University Press.

up to 15 min after injection, density distributions peak similarly at low densities, the increasing sedimentation coefficient, indicated by the shift from fraction P to fraction L, probably reflects an increase of particle size. The further increase of sedimentation coefficient at later times probably results from an increase in density from that of endosomes to that of secondary lysosomes.

The irregular shape of tubulospherical endosomes (CURL) can however not be neglected and is expected to decrease their sedimentation rate, possibly resulting in a shift from fraction L to fraction F. Interestingly, by gel filtration of hepatocyte extracts solubilized in Triton X-100, internalized asialofetuin moves as a ligand-receptor complex at 2 min, is being uncoupled around 5 min, and is almost completely dissociated at 15 min (Kindberg *et al.*, 1984). Hence, the fluffy layer, which contains most of galactose-exposing proteins at times corresponding to ligand-receptor uncoupling, can thus be enriched in tubulospherical (sorting) endosomes.

In contrast to galactose-exposing proteins, asialotransferrin type III (used to label the recycling/regurgitation pathway; Debanne *et al.*, 1984), as well as pIgA and $F(ab')_2$ antirat secretory component (tracers of the transcytotic pathway; Limet *et al.*, 1985), remain mostly associated with "restricted" microsomes. Hence, if sorting of these ligands also involves the CURL (Geuze *et al.*, 1984), either the transit in tubulospherical endosomes is brief and tubular extensions containing membrane-bound recycled or transcytosed ligands are rapidly cleaved off *in vivo,* or a large proportion of them are ruptured during homogenization.

2.3. Isopycnic Centrifugation

The major common characteristic of endosomes in rat hepatocytes is their low equilibrium density, which is most convenient to distinguish their distribution in density gradients from that of plasma membrane and especially secondary lysosomes. However, even low-density preparations purified 80-fold from the rat liver homogenate remain largely contaminated by Golgi, plasma membrane, and especially endoplasmic reticulum vesicles (Quintart *et al.*, 1983; Kay *et al.*, 1984), because endosomes are only a minor cell compartment in rat hepatocytes (Quintart *et al.*, 1984). Fortunately, preliminary results indicate that low-density preparations from rat liver homogenates are actually depleted in endosomes from sinusoidal cells (Quintart *et al.*, 1983). Hence, whereas conventional density equilibration is appropriate to compare the fate of various ligands, by following the time course of the distributions, one sees that additional steps are clearly required for purification purposes.

Differences in equilibrium density between endosomal compartments are more subtle. They can be difficult to detect in the hypertonic linear sucrose gradients, which nevertheless provide highly reproducible results. A better resolution has been reported with Nycodenz gradients, or when using flat isotonic gradients made of Ficoll or Percoll.*

In sucrose gradients, the equilibrium density of GalBSA is first shifted from an unimodal peak at 1.13 g/ml in density in the P fraction isolated at 2 min after injection to an even lower density (1.09–1.11 g/ml) in the L fraction isolated at 10 min after injection (Figure 3; Limet *et al.*, 1985; Wall and Hubbard, 1985). The initial displacement from early-light, to intermediate–very light particles is reproduced in Nycodenz gradients when using rat liver (Evans and Flint, 1985) but not isolated rat hepatocytes (Kindberg *et al.*, 1984), and can also be observed with Ficoll (Branch *et al.*, 1987) and Percoll gradients (Kennedy and Cooper, 1988).

If there is no inward movement of fluid, a lighter equilibrium density of a particle may indicate a relatively higher contribution of its low-density content (the density of plasma is 1.027 g/ml) compared to its membrane (the equilibrium density of plasma membrane, indicated by 5′-nucleotidase is around 1.17 g/ml in sucrose gradients; see Figure 4). This is compatible with the preferential removal of membrane by the tubular extensions of secondary endosomes. Indeed, most endosomes of the L fraction isolated in the very low-density region are large vesicles.

Between 10 and 20 min after injection and at variable kinetics from rat to rat, the distribution of GalBSA becomes bimodal. Ligand is rapidly transferred from the very-low-density region to the high-density region, where it largely overlaps with lysosomal enzymes (1.18–1.20 g/ml, see upper panel of Figure 6). Finally, the distribution of GalBSA again becomes unimodal and its peak is slightly displaced to a higher density, which now fully coincides with that of lysosomal markers (1.21 g/ml; Limet *et al.*, 1985). This transient accumulation in dense lysosomes is easier to demonstrate with GalBSA than with asialofetuin, presumably because BSA is more resistant to proteolysis.

In our estimation, the abrupt displacement from 1.11 to 1.18 g/ml is best explained by the fusion of a low-density multivesicular endosome with a preexisting high-density secondary lysosome, generating a particle of intermediate density. This is then followed by randomization of the lysosomal compartment.

*The curvilinear shape of self-generating Percoll gradients is however not fully reproducible and may lead to the artificial separation of a homogenous population of particles in two fractions of continuous properties (discussed in Draye *et al.*, 1987). The equilibrium density of a similar type of particle also varies in different density media, because of their unequal effects on the osmotic space of the particles (de Duve *et al.*, 1959; Beaufay and Amar-Costesec, 1976). In rat-liver subcellular fractions, the plasma membrane marker 5′-nucleotidase equilibrates at higher density than endosomes in hypertonic sucrose gradients, but at lower density in isotonic Percoll gradients (Kennedy and Cooper, 1988). The sequential endosomes labeled by injected cholera toxin also show opposite density changes in sucrose and Percoll gradients (Janicot and Desbuquois, 1987).

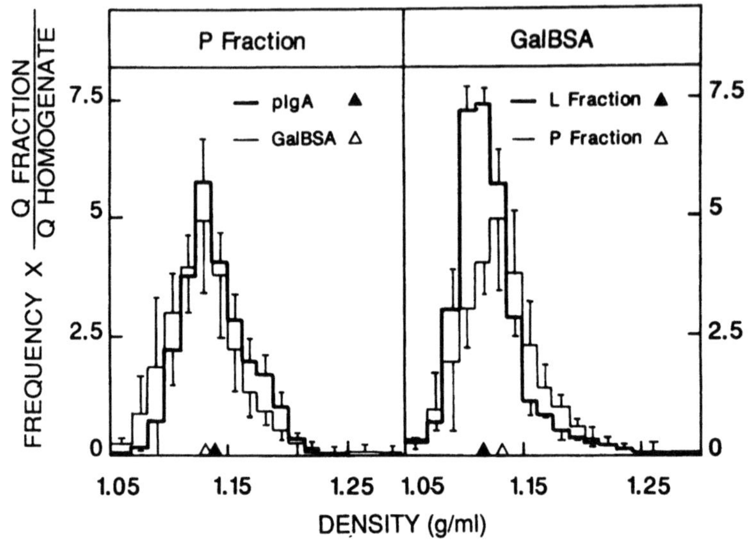

FIGURE 3. Resolution of endosomes in linear sucrose gradients. The left panel shows that the density distributions of GalBSA and pIgA in the P fraction isolated at 3 min after injection are not distinguished in linear sucrose gradients, indicating that the two ligands are associated with physically similar, if not the same, primary endosomes. The right panel demonstrates that between 3 and 10 min after injection, the distribution of injected GalBSA is shifted from early, low-density endosomes (P fraction, around 1.13 g/ml in density) to late, very-low density endosomes (L fraction including the fluffy layer, around 1.11 g/ml in density). In contrast, the distributions of pIgA and $F(ab')_2$ antirat secretory component remain essentially identical from 3 to 20 min after injection (P fraction, 1.13 g/ml in density; not shown). Typical recoveries in the gradients ranged between 96 and 103%. Distributions are means ± standard deviations of five experiments, normalized according to Leighton *et al.* (1968), and are expressed with respect to the homogenate. Arrows indicate the median density of distributions. Adapted from Limet *et al.* (1985), by copyright permission of the *Biochemical Journal of Biochemistry*.

Conventional density gradients thus provide useful information. Because of their limited potential in resolving the various endosomal compartments, however, alternative approaches are being explored, based on *in vivo* or *in vitro* density perturbations.

2.4. *In Vivo* Density Modifications

In vivo density modifications take advantage of three classes of agents or procedures, aiming at producing controlled perturbations at selected organelles of the endocytic apparatus in the living cell. A first possibility is to delay the transfer between endosomes of different equilibrium densities. The class I of perturbants includes intermediate temperatures (15–20°C block; Dunn *et al.*, 1980), microtubule-depolymerizing agents (Kolset *et al.*, 1979; Robert *et al.*,

1985; Bergeron *et al.*, 1986; Perez *et al.*, 1988; Clarke and Weigel, 1989), and local anesthetics (Tolleshaug *et al.*, 1982).

A second possibility is to alter the density of the organelle itself. This is exemplified by the overload of multivesicular bodies by lipoproteins upon estrogen treatment (Hornick *et al.*, 1985), the accumulation of poorly digested or nondigestible compounds like Triton WR 1339–lipoprotein complexes in the lysosomes of hepatocytes (Limet *et al.*, 1985), or colloidal gold and invertase in the lysosomes of sinusoidal cells (Henning and Plattner, 1974; Jadot *et al.*, 1986). Endocytic tracers still reach lysosomes preloaded with invertase or Triton WR 1339 (Jadot *et al.*, 1986; Janicot and Desbuquois, 1987). In rat liver, the class II of perturbants appears to be more generally applicable to lysosomes than to endosomes. In particular, *in vivo* density shift by colloidal gold, which offers great potentials for the study of endosomal compartments from cultured cells (Beaumelle *et al.*, 1990), seems less promising for rat hepatocytes, because a moderate uptake of gold probes does not result in appreciable displacement of endosomes (Wall and Hubbard, 1985) and possible misaddressing due to colliodoplexy of sinusoidal cells.

A third possibility is to exploit differences in vacuolar pH, so as to affect traffic like the class I of perturbants, and/or to alter the density of the organelles like the class II of perturbants, by producing their osmotic swelling. The class III of perturbants includes the acidotropic weak bases like ammonium chloride (Harford *et al.*, 1983) and chloroquine (Limet *et al.*, 1985; Bergeron *et al.*, 1986; Clarke and Weigel, 1989), and the proton ionophore monensin (Clarke and Weigel, 1989). The effects of the class II (Triton WR 1339) and the class III of perturbants (chloroquine) are exemplified in Figure 4.

In retrospect, interpretation of experiments based on perturbants is difficult because most of them have pleiotropic effects, and because physical alterations of particles may also change their functional properties. For example, the acidotropic weak bases also decrease the number of surface receptors by increasing their intracellular sequestration, delay vesicular traffic (as illustrated in Figure 4), and inhibit lysosomal proteolysis (reviewed by Dean *et al.*, 1984; Zijderhand-Bleekemolen *et al.*, 1987). Furthermore, the protease inhibitor leupeptin, expected to only block proteolysis in lysosomes, also acts upstream by retarding transport to these organelles (Tolleshaug and Berg, 1981). Hence, to provide less ambiguous results, alternative procedures have been designed in which the organelles are unaffected *in vivo*, but particles are subsequently submitted to density modifications *in vitro*.

2.5. *In Vitro* Density Modifications

In vitro density modifications are based on the endocytic uptake of enzymes by the living cell, followed by a cytochemical reaction on isolated subcellular

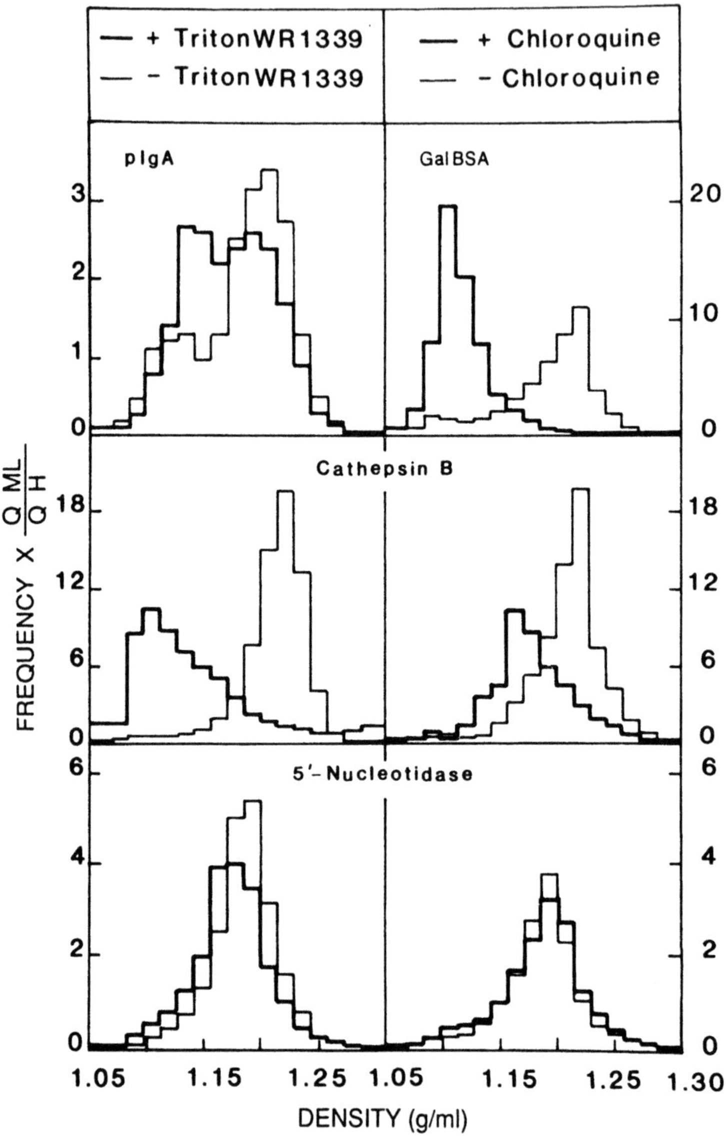

FIGURE 4. *In vivo* density modifications of rat hepatocyte endocytic organelles: effect of drugs. Rats were pretreated either with Triton WR 1339 at left (0.9 mg/g body mass intraperitoneally, 4 days before the experiment), chloroquine at right (50 μg/g body mass intraperitoneally, 30 min before the experiment), or received saline for control. In each group, rats were sacrificed at 20 min after the intravenous injection of pIgA at left (5 μg/g body mass), or GalBSA at right (0.5 μg/g body mass). Combined ML fractions (including the fluffy layer) were equilibrated in linear sucrose gradients. Distributions show representative experiments, normalized after Leighton *et al.* (1968), and are expressed with respect to the homogenate.

fractions. Cytochemistry leads to the accumulation of a dense reaction product in the particles containing the enzyme and allows their retrieval based on the selective density shift.

This approach offers several advantages (Courtoy *et al.*, 1988): (1) the enzyme can be introduced into predefined endocytic organelles, usually by its conjugation to an appropriate ligand; (2) under correct conditions, enzyme-tracer conjugates behave as the native tracer (see Figure 1 and left panel of Figure 5); (3) the labeled organelles and particles can be identified by ultrastructural cytochemistry in the intact cells and subcellular fractions, respectively; (4) before the cytochemical reaction is applied to the isolated fraction, the equilibrium density of the particles is unaffected, so that they can first be recovered at their expected natural position; (5) after cytochemistry, particles are placed in a second gradient at much heavier densities, in a region that had been cleared of contaminants by the first gradient; and (6) the catalytic activity of the enzyme combined with the limited volume to be filled by the reaction product result in a sensitive procedure; in our experience, about ten copies of conjugate are sufficient to displace an endosome.

In vitro density modifications have used peroxidase (Figure 5, right panel; Courtoy *et al.*, 1984) or cholinesterase (Helmy *et al.*, 1986). In the study of the rat liver, they have been applied to the endocytic coated vesicles (Helmy *et al.*, 1986), to the endosomal compartments of the degradation pathway (Quintart *et al.*, 1984; Kay *et al.*, 1984; Janicot and Desbuquois, 1987; Kennedy and Cooper, 1988), and to the endosomal compartments of the transcytotic pathway (Quintart *et al.*, 1989).

Density-shift procedures are specific as long as the enzyme is limited to the interior of the endocytic vesicles (Courtoy *et al.*, 1984). If peroxidase contaminates the medium, specificity can nevertheless be restored with thimerosal (Courtoy, 1989). Their greatest potential is to resolve both particles of indistinguishable physical properties, such as the various transcytotic compartments (Quintart *et al.*, 1989) and endocytic from exocytic coated vesicles (Helmy *et al.*, 1986). The cytochemical reaction product may limit the subsequent bio-

◄

In the absence of perturbation, ligand recovered with the ML fraction (27% of pIgA, but 66–80% of GalBSA of the homogenate) largely codistribute with lysosomes (cathepsin B). Upon Triton WR 1339 accumulation, there is a marked decrease of the equilibrium density of lysosomes, but only a small and partial displacement of pIgA, indicating that this minor component of the distribution of pIgA at late times after injection is distinct from lysosomes. Chloroquine accumulation causes the osmotic swelling of lysosomes and produces a moderate decrease of their equilibrium density (of about 0.050 g/ml), but causes a major decrease in the density distribution of GalBSA (about 0.100 g/ml). Since the distribution of ligand and lysosomes are dissociated by chloroquine, the effect of this compound cannot be simply ascribed to a displacement of lysosomes, and is attributed to a delay in vesicular traffic. Adapted from Limet *et al.* (1985), by copyright permission of the *European Journal of Biochemistry*.

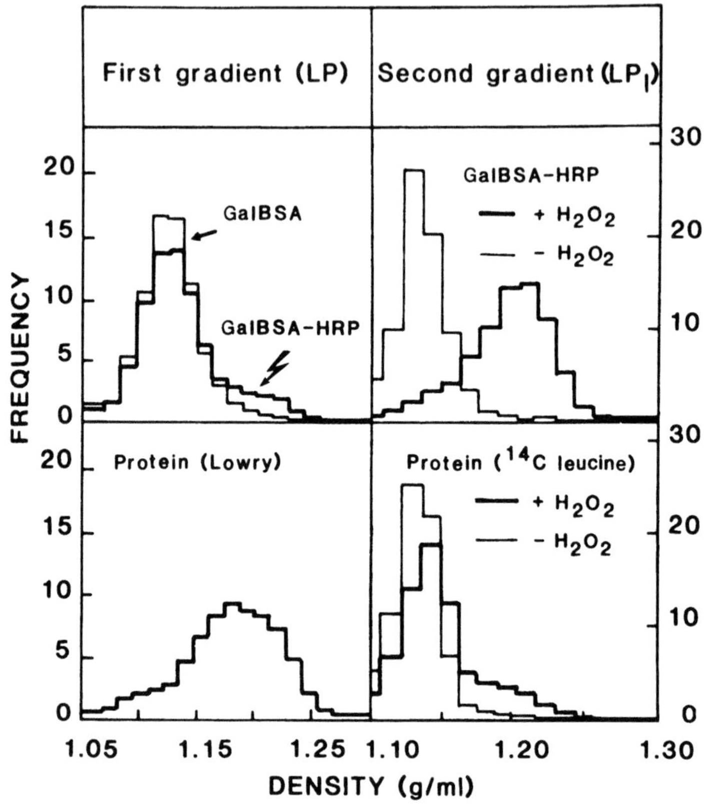

FIGURE 5. *In vitro* density modification of rat hepatocyte endosomes: the DAB-induced density shift. At left, a rat was sacrificed at 10 min after the intravenous injection of GalBSA-HRP (1 μg/g body mass) and the LP fraction was equilibrated in a first linear sucrose gradient (from 1.05 to 1.30 ml in density). Particles recovered between 1.11 and 1.13 g/ml in density (LP_1 pool) were incubated in 5.5 mM diaminobenzidine, with 11 mM H_2O_2 (test, thick lines) or without H_2O_2 (control; thin lines) and equilibrated again in a second linear sucrose gradient (from 1.10 to 1.30 ml in density). In the presence of H_2O_2, a major shift of the ligand distribution is observed, with the concomitant displacement of a minor fraction of liver protein. Reproduced from Courtoy *et al.* (1988), by copyright permission of Alan R. Liss.

chemical analysis and could especially inhibit the activity of marker enzymes present within shifted particles (Ajioka and Kaplan, 1987). In our biochemical studies, enzyme inactivation only required minor corrections (Courtoy *et al.*, 1988). Inactivation can, however, be purposely amplified and used to measure quenched activities attributable to endosomes, insofar as quenching is complete, reproducible, and specific to the enzyme-containing particles (Stoorvogel *et al.*, 1989, 1991).

Application of the density-shift technique to rat-liver particles has indicated that apical plasma-membrane markers are largely absent along the degradation pathway (Figure 6; Quintart *et al.*, 1984), but are acquired by endosomes at the late stages of transcytosis (Quintart *et al.*, 1989). The procedure also proved quite useful to resolve endosomes from Golgi-derived particles (Kay *et al.*, 1984).

2.6. Use of Surface Properties

Besides using differences in size and density (natural or modified), it is possible to resolve the various endosomal compartments of rat hepatocytes on the basis of surface properties. Nonspecific differences in electrical charge at the particle surface can be exploited by free flow electrophoresis. Indeed, endosomes share with lysosomes a faster movement toward the anode in an electrical field, as compared with the other subcellular particles including Golgi-derived elements (Evans and Flint, 1985; Fuchs *et al.*, 1989a). After retrieval of a mixed population of endosomes and lysosomes by free flow electrophoresis, the two types of particles are easily separated by isopycnic centrifugation. However, at the time this review is completed, the potential of free flow electrophoresis to resolve the rat hepatocyte endosomal compartments from one another remains to be fully explored. In view of the complexity of the problem, alternative methods based on nonspecific properties such as phase partition (Gierow *et al.*, 1986) should not be overlooked.

Specific differences in surface properties are based on the immunological recognition of unique antigens at the cytoplasmic aspect of isolated particles. Immunoisolation has brought major achievements in cell-free fusion assays involving particles isolated from cultured cells, because appropriate antigens can be selectively introduced in predefined endosomes (reviewed by Gruenberg and Howell, 1989). With regard to rat-liver particles, immunoisolation has permitted researchers to distinguish endosomes that do or do not carry the galactose-specific receptor (Mueller and Hubbard, 1986). The former are small particles (100 nm in diameter and thus expected to sediment in the P fraction) and have a low equilibrium density (1.12 g/ml in sucrose gradients). The latter are large particles (400 nm in diameter, expected to sediment in the L fraction), have a very, very low equilibrium density (down to 1.09 g/ml), and include multivesicular endosomes.

3. COMPOSITION AND FUNCTIONS

3.1. Intrinsic Membrane Polypeptides

We still know very little about the polypeptide membrane composition of the various endosomal compartments of rat hepatocytes and no specific marker of endosomes has been identified so far in these cells. It is clear that rat hepatocyte endosomes contain receptors (Weigel and Oka, 1983), whose numbers vary depending on exposure to ligands (Desbuquois *et al.*, 1982; Dunn and Hubbard, 1984; Lai *et al.*, 1989a,b), and which are not equally distributed within all endosomal compartments (Mueller and Hubbard, 1986; Stoorvogel *et al.*, 1989). Plasma membrane marker enzymes are also unequally distributed in all of them (see section 2.5). It is, however, likely that the general membrane composition of at least the primary and secondary endosomes accessible from the basolateral and apical surface largely resembles that of corresponding plasma membrane domains, and that important overlaps exist between all endosomal compartments (Evans and Hardison, 1985).

Endosomal preparations contain an electrogenic proton pump of the vacuolar type, operationally defined as the NEM-sensitive monensin-stimulated Mg^{2+}-ATPase (Hornick *et al.*, 1985; Evans and Flint, 1985; Fuchs *et al.*, 1989a). Along the degradation pathway in rat hepatocytes, late endosomes are more acidified than early endosomes (Fuchs *et al.*, 1992). To what extent possible differences in pH along the degradation pathway are accounted for by the counteraction of the electrogenic $3Na^{+}/2K^{+}$,ATPase, by a chloride conductance, or still another mechanism is not clear so far (Cain *et al.*, 1989; Fuchs *et al.*, 1989b). Interestingly, transcytotic endosomes are not acidified (Fuchs *et al.*, 1992). Other probable polypeptides of the endosomal membrane include pumps and channels for other ions such as Cl^{-} and Ca^{2+} (Fuchs *et al.*, 1989a; Diaz *et al.*, 1989), and proteins interacting with microtubules (Vergés *et al.*, 1992).

It remains to identify, in each set of the various endosomal compartments, the unique intrinsic membrane polypeptides that could be used as molecular markers of their maturation and act as recognition signals for their specific addressing among the complex system of distinct acceptor organelles. Indeed, in the polarized hepatocyte, various types of endosomes converge along the microtubular network into the pericanalicular region, which concentrates all types of acceptor particles, namely lysosomes, (specific target of the multivesicular endosomes), the trans-Golgi reticulum (acceptor of some recycling endosomes), and bile canaliculi (the ultimate station of transcytotic endosomes). The specific fusion with lysosomes of the endosomes of the degradation pathway has been

FIGURE 6. Analysis of the endosomes of the degradation pathway by the DAB-induced density shift. The upper panel compares the unmodified density distribution of GalBSA-HRP in the LP fraction from a rat sacrificed at 10 or 20 min after ligand injection. At 20 min, no intermediate can be detected between the low-density peak (endosomes) and the high-density peak which largely overlaps with N-acetyl-β-glucosaminidase, indicated by the broken line (lysosomes). The lower panel shows the effect of DAB cytochemistry (thick line), or a mock treatment (thin line), on the density distributions of the ligand and the small amount of lysosomal marker enzyme present in the low-density peak. DAB cytochemistry causes the almost complete displacement of the tracer but has no detectable effect on the lysosomal marker, at either 10 or 20 min after injection. Differences in ligand displacement result from unequal enzymatic activities between GalBSA-HRP preparations. Partially reproduced from Courtoy *et al.* (1984) and Quintart *et al.* (1984), by copyright permission of the Rockefeller University Press.

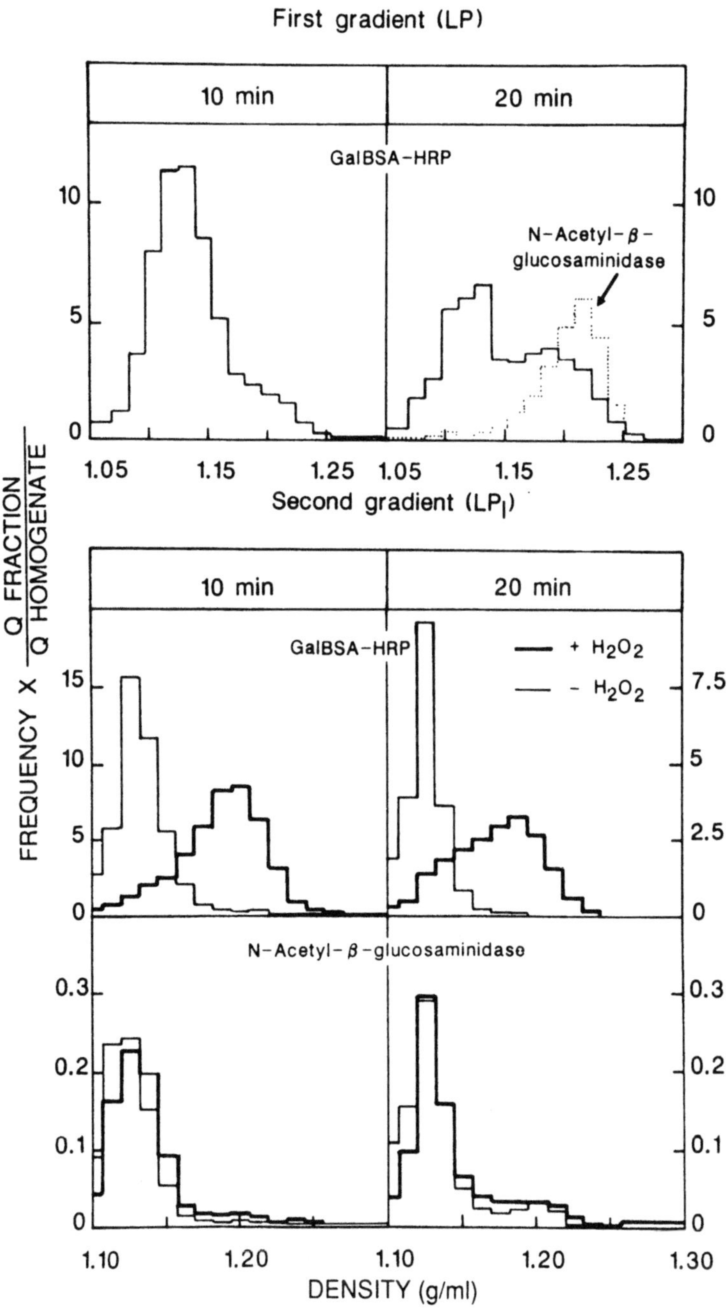
First gradient (LP)
10 min
20 min
GalBSA-HRP
N-Acetyl-β-glucosaminidase
1.05
1.15
1.25
Second gradient (LPl)
FREQUENCY X Q FRACTION / Q HOMOGENATE
GalBSA-HRP
+ H2O2
- H2O2
N-Acetyl-β-glucosaminidase
DENSITY (g/ml)

demonstrated in a cell-free system established with rat-liver fractions (Mullock *et al.*, 1989). *In vitro* fusion of transcytotic endosomes with isolated canalicular membranes has also been reported, but its specificity was not demonstrated (Musil and Baenziger, 1988). A 108 kD antigen has recently been localized to late transcytotic endosomes of rat liver (Sztul *et al.*, 1991).

There is currently great interest in the regulation of the endocytic apparatus by GTP-binding proteins (Mayorga *et al.*, 1989). Small isoprenylated GTPases of the rab family reversibly bind to and control the fusion of selected endosomal compartments of cultured cells (Chavrier *et al.*, 1990; Gorvel *et al.*, 1991; van der Sluijs *et al.*, 1991). Work in progress also shows involvement of large trimeric GTP-binding proteins in the degradation or the transcytotic pathway, including in rat liver. It is to be hoped that the corresponding "receptors" will be identified in the near future at the cytoplasmic surface of specific endosomal membrane domains of rat hepatocytes.

3.2. Membrane Lipids

The first evidence that the membrane of coated pits and endosomes is rich in cholesterol came from the use of cholesterol-complexing agents. At the electron microscope, the antibiotic filipin produces a corrugated membrane deformation of rat-liver coated pits, which is characteristic of a high cholesterol content, provided the stabilizing coat had been previously removed (Steer *et al.*, 1984). Another approach is the treatment of particles with appropriate concentrations of the detergent digitonin, followed by isopycnic centrifugation. This procedure has little effect on the particles derived from the secretory pathway, but causes an appreciable density shift of all plasma-membrane markers (Amar-Costesec *et al.*, 1974b), as well as of endosomes containing insulin (Smith and Peeters, 1982), asialoglycoprotein and asialoglycoprotein receptors (Wall and Hubbard, 1985), transferrin (Sibille *et al.*, 1986), and pIgA (Limet *et al.*, 1982).

The cholesterol/phospholipid molar ratio of endocytic coated pits and early endosomes purified from rat liver is around 0.80 (Evans and Hardison, 1985; Helmy *et al.*, 1986). This value is comparable to that obtained with purified plasma membrane (Thinès-Sempoux, 1972; Wibo *et al.*, 1981), but much higher than in particles derived from the Golgi (around 0.30; Wibo *et al.*, 1981; Helmy *et al.*, 1986), and particularly than in the outer mitochondrial membrane (less than 0.10; Henning, 1974; Wibo *et al.*, 1981). Conceivably, a similar cholesterol content might facilitate the mixing of early endosomes and the fusion of recycling endosomes with the plasma membrane, while preventing mixing with unappropriate organelles.

On the other hand, data are compatible with a decreasing gradient of cholesterol membrane concentration along the degradation pathway, from 0.80 in early endosomes (Evans and Hardison, 1985), to 0.69 in secondary endosomes (Evans and Hardison, 1985), 0.46 in multivesicular endosomes (Hornick *et al.*, 1985), and down to 0.27 at the membrane of lysosomes purified without overload of content (Henning, 1974).

3.3. Soluble Content

Due to the absence of ultrafiltration by the sinusoidal wall, the soluble content of primary endocytic basolateral vesicles in rat hepatocytes is essentially identical to the plasma, but it rapidly undergoes several modifications in subsequent compartments. First, the ionic composition is altered as a result of transmembrane fluxes or exchanges, exemplified by the secretion of protons into the vacuolar lumen. It is likely that several other ionic alterations do occur, insofar as pumps and channels are internalized from the cell surface (where they interact with an almost infinite space, containing a permanently replaced extracellular fluid) into the small endocytic organelles (where they face a very limited closed compartment, with no readily exchangeable content, except during the brief intervals between fusion and fission of the vesicles).

Second, a large number of constituents initially adsorbed at the cell surface to receptors (or to less specific binding sites) are released into the endosomal lumen, because of the generation of a dissociative microenvironment. Noticeable exceptions are transferrin (Dautry-Varsat *et al.*, 1983), pIgA (Limet *et al.*, 1982; Schiff *et al.*, 1986), and a fraction of EGF (Dunn and Hubbard, 1984; Lai *et al.*, 1989a). It should be reminded that the actual kinetics of dissociation in endosomes *in situ* cannot be simply extrapolated from experiments carried out at the cell surface in environments mimicking the presumptive endosomal milieu, since dissociation should be opposed by the rapid increase of free concentration in the limited vacuolar space.

Third, if dissociation occurs at stages posterior to the divergence of the transcytotic pathway, acid-sensitive tracers like asialoglycoproteins will be almost absent from the distal transcytotic endosomes, but relatively enriched in lumen of the later endosomal compartments of the degradation pathway. This difference in content is exploited to selectively retrieve the latter particles by density shift procedures using GalBSA-HRP.

Finally, although newly synthesized lysosomal enzymes are delivered or activated in late endosomes (Brown *et al.*, 1986; Geuze *et al.*, 1988; Casciola and Hubbard, 1991; Rijnboutt *et al.*, 1991), the actual proportion of acid hydrolase activity in all preparations of rat liver low-density endosomes is extremely low (Quintart *et al.*, 1983, 1984; Hornick *et al.*, 1985; Fuchs *et al.*, 1989a), because half-life of hydrolases in secondary lysosomes is very long.

3.4. Posttranslational Modifications

Three posttranslational modifications occur in rat-liver endosomes: (1) phosphorylation (and dephosphorylation?); (2) proteolysis and (3) reduction (and interchange?) of disulfide bridges.

3.4.1. Phosphorylation (and Dephosphorylation?)

Several receptors at the basolateral membrane of rat hepatocytes are phosphorylated on their cytoplasmic domain. This involves either serine residues, in the galactose-specific lectin (Takahashi *et al.*, 1985) and the secretory component (Larkin *et al.*, 1986; Casanova *et al.*, 1990), or tyrosine residues, in the receptor for insulin (Khan *et al.*, 1989) or EGF (Lai *et al.*, 1989b).

The injection of EGF triggers EGF receptor autophosphorylation activity in endosomal preparations, and its kinetics is coincident with that of downregulation of surface receptor activity (Lai *et al.*, 1989b). Hence, phosphorylation could be involved in the specific regulation of endocytosis and/or intracellular retention of plasma-membrane constituents. Conversely, dephosphorylation would permit recycling. In addition, the translocation to the perinuclear region of endosomes carrying activated receptors lends support to the hypothesis that endocytosis participates in the intracellular signaling of polypeptide hormones and growth factors (Bergeron *et al.*, 1985; Wada *et al.*, 1992).

3.4.2. Proteolysis in Endosomes

Endosomes of rat hepatocytes do not degrade most proteins to completion (Debanne *et al.*, 1984; Limet *et al.*, 1985; Casciola and Hubbard, 1991), but are capable of fully degrading internalized insulin, and this activity could entirely account for the termination of the hormonal signal. The activity of the endosomal insulin-glucagon protease(s) is optimal at mildly acidic pH, is inhibited by bacitracin but insensitive to pepstatin and leupeptin (Pease *et al.*, 1987; Doherty *et al.*, 1990) and is most active at a pH of 6.0 (Desbuquois *et al.*, 1992). Because of the brief delay between endocytosis and insulin degradation (Bergeron *et al.*, 1986), this protease activity is tentatively attributed to the secondary endosomes (see section 4.2). In contrast to other cell types, rat-liver endosomes do not degrade EGF (Lai *et al.*, 1989a, compared to Schaudies *et al.*, 1987).

Partial proteolysis in rat-hepatocyte endosomes probably participates in the maturation of the lysosomal matrix enzymes, which are first synthesized as transmembrane (lysosomal acid phosphatase) or membrane-bound (cathepsin D) precursors, transiently integrated to the cell surface, and finally addressed via endosomes to the lysosomes (for a review, see Von Figura and Hasilik, 1986; Lippincott-Schwartz and Fambrough, 1986). Activation of cathepsin D involves two referential proteolytic steps, one in endosomes and the other in lysosomes (Diment *et al.*, 1988; Rijnboutt *et al.*, 1991).

3.4.3. Reduction of Disulfide Bridges in Endosomes

Reduction of disulfide bridges has been demonstrated in the endosomes of rat hepatocytes. As already mentioned, injected antisecretory component $F(ab')_2$

antibodies are efficiently transferred into bile. When recovered after transcytosis, the probe is identical to Fab′ produced by mild *in vitro* reduction, but remains as $F(ab')_2$ when simply incubated with bile (Limet *et al.*, 1985). During transcytosis, an increasing proportion of the tracer isolated with low-density microsomes correspond to Fab′, indicating that reduction can be initiated in the transcytotic endosomes and completed at the canalicular membrane (Courtoy *et al.*, 1992; Figures 7 and 8).

Injected cholera toxin is also efficiently taken up by the liver, apparently via clathrin-independent invaginations, and is transferred at 10 min after injection to endosomes that are also accessible to asialoglycoproteins. There, its A subunit is slowly and marginally reduced into the A_1 fragment (Janicot and Desbuquois, 1987). This indicates that reduction of disulfide bridges can also occur in late

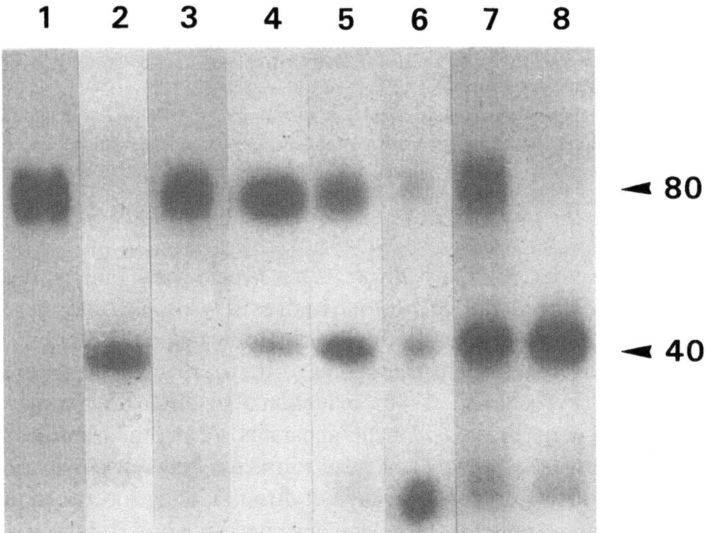

FIGURE 7. Reduction of $F(ab')_2$ antirat secretory component during transcytosis. Tracer was analyzed by SDS-PAGE carried out under nonreducing conditions, followed by autoradiography. Lane 1: injected ($F(ab')_2$ (80 kD); lane 2: tracer recovered in bile after transcytosis; lane 3: tracer incubated *in vitro* in undiluted bile; lane 4: tracer recovered at 2 min after injection in the low-density pool (1.095–1.135 g/ml in density) from microsomes of the liver homogenate (preparations corresponding to 23%, 38% and 70% of injected dose, respectively); lane 5: tracer at 10 min after injection in the low-density pool from microsomes of the liver homogenate (corresponding to 30%, 45%, and 80% of injected dose, respectively); lane 6: tracer in a total extract of cultured rat hepatocytes, after a 2 hr uptake at 37°C; lane 7: partial *in vitro* reduction of $F(ab')_2$ by 1 hr incubation at 37°C in 0.8 mM mercaptoethylamine, in 10 mM acetate, pH 5; lane 8: complete *in vitro* reduction into Fab′ (40 kD) by a similar incubation as at lane 7, except for 4 mM mercaptoethylamine.

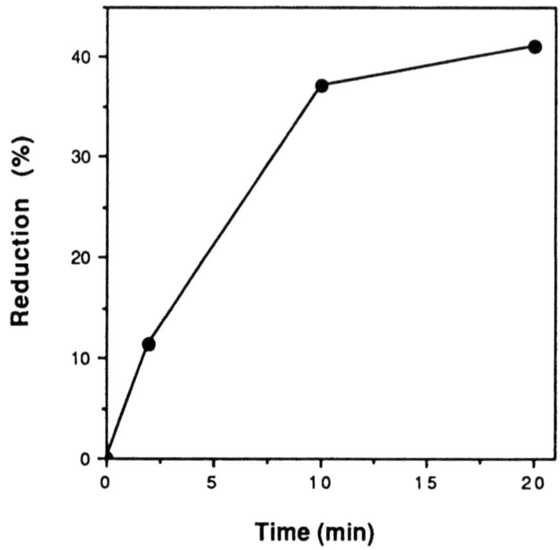

FIGURE 8. Quantitation of reduction of F(ab')$_2$ in rat endosomes. Gels similar to those illustrated in Figure 7, lanes 4 and 5, have been scanned by laser densitometry and the fraction reduced into Fab' was expressed as percentage of total radioactivity. Values are means of three experiments at each interval after injection.

endosomes of the degradation pathway of rat hepatocytes. Finally, recent evidence suggests that reduction of disulfide bridges in lysosomes play an important role in antigen presentation by macrophages (Collins *et al.*, 1991).

Although divergent interpretations have been proposed on its exact subcellular location (Feener *et al.*, 1990), reduction of disulfide bridges may thus be an important property of the endocytic apparatus. Along the transcytotic pathway, it may be essential in covalent bond formation between polymeric immunoglobulins and their receptor. Exchange of disulfide bridges between pIgA and the secretory component has been demonstrated *in vitro* (Lindh and Björk, 1976), and is promoted by crude microsomal fractions (probably including endosomes; Murkofsky and Lamm, 1979). Because no free thiol is available on the native secretory component (Lindh and Björk, 1976), reduction of some intramolecular disulfide bridges of this molecule may be a prerequisite for the establishment of intermolecular disulfide bridges with pIgA. The formation of a covalent bond between pIgA and its receptor would further stabilize the complex for further transcytosis and after secretion.

3.5. Sorting, Segregation, and Addressing

Sorting in endosomes includes the separation of membrane constituents with different destinations (membrane sorting), as well as separation of membrane and content (phase sorting; Courtoy *et al.*, 1985). It is likely, but not demonstrated, that the major endocytic activity in rat hepatocytes *in situ* occurs at the basolateral membrane. Subsequent membrane sorting from basolateral endosomes is required to allow for the specific retrieval of recycled receptors to the basolateral membrane, as well as for the correct endocytic addressing of apical membrane constituents including the secretory component (Bartles *et al.*, 1987).

Immunoelectron microscopic studies provide clear evidence that constituents are clustered in different protruding extensions of the tubulospherical endosomes, indicating that after mixing, they are selectively included into either recycling or transcytotic vesicles (Geuze *et al.*, 1984; 1987). If the flux of membrane to the recycling pathway overwhelms that following transcytosis (see section 4.4), an apical signal operating at the level of tubulospherical endosomes would provide sufficient information, since the bulk membrane flow could then be carried by default back to the basolateral domain. This attractive hypothesis and its underlying molecular mechanisms remain to be demonstrated.

For the lysosomal membrane constituents that transit at the cell surface, addressing to the lysosomes could also result from their selective insertion in the endosomes of the degradation pathway, although a net accumulation in the lysosomal membrane could be explained by other mechanisms (discussed in Draye *et al.*, 1988).

Phase sorting is easily achieved by the differential dissociation of some membrane-bound constituents. This is probably not an all-or-none phenomenon, as adsorbed constituents may exhibit different vulnerabilities to the endosomal environment, which may itself vary along the various endocytic pathways (for a multistage sorting model, see Stoorvogel *et al.*, 1989). A quantitative model in which the segregation of fluid and membrane in rat hepatocyte endosomes can be simply accounted for by the geometry and dynamics of transient tubulospherical endosomal units has been presented elsewhere (Courtoy, 1991).

4. A STRUCTURAL MODEL OF RAT HEPATOCYTE ENDOSOMES

It is useful at this stage to integrate morphological and cell fractionation data on the endocytic apparatus of rat hepatocytes, in light of the current knowledge on endocytosis, to propose a structural model, and to mention some unsolved issues. This is the purpose of Figure 9, which (1) describes in realistic respective dimensions the complexity of endosomes in rat hepatocytes; (2) provides a clear definition of the functional compartments, based on appropriate tracers, stage of processing, and physical properties; and (3) indicates their possible relations with each other, as well as with other subcellular organelles and plasmalemmal domains. We will start by the sequential endosomal compartments of the degradation pathway, not necessarily because it is the most important, but because it has

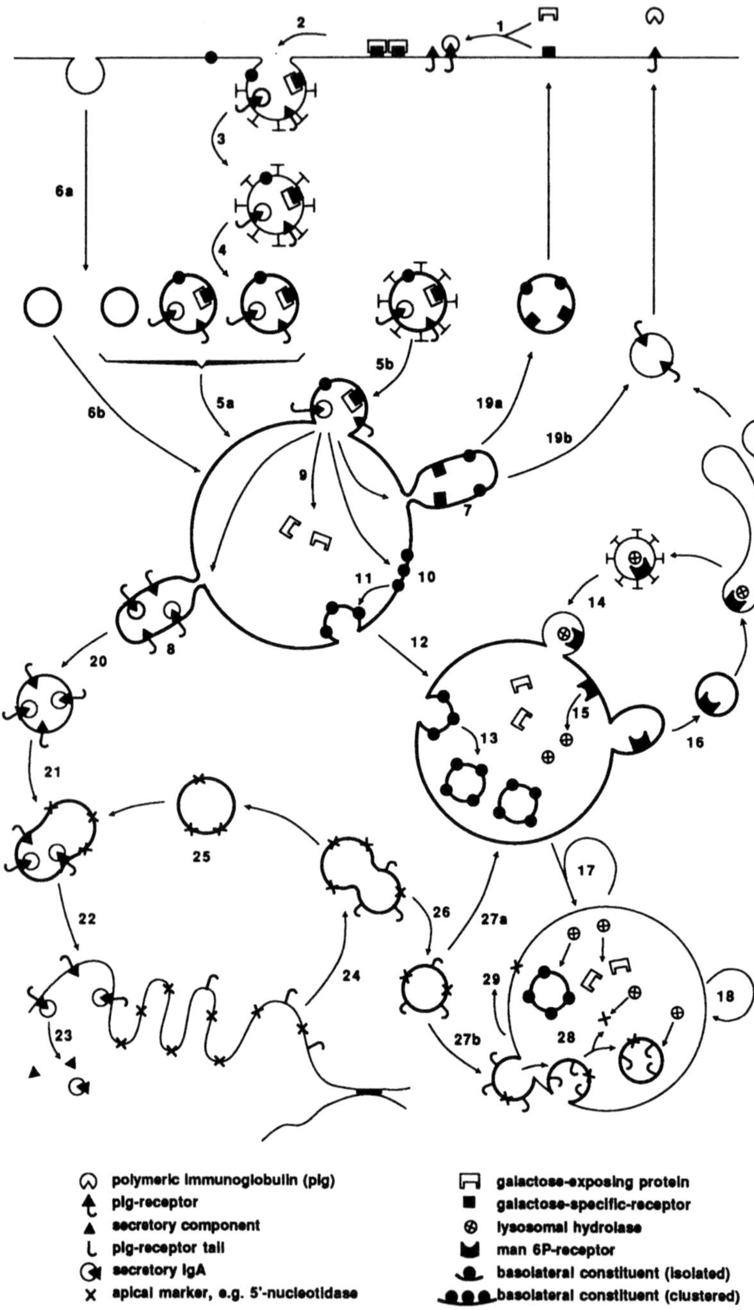

FIGURE 9. A structural model of endosomes in rat hepatocytes. Soluble tracers are indicated by open symbols, and membrane receptors or markers by filled symbols. Endosomes are drawn with thick lines. In rat hepatocytes, the Golgi complex, multivesicular endosomes and lysosomes are all concentrated around bile canaliculi. For further explanation, see text.

been the most studied, for the convenience of its labeling by fluid-phase tracers and most ligands.

4.1. Primary Basolateral Endosomes

Patches of basolateral (sinusoidal) membrane, including the various ligand-receptor complexes formed and clustered at the cell surface (step 1 of Figure 9; all step numbers refer to this figure) and some unoccupied receptors, are internalized through coated pits (step 2), either by selective inclusion in preformed clathrin-covered domains or by local clathrin accretion on clustered receptors (Brodsky, 1988). Coated pits generate coated vesicles (step 3), the vast majority of which uncoat (step 4). We here define both uncoated vesicles derived from clathrin-coated pits and vesicles derived from clathrin-independent endocytic pits as primary endosomes.

Primary endosomes of basolateral origin are about 100 nm in diameter and are located at the cell periphery, preferentially around the junction of the sinusoidal domain with the lateral domain (Geuze *et al.*, 1987). The corresponding particles are essentially recovered in the microsomal fraction and equilibrate around 1.13 g/ml in sucrose gradients. They are probably equivalent to the particles enriched in fraction DN2 of Evans and Flint (1985). Part of the receptor-positive endosomes should also be recovered in these preparations (Mueller and Hubbard, 1986).

Primary endosomes contain a mixture of ligands and receptors destined to the various subsequent pathways, and ligand-receptor complexes remain mostly, if not entirely, coupled at this stage (Kindberg *et al.*, 1984). The stability of complexes in primary endosomes could result from the counteracting effect of free ligand released in their very small volume (about 0.5×10^{-18} liters) and from a lack of acidification (Fuchs *et al.*, 1992). It is unknown whether the absence of acidification reflects the lack of proton pumps, antagonism by the electrogenic $3Na^+/2K^+$,ATPase, an unusually high membrane conductance to protons (Fuchs *et al.*, 1989a), the large initial buffering capacity of membrane glycocalyx and internalized fluid (identical to that of plasma, i.e. 25–35 meq/liter, in the case of hepatocytes), or another unsuspected reason.

4.2. Secondary (Tubulospherical) Endosomes

Uncoated vesicles rapidly coalesce into larger endocytic vesicles, referred to hereafter as secondary endosomes (step 5a). Alternatively, uncoated vesicles, together with a few partially coated vesicles, fuse with preexisting large endosomes (step 5b) (Griffiths *et al.*, 1989). Vesicles derived from clathrin-independent endocytic pits (reviewed by Van Deurs *et al.*, 1989) are also believed to mix with the large endosomes (steps 6a and 6b) (Janicot and Desbuquois, 1987; Tran *et al.*, 1987). Secondary endosomes thus represent a station of intracellular mixing of all internalized constituents, whatever their nature or their receptor, their kinetics of internalization, and their port of entry. These particles are referred to by others as early endosomes (Hubbard, 1989), CURL (Geuze *et al.*, 1983) or simply endosomes (de Chastellier *et al.*, 1987).

Secondary endosomes sediment mostly in the light mitochondrial fraction studied by Quintart *et al.* (1983) and are probably enriched in the fluffy layer. They equilibrate around the density of 1.11 g/ml in linear sucrose gradients (Quintart *et al.*, 1983, 1984; Limet *et al.*, 1985). These particles probably correspond to fraction DN1 of Evans and Flint (1985) and to peak B of Kennedy and Cooper (1988). Another set of receptor-positive endosomes should be found among secondary endosomes.

The geometries involved mean that coalescence of numerous small vesicles into larger structures generates excess membrane, if there is no major inflow of water from the cytosol into the lumen (Duncan and Pratten, 1977). Excess membrane is accommodated by its retrieval into narrow tubular extensions at the expense of a comparatively smaller fraction of internalized fluid. Tubulospherical endocytic vesicles, made of a central vesicle of about 500 nm in diameter connected to several long and narrow tubules of about 50 nm in diameter, are evident in fixed tissue and are maintained throughout subcellular fractionation (Quintart *et al.*, 1983).

Kinetics of transfer of ligand into tubulospherical endosomes (from 2 to 10 min) correspond with that of uncoupling of galactose-exposing proteins from their corresponding liver lectin (step 9) (Harford *et al.*, 1983; Kindberg *et al.*, 1984). Immunogold labeling is consistent with this interpretation (Geuze *et al.*, 1987). Uncoupling of various ligand-receptor complexes is attributed to a pH drop (Harford *et al.*, 1983; Mellman *et al.*, 1986) and possibly changes in the concentration of divalent cations (Diaz *et al.*, 1989). Acidification could enhance the degradation of dissociated insulin, which starts after a delay of a few minutes (Bergeron *et al.*, 1986; Doherty *et al.*, 1990).

Once internalized, membrane-bound constituents will thus be discriminated according to their differential response to the environment of secondary endosomes (phase partition). Furthermore, the tubulospherical geometry is particularly well suited for the subsequent segregation of membrane and content. Thus, secondary endosomes that collect, sort, and segregate virtually all kinds of internalized material serve as the central sorting station along the endocytic apparatus, for which we coined the term *dianemosome* (Courtoy *et al.*, 1985).

4.3. Tertiary (Multivesicular, Late) Endosomes

The central portions of tubulospherical endosomes will mature on their own to become multivesicular endosomes, the tertiary (late) endosomes committed to the degradation pathway (step 12). The generation of a multivesicular endosome probably involves several changes. First, as in other cell types (Roederer *et al.*, 1987; Schmid *et al.*, 1988), internal pH drops further, possibly because recycling endosomes have removed the $3Na^+/2K^+$,ATPase, which counteracted the electrogenic H^+,ATPase in the earlier endosomes (Fuchs *et al.*, 1992). Multivesicular endosomes of rat liver have been shown to possess an electrogenic proton pump (Van Dyke *et al.*, 1985).

Second, the endosomal lumen seems to become more electron-dense. The apparent concentration of content would reflect the exclusion of water and solute, a process expected to cause a progressive increase of the density of the particle. This has been observed *in vitro* with endosomes isolated from cultured fibroblasts (Murphy, 1989), but not with late endosomes of rat liver (Mullock *et al.*, 1989).

Third, there is a concomitant generation of intraparticular vesicles by inward budding and vesiculation of the limiting membrane, which creates the typical multivesicular endosomes. This represents another way for the disposal of excess membrane resulting from the loss of water (Duncan and Pratten, 1977). The behavior of EGF-receptor complexes suggests that membrane constituents destined for degradation are preferentially clustered on the inward budding patches (steps 10 and 11), which subsequently generate the intraparticular vesicles (step 13) (McKanna *et al.*, 1979).

Fourth, late endosomes of the degradation pathway are supplemented by newly synthesized lysosomal enzymes carried with mannose-6-phosphate receptors by Golgi-derived coated vesicles (step 14) (Lemansky *et al.*, 1987). Enzyme-receptor complexes are believed to dissociate in multivesicular endosomes or prelysosomes (step 15), and mannose-6-phosphate receptors to recycle back to the Golgi apparatus (step 16). The latter route might correspond to the pathway by which a small proportion of internalized membrane constituents or adsorbed tracers reach the Golgi (Ottosen *et al.*, 1980; Regoeczi *et al.*, 1982; Snider and Rogers, 1985; Van Deurs *et al.*, 1988).

Late endosomes of the degradation pathway thus contain substrates, acidic pH, and newly synthesized lysosomal enzymes (prelysosomes). However, as discussed above, general proteolysis in those late endosomes of the growth-arrested rat hepatocytes is not significant, because of their very low content in lysosomal enzymes. It will really start only after the fusion with secondary lysosomes (step 17), resulting in a structure of intermediate density (Berg *et al.*, 1985). Endocytosed material is finally recovered at high density, following randomization of the lysosomal compartment (step 18) (Limet *et al.*, 1985; Ferris *et al.*, 1987; Deng and Storrie, 1988).

4.4. Recycling Basolateral Endosomes

In the secondary endosomes accessible from the sinusoidal domain, segregation of membrane constituents in the tubular extensions or the central body has been documented (Geuze *et al.*, 1987), but the molecular mechanisms involved are unknown. The contribution of the tubular extensions that generate vesicles recycling from the tubulospherical endosomes to unknown sites at the basolateral membrane (defined as recycling basolateral endosomes; step 7) probably overwhelms that of transcytotic vesicles (step 8). The extensive tubular endosomal network has been demonstrated in other cell types (Hopkins *et al.*, 1990; Tooze and Hollinshead, 1991). This seems necessary to account for the regurgitation of about 80%, compared to the transcytosis of only 2% of internalized fluid-phase tracers (Limet *et al.*, 1985; Scharschmidt *et al.*, 1986; Blomhoff *et al.*, 1989). A minor transcytotic pathway would also prevent randomization of basolateral and apical resident constituents.

Recycling vesicles constitute the third population of receptor-positive endosomes (Mueller and Hubbard, 1986). Recycling endosomes, which contain rapidly regurgitated fluid-phase tracers and recycled receptors (Oka and Weigel, 1983), return to unknown sites of the basolateral domain, either directly (step 19a), or after fusion with the trans-Golgi reticulum and incorporation in secretory vesicles (step 19b). In the rat liver, secretory vesicles constitutively carry plasma proteins (Kelly, 1985), together with newly synthesized basolateral and apical membrane constituents (Sztul *et al.*, 1983; Bartles *et al.*, 1987).

4.5. Transcytotic Endosomes

The complementary set of tubular extensions, derived from secondary endosomes accessible from the basolateral domain, concentrate newly synthesized apical membrane constituents into transcytotic endosomes (step 8). The receptor for polymeric immunoglobulins, is carried by transcytosis (steps 20–22) and becomes transiently associated with the canalicular membrane. After being cleaved off at the canalicular membrane, the ectoplasmic domain continues its journey into bile (step 23) (Musil and Baenziger, 1988). Other constituents use the transcellular route to become stable residents of the canalicular membrane (Bartles *et al.*, 1987) and possibly of an exchangeable pericanalicular membrane pool (steps 24 and 25) (Quintart *et al.*, 1989).

Early endosomes committed to transcytosis (early transcytotic endosomes, step 20) are labeled by polymeric pIgA at 5–10 min, are sedimenting in the microsomal fraction, and are indistinguishable in equilibrium density in sucrose gradients from primary endosomes (Limet *et al.*, 1985). The canalicular membrane markers 5′-nucleotidase and alkaline phosphodiesterase I are remarkably absent from the early transcytotic endosomes, but appear to become significantly associated with the late transcytotic endosomes, labeled at 20 min (step 22) (Quintart *et al.*, 1989).

A possible explanation is that early transcytotic vesicles fuse with a pool of vesicles derived from the canalicular membrane and rapidly recycle to it (steps

24 and 25). The delay between the arrival of pIgA in the pericanalicular region (at about 10 min after injection), and its demonstration in the canalicular lumen (after 20 min or more), may correspond to some unidentified maturation process and indicate that only the late transcytotic endosomes have acquired the competence to fuse with the canalicular membrane. Whether the 108 kD-antigen (Sztul *et al.*, 1991) is limited to a subset of transcytotic endosomes of rat hepatocytes has not been clarified so far.

4.6. Apical Endosomes

Cleavage of the receptor for polymeric immunoglobulins allows for the secretion of the free secretory component and of secretory IgA but leaves behind the 32–34-kD tail fragment in the canalicular membrane. If the tail were not disposed of, one might calculate that its accumulation would occupy a membrane area equivalent to the entire canalicular membrane after about one day.* Although there is evidence for disposal of the tail by selective vesiculation into the canalicular lumen (Solari *et al.*, 1989), one can as well envisage that it is removed with the primary apical endosomes (step 24) which generate the recycling apical pool (step 25), then segregated into vesicles destined for degradation (step 26) either via multivesicular endosomes (step 27a) or by direct fusion with secondary lysosomes (step 27b), discharged in the lysosomal lumen, and finally degraded (step 28). Indeed, the retrograde injection of peroxidase in the bile duct results in the labeling of multivesicular endosomes and lysosomes and remains essentially in this pericanalicular location (Matter *et al.*, 1969). Whereas transcytosis occurs in both directions in MDCK cells (Bomsel *et al.*, 1989), the existence of a reverse vesicular pathway, connecting the canalicular membrane to the basolateral domain (reverse transcytosis), has not been demonstrated so far, with the exception of pathological conditions increasing canalicular pressure (Matter *et al.*, 1969; Chen *et al.*, 1987).

ACKNOWLEDGMENTS. This work was supported by grants 2.4549.84-88 and 3.4570.84-88 of the Belgian Fonds National de la Recherche Scientifique, at which the author was senior research associate, as well as by the Belgian State—

*In the rat bile, the secretory component occurs mostly in the free form (80 kD), whose excretion amounts to 9 mg/day in the average 250 g-Wistar rat (Lemaître-Coelho *et al.*, 1977). Taking into account the total population of hepatocytes in the liver (1.33×10^9 cells; Weibel *et al.*, 1969), the excretion of free secretory component corresponds to 50×10^6 molecules/hepatocyte/day. Assuming that a globular tail of 32–34 kD occupies a membrane area of 10 nm^2 (De Roe *et al.*, 1987), its production would correspond to the addition of 500 μm^2/hepatocyte/day, a value comparable to the total canalicular membrane area (700–850 μm^2/hepatocyte) (Weibel *et al.*, 1969; Quintart *et al.*, 1984).

Prime Minister's office—Science Policy Programming (concerted actions, grant 82/87-39, and framework of interuniversity attraction poles, grant 7). It largely reflects stimulating discussions with Pierre Baudhuin and the close collaboration with the previous collaborators of our research unit. The outstanding technical help of Michèle Leruth-Deridder, Nicole Delflasse and Francisca N'Kuli-Pyrrhon, the typing assistance of Michèle Geist and the graphic art of Claude De Meyere are gratefully acknowledged. This work is dedicated to my family, as well as to ML and CF.

5. REFERENCES

Ajioka, R. S., and Kaplan, J., 1987, Characterization of endocytic compartments using the horseradish peroxidase-diaminobenzidine density shift technique, *J. Cell Biol.* **104:**77–85.

Amar-Costesec, A., Beaufay, H., Wibo, M., Thinès-Sempoux, D., Feytmans, E., Robbi, M., and Berthet, J., 1974a, Analytical study of microsomes and isolated subcellular membranes from rat liver. II. Preparation and composition of the microsomal fraction, *J. Cell Biol.* **61:**201–212.

Amar-Costesec, A., Wibo, M., Thinès-Sempoux, D., Beaufay, H., and Berthet, J., 1974b, Analytical study of microsomes and isolated subcellular membranes from rat liver. IV. Biochemical, physical and morphological modifications of microsomal components induced by digitonin, EDTA, and pyrophosphate, *J. Cell Biol.* **62:**717–745.

Bartles, J. R., Feracci, H. M., Stieger, B., and Hubbard, A. L., 1987, Biogenesis of the rat hepatocyte plasma membrane in vivo: comparison of the pathways taken by apical and basolateral proteins using subcellular fractionation, *J. Cell Biol.* **105:**1241–1251.

Beaumelle, B. D., Gibson, A., and Hopkins, C. R., 1990, Isolation and preliminary characterization of the major membrane boundaries of the endocytic pathway in lymphocytes, *J. Cell Biol.* **111:**1811–1823.

Beaufay, H. and Amar-Costesec, A., 1976, Cell fractionation techniques, in *Methods in Membrane Biology, vol. 6* (E. D. Korn, ed.), pp. 1–100, Plenum Press, New York.

Berg, T., Ford, T., Kindberg, G. M., Blomhoff, R., and Drevon, C. A., 1985, Intracellular degradation of asialoglycoproteins in hepatocytes starts in a subgroup of lysosomes, *Exp. Cell Res.* **156:**570–574.

Bergeron, J. J. M., Cruz, J., Khan, M. N., and Posner, B. I., 1985, Uptake of insulin and other ligands into receptor-rich endocytic components of target cells: the endosomal apparatus, *Ann. Rev. Physiol.* **47:**383–403.

Bergeron, J. J. M., Searle, N., Khan, M. S., and Posner, B. I., 1986, Differential and analytical subfractionation of rat-liver components internalizing insulin and prolactin, *Biochemistry,* **25:**1756–1764.

Besterman, J. M., Airhart, J. A., Woodworth, R. C., and Low, R. B., 1981, Exocytosis of pinocytosed fluid in cultured cells; kinetic evidence for rapid turnover and compartmentation, *J. Cell Biol.* **91:**716–727.

Blomhoff, R., Nenseter, M. S., Green, M. H., and Berg, T., 1989, A multicompartmental model of fluid-phase endocytosis in rabbit-liver parenchymal cells, *Biochem. J.* **262:**605–610.

Blouin, A., Bolender, R. P., and Weibel, E. R., 1977, Distribution of organelles and membranes between hepatocytes and nonhepatocytes in the rat-liver parenchyma: a stereological study, *J. Cell Biol.* **72:**441–455.

Bomsel, M., Prydz, K., Parton, R. G., Gruenberg, J., and Simons, K., 1989, Endocytosis in filter-grown Madin-Darby Canine Kidney cells, *J. Cell Biol.* **109:**3243–3258.

Branch, W. J., Mullock, B. M., and Luzio, J. P., 1987, Rapid subcellular fractionation of the rat-liver endocytic compartments involved in transcytosis of polymeric immunoglobulin A and endocytosis of asialofetuin, *Biochem. J.* **244**:311–315.

Brodsky, F. M., 1988, Living with clathrin: its role in intracellular membrane traffic, *Science* **242**:1396–1402.

Brown, W. J., Goodhouse, J., and Farquhar, M. G., 1986, Mannose-6-phosphate receptors for lysosomal enzymes cycle between the Golgi complex and endosomes, *J. Cell Biol.* **103**:1235–1247.

Cain, C. C., Sipe, D. M., and Murphy, R. F., 1989, Regulation of endocytic pH by the Na^+,K^+-ATPase in living cells, *Proc. Natl. Acad. Sci. USA* **86**:544–548.

Casanova, J. E., Breitfeld, P. P., Ross, S. A., and Mostov, K. E., 1990, Phosphorylation of the polymeric immunoglobulin receptor is required for its efficient transcytosis, *Science* **248**:742–745.

Casciola, R. and Hubbard, A. L., 1991, Hydrolases in intracellular compartments of rat liver cell: evidence for selective activation and/or delivery, *J. Biol. Chem.* **266**:4341–4347.

Chavrier, P., Parton, R. G., Hauri, H. P., Simons, K., and Zerial, M., 1990, Localization of low molecular weight GTP-binding proteins to exocytic and endocytic compartments, *Cell* **62**:317–329.

Chen, S. D., Widnell, C., Vaerman, J. P., Baudhuin, P., and Courtoy, P. J., 1987, Ultrastructural localization of 5′-nucleotidase in the normal rat liver, and after bile duct ligation, *J. Cell Biol.* **105**:233a.

Clarke, B. J., and Weigel, P. H., 1989, Differential effects of leupeptin, monensin and colchicine on ligand degradation by the two asialoglycoprotein receptor pathways in isolated rat hepatocytes, *Biochem. J.* **262**:277–284.

Collins, D. S., Unanue, E. R., and Harding, C. V., 1991, Reduction of disulfide bonds within lysosomes is a key step in antigen processing, *J. Immunol.* **147**:4054–4059.

Courtoy, P. J., 1984, Receptor-mediated endocytosis in rat hepatocytes with special reference to the pathways of polymeric IgA and Galactose-exposing proteins, (*Thèse d'Agrégation de l'Enseignement Supérieur*), Bruxelles, pp. 1–156.

Courtoy, P. J., 1989, Use of thimerosal in diaminobenzidine-induced density shift, *J. Cell Biol.* **109**:325a.

Courtoy, P. J., 1991, Dissection of endosomes, *in* "Intracellular trafficking of proteins" (C. Steer and J. Hanover, eds.), pp. 103–156, Cambridge University Press.

Courtoy, P. J., Quintart, J., and Baudhuin, P., 1984, Shift of equilibrium density induced by 3,3′-diaminobenzidine cytochemistry: a new procedure for the analysis and purification of peroxidase-containing organelles, *J. Cell Biol.* **98**:870–876.

Courtoy, P. J., Quintart, J., Limet, J. N., De Roe, C., and Baudhuin, P., 1985, Polymeric IgA and galactose-specific pathways in rat hepatocytes: evidence for intracellular ligand sorting, in *Endocytosis* (I. Pastan and M. C. Willingham, eds.), pp. 163–194, Plenum Press, New York.

Courtoy, P. J., Quintart, J., Draye, J. P., and Baudhuin, P., 1988, The DAB-induced density shift: principle, validity, and application to endosomes, *in* "Cell-free analysis of membrane traffic", Progress in Clinical and Biological Research, vol. 270, (D. J. Morré, K. E. Howell, G. M. W. Cook, and W. H. Evans, eds.), pp. 169–183, Alan R. Liss, New York.

Courtoy, P. J., Leruth-Deridder, M., Vaerman, J. P., and Baudhuin, P., 1992, Analytical subcellular fractionation of receptor-mediated transcytosis in rat hepatocytes, in "*Endocytosis, from cell biology to health, disease and therapy*" (P. J. Courtoy, ed.), NATO ASI Series H62, Springer Verlag, pp. 291–300.

Dautry-Varsat, A., Ciechanover, A., and Lodish, H. F., 1983, pH and the recycling of transferrin during receptor-mediated endocytosis, *Proc. Natl. Acad. Sci. USA* **80**:2258–2262.

Dean, R. T., Jessup, W., and Roberts, C. R., 1984, Effects of exogenous amines on mammalian cells, with particular reference to membrane flow, *Biochem. J.* **217**:27–40.

Debanne, M. T., Bolyos, M., Gauldie, J., and Regoeczi, E., 1984, Two populations of prelysosomal structures transporting asialoglycoproteins in rat liver, *Proc. Natl. Acad. Sci. USA* **81**:2995–2999.

de Brabander, M., Nuydens, R., Geerts, H., and Hopkins, C. R., 1988, Dynamic behavior of the transferrin receptor followed in living epidermoid carcinoma (A431) cells with nanovid microscopy, *Cell Motility and the Cytoskeleton* **9**:30–47.

de Chastellier, C., Lang, T., Ryter, A., and Thilo, L., 1987, Exchange kinetics and composition of endocytic membranes in terms of plasma membrane constituents: a morphometric study in macrophages, *Eur. J. Cell Biol.* **44**:112–123.

de Duve, 1964, Principles of tissue fractionation, *J. Theoret. Biol.* **6**:33–59.

de Duve, C., Berthet, J., and Beaufay, H., 1959, Gradient centrifugation of cell particles: theory and applications, *Prog. Biophys. Biol. Chem.* **9**:325–369.

de Duve, C., Pressman, B. C., Gianetto, R., Wattiaux, R., and Appelmans, F., 1955, Tissue fractionation studies. VI. Intracellular distribution patterns of enzymes in rat-liver tissue, *Biochem. J.* **60**:604–617.

Deng, Y. and Storrie, B., 1988, Animal cell lysosomes rapidly exchange membrane proteins, *Proc. Natl. Acad. Sci. USA* **85**:3860–3864.

De Roe, C., Courtoy, P. J., and Baudhuin, P., 1987, A model of protein-colloidal gold interactions, *J. Histochem. Cytochem.* **35**:1191–1198.

Desbuquois, B., Lopez, S., and Burlet, H., 1982, Ligand-induced translocation of insulin receptors in intact rat liver, *J. Biol. Chem.* **257**:10852–10860.

Desbuquois, B., Authier, F., Clot, J. P., Janicot, M., and Fouque, F., 1992, Degradation of insulin and glucagon in isolated liver endosomes: functional relationships with ATP-dependent endosomal acidification and partial characterization of degradation products, in "*Endocytosis, from cell biology to health, disease and therapy*" (Courtoy P. J. ed.), NATO ASI Series H62, Springer, Heidelberg, pp. 141–149.

Diaz, R., Wileman, T. E., Anderson, S. J., and Stahl, P., 1989, The use of permeabilized cells to study the ion requirements of receptor-ligand dissociation in endosomes, *Biochem. J.* **260**:127–134.

Diment, S. and Stahl, P., 1985, Macrophage endosomes contain proteases which degrade endocytosed protein ligands, *J. Biol. Chem.* **260**:15311–15317.

Diment, S., Leech, M. S., and Stahl, P. D., 1988, Cathepsin D is membrane-associated in macrophage endosomes, *J. Biol. Chem.* **263**:6901–6907.

Doherty, J. J. II, Kay, D. G., Lai, W. H., Posner, B. I., and Bergeron, J. J. M., 1990, Selective degradation of insulin within rat-liver endosomes, *J. Cell Biol.* **110**:35–42.

Draye, J. P., Courtoy, P. J., Quintart, J., and Baudhuin, P., 1987, Relations between plasma membrane and lysosomal membrane. 2. Quantitative evaluation of plasma membrane marker enzymes in the lysosomes, *Eur. J. Biochem.* **170**:405–411.

Draye, J. P., Quintart, J., Courtoy, P. J., and Baudhuin, P., 1988, A quantitative model of traffic between plasma membrane and secondary lysosomes: evaluation of inflow, lateral diffusion, and degradation, *J. Cell Biol.* **107**:2109–2115.

Duncan, R., and Pratten, M. K., 1977, Membrane economics in endocytic systems, *J. Theoret. Biol.* **66**:727–735.

Dunn, W. A. and Hubbard, A. L., 1984, Receptor-mediated endocytosis of epidermal growth factor by hepatocytes in the perfused rat liver: ligand and receptor dynamics, *J. Cell Biol.* **98**:2148–2159.

Dunn, W. A., Hubbard, A. L., and Aronson, N. N. Jr., 1980, Low temperature selectively inhibits fusion between pinocytic vesicles and lysosomes during heterophagy of ^{125}I-asialofetuin by the perfused rat liver, *J. Biol. Chem.* **255**:5971–5978.

Evans, W. H., and Flint, N., 1985, Subfractionation of hepatic endosomes in Nycodenz gradients and by free flow electrophoresis, *Biochem. J.* **232**:25–32.

Evans, W. H., and Hardison, W. G. M., 1985, Phospholipid, cholesterol, polypeptide, and glycoprotein composition of hepatic endosome subfractions, *Biochem. J.* **232**:33–36.

Feener, E. P., Shen, W. C., and Ryser, H. J. P., 1990, Cleavage of disulfide bonds in endocytosed macromolecules: a processing not associated with lysosomes or endosomes. *J. Biol. Chem.* **265**:18780–18785.

Ferris, A. L., Brown, J. C., Park, R. D., and Storrie, B., 1987, Chinese hamster ovary-cell lysosomes rapidly exchange contents, *J. Cell Biol.* **105**:2703–2712.

Fuchs, R., Mâle, Ph., and Mellman, I., 1989a, Acidification and ion permeabilities of highly purified rat-liver endosomes, *J. Biol. Chem.* **264**:2212–2220.

Fuchs, R., Schmid, S., and Mellman, I., 1989b, A possible role for Na^+·K^+-ATPase in regulating ATP-dependent endosome acidification, *Proc. Natl. Acad. Sci. USA.* **86**:539–543.

Fuchs, R., Schmid, S., Mellman, I., and Klapper, H., 1992, Regulation of ATP-dependent endosome acidification, in "*Endocytosis from cell biology to health, disease and therapy*" (Courtoy P. J., ed.). NATO ASI series H62, Springer, Heidelberg, pp. 135–140.

Geuze, H. J., Slot, J. W., Strous, G. J. A. M., Lodish, H. F., and Schwartz, A. L., 1983, Intracellular site of asialoglycoprotein receptor-ligand uncoupling: double label immunoelectron microscopy during receptor-mediated endocytosis, *Cell* **32**:277–287.

Geuze, H. J., Slot, J. W., Strous, G. J. A. M., Peppard, J., Von Figura, J., Hasilik, A., and Schwartz, A. L., 1984, Intracellular receptor sorting during endocytosis: comparative immunoelectron microscopy of multiple receptors in rat liver, *Cell* **37**:195–204.

Geuze, H. J., Slot, J. W., and Schwartz, A. L., 1987, Membranes of sorting organelles display lateral heterogeneity in receptor distribution, *J. Cell Biol.* **104**:1715–1723.

Geuze, H. J., Stoorvogel, W., Strous, G. J., Slot, J.W., Bleekemolen, J. D., and Mellman, I., 1988, Sorting of mannose-6-phosphate receptors and lysosomal membrane proteins in endocytic vesicles, *J. Cell Biol.* **107**:2491–2501.

Gierow, P., Sommarin, M., Larsson, C., and Jergil, B., 1986, Fractionation of rat-liver plasma membrane region by two-phase partitioning, *Biochem. J.* **235**:681–685.

Gordon, P. B. and Seglen, P. O., 1988, Prelysosomal convergence of autophagic and endocytic pathways, *Biochem. Biophys. Res. Comm.* **151**:40–47.

Gorvel, J. P., Chavrier, P., Zerial, M., and Gruenberg, I., 1991, rab 5 controls early endosome fusion in vitro, *Cell* **64**:915–925.

Grant, K. I., Casciola, L. A. F., Coetzee, G. A., Sanan, D. A., Gevers, W., and van der Westhuyzen, 1990, Ammonium chloride causes reversible inhibition of low-density lipoprotein receptor recycling and accelerates receptor degradation, *J. Biol. Chem.* **265**:4041–4047.

Griffiths, G., and Gruenberg, J., 1991, The arguments for pre-existing early and late endosomes, *Trends Cell Biol.* **1**:5–9.

Griffiths, G., Hoflack, B., Simons, K., Mellman, I., and Kornfeld, K., 1988, The mannose-6-phosphate receptor and the biogenesis of lysosomes, *Cell* **52**:329–341.

Griffiths, G., Back, R., and Marsh, M., 1989, A quantitative analysis of the endocytic pathway in baby hamster kidney cells, *J. Cell Biol.* **109**:2703–2720.

Gruenberg, J. and Howell, K. E., 1989, Membrane traffic in endocytosis: insights from cell-free assays, *Ann. Rev. Cell Biol.* **5**:453–481.

Harford, J., Bridges, K., Ashwell, G., and Klausner, R. D., 1983, Intracellular dissociation of receptor-bound asialoglycoproteins in cultured hepatocytes: a pH-mediated nonlysosomal event, *J. Biol. Chem.* **258**:3191–3197.

Helmy, S., Porter-Jordan, K., Dawidowicz, E. A., Pilch, P., Schwartz, A. L., and Fine, R. E., 1986, Separation of endocytic from exocytic coated vesicles using a novel cholinesterase-mediated density-shift technique, *Cell* **44**:497–506.

Henning, R., 1974, The lysosomal membrane: characteristic features of composition and function, in *Methodological Developments in Biochemistry* (E. Reid, ed.), pp. 187–194, Longman, London.

Henning, R., and Plattner, H., 1974, Isolation of rat-liver lysosomes by loading with colloidal gold, *Biochim. Biophys. Acta* **354:**114–120.

Hopkins, C. R., Gibson, A., Shipman, M., and Miller, K., 1990, Movement of internalized ligand-receptor complexes along a continuous endosomal reticulum, *Nature* **346:**335–339.

Hoppe, C. A., Connolly, T. P., and Hubbard, A. L., 1985, Transcellular transport of polymeric IgA in the rat hepatocyte: biochemical and morphological characterization of the transport pathway, *J. Cell Biol.* **101:**2113–2123.

Hornick, C., Hamilton, R. L., Spaziani, E., Enders, G. H., and Havel, R. J., 1985, Isolation and characterization of multivesicular bodies from rat hepatocytes: an organelle distinct from secretory vesicles of the Golgi apparatus, *J. Cell Biol.* **100:**1558–1569.

Hubbard, A. L., 1989, Endocytosis. *Current opinion in Cell Biology* **1:**675–683.

Jacques, P., 1968, Epuration plasmatique des protéines étrangères, leur capture et leur destinée dans l'appareil vacuolaire du foie, (*Thèse d'Agrégation de l'Enseignement Supérieur*), Librairie Universitaire, Louvain, pp. 1–149.

Jadot, M., Misquith, S., Dubois, F., Wattiaux-De Coninck, S., and Wattiaux, R., 1986, Intracellular pathway followed by invertase endocytosed by rat liver, *Eur. J. Biochem.* **161:**695–700.

Janicot, M. and Desbuquois, B., 1987, Fate of injected ^{125}I-labeled cholera toxin taken up by rat liver *in vivo.* Generation of the active A_1 peptide in the endosomal compartment, *Eur. J. Biochem.* **163:**433–442.

Kay, D. G., Khan, M. N., Posner, B. I., and Bergeron, J. J. M., 1984, *In vivo* uptake of insulin into hepatic Golgi fractions: application of the diaminobenzidine-shift protocol. *Biochem. Biophys. Res. Comm.* **123:**1144–1148.

Kelly, R. B., 1985, Pathways of protein secretion in eukaryotes, *Science* **230:**25–32.

Kennedy, J. and Cooper, C., 1988, The time-dependent distribution of ^{125}I-asialo-orosomucoid-horseradish peroxidase and ^{131}I-immunoglobulin A among three endosomal subfractions isolated from rat liver, *Biochem. J.* **252:**739–752.

Kielian, M. C., Marsh, M., and Helenius, A., 1986, Kinetics of endosome acidification detected by mutant and wild-type Semliki Forest Virus, *EMBO J.* **5:**3103–3109.

Kindberg, G. M., Ford, T., Blomhoff, R., Rickwood, D., and Berg, T., 1984, Separation of endocytic vesicles in Nycodenz gradients, *Anal. Biochem.* **142:**455–462.

Khan, M. N., Baquiran, G., Brule, C., Burgess, J., Foster, B., Bergeron, J. J. M., and Posner, B. I., 1989, Internalization and activation of the rat-liver insulin receptor kinase *in vivo, J. Biol. Chem.* **264:**12931–12940.

Klausner, R. D., Ashwell, G., Van Renswoude, J., Harford, J. B., and Bridges, K. R., 1983, Binding of apotransferrin to K562 cells: explanation of the transferrin cycle, *Proc. Natl. Acad. Sci. USA* **80:**2263–2266.

Kolset, S. O., Tolleshaug, H., and Berg, T., 1979, The effects of colchicine and cytochalasin B on uptake and degradation of asialoglycoproteins in isolated rat hepatocytes, *Exp. Cell Res.* **122:**159–167.

Lai, W. H., Cameron, P. H., Wada, I., Dohert, J. J. II, Kay, D. G., Posner, B. I., and Bergeron, J. J. M., 1989a, Ligand-mediated internalization, recycling, and downregulation of the epidermal growth factor receptor *in vivo, J. Cell Biol.* **1989:**2741–2749a.

Lai, W. H., Cameron, P. H., Doherty, J. J. II, Posner, B. I., and Bergeron, J. J. M., 1989b, Ligand-mediated autophosphorylation activity of the epidermal growth factor during internalization, *J. Cell Biol.* **109:**2751–2760.

Larkin, J. M., Sztul, E. S., and Palade, G. E., 1986, Phosphorylation of the rat hepatic polymeric IgA receptor, *Proc. Natl. Acad. Sci. USA* **83:**4759–4763.

Leighton, F., Poole, B., Beaufay, H., Baudhuin, P., Coffey, J. W., Fowler, S., and de Duve, C., 1968, The large-scale separation of peroxisomes, mitochondria and lysosomes from the livers of rats injected with Trition WR-1339. Improved isolation procedures, automated analysis, biochemical and morphological properties of fractions, *J. Cell Biol.* **37:**482–513.

Lemaître-Coelho, I., Jackson, G. D. F., and Vaerman, J. P., 1977, Rat bile as a convenient source of secretory IgA and free secretory component, *Eur. J. Immunol.* **8**:588–590.
Lemansky, P., Hasilik, A., Von Figura, K., Helmy, S., Fishman, J., Fine, R. E., Kedersha, N. L., and Rome, L. N., 1987, Lysosomal enzyme precursors in coated vesicles derived from the exocytic and endocytic pathways, *J. Cell Biol.* **104**:1743–1748.
Limet, J. N., Schneider, Y. J., Vaerman, J. P., and Trouet, A., 1982, Binding, uptake and intracellular processing of polymeric rat IgA by cultured rat hepatocytes, *Eur. J. Biochem.* **125**:437–443.
Limet, J. N., Quintart, J., Schneider, Y-J., and Courtoy, P. J., 1985, Receptor-mediated endocytosis of polymeric IgA and galactosylated serum albumin in rat liver. Evidence for intracellular ligand sorting and identification of distinct endosomal compartments, *Eur. J. Biochem,* **146**:539–548.
Lindh, E. and Björk, I., 1976, Binding of secretory component to dimers of immunoglobulin A *in vitro:* mechanism of the covalent bond formation, *Eur. J. Biochem.* **62**:263–270.
Lippincott-Schwartz, J. and Fambrough, D. M., 1986, Lysosomal membrane dynamics: structure and interorganellar movement of a major lysosomal membrane glycoprotein, *J. Cell Biol.* **102**:1593–1605.
Magnusson, S. and Berg, T., 1989, Extremely rapid endocytosis mediated by the mannose receptor of sinusoidal endothelial rat-liver cells, *Biochem. J.* **257**:651–656.
Marsh, M. and Helenius, A., 1989, Virus entry into animal cells, *Adv. in Virus Res.* **36**:107–151.
Matsuura, S., Eto, S., Kato, K., and Tashiro, Y., 1984, Ferritin immunoelectron microscopic localization of 5′-nucleotidase on rat-liver cell surface, *J. Cell Biol.*, **99**:166–173.
Matter, A., Orci, L., and Rouiller, C., 1969, A study on the permeability barriers between Disse's space and the bile canaliculus. *J. Ultrastruct. Res. (Supp)* **11**:1–71.
Mayorga, L. S., Diaz, R., Stahl, P. D., 1989, Regulatory role for GTP-binding proteins in endocytosis, *Science* **244**:1475–1477.
McKanna, J. A., Haigler, H. T., and Cohen, S., 1979, Hormone receptor topology and dynamics: morphological analysis using ferritin-labeled epidermal growth factor, *Proc. Natl. Acad. Sci. USA,* **76**:5689–5693.
Mellman, I., Fuchs, R., and Helenius, A., 1986, Acidification of the endocytic and exocytic pathways, *Ann. Rev. Biochem.* **55**:663–700.
Mueller, S. C. and Hubbard, A. L., 1986, Receptor-mediated endocytosis of asialoglycoproteins by rat hepatocytes: receptor-positive and receptor-negative endosomes, *J. Cell Biol.* **102**:932–942.
Mullock, B. M., Branch, W. J., Van Schaik, M., Gilbert, L. K., and Luzio, J. P., 1989, Reconstitution of an endosome-lysosome interaction in a cell-free system, *J. Cell Biol.* **108**:2093–2099.
Munniksma, J., Noteborn, M., Kooistra, T., Stienstra, S., Bouma, J. M. W., Gruber, M., Brouwer, A., Praaning-Van Dalen, D., and Knook, D. L., 1980, Fluid endocytosis by rat liver and spleen, *Biochem. J.* **192**:613–621.
Murkofsky, N. A. and Lamm, M. E., 1979, Effect of a disulfide interchange enzyme on the assembly of human secretory immunoglobulin A from immunoglobulin A and free secretory component, *J. Biol. Chem.* **254**:12181–12184.
Murphy, R. F., 1989, Endosomal pH regulation and lysosome biogenesis, *J. Cell Biol.* **109**:6a.
Murphy, R. F., 1991, Maturation models for endosome and lysosome biogenesis, *Trends Cell Biol.* **1**:77–82.
Musil, L. S. and Baenziger, J. U., 1988, Proteolytic processing of rat-liver membrane secretory component: cleavage activity is localized to bile canalicular membranes, *J. Biol. Chem.* **263**:15799–15808.
Oka, J. A. and Weigel, P. H., 1983, Recycling of the asialoglycoprotein receptor in isolated rat hepatocytes: dissociation of internalized ligand from receptor occurs in two kinetically and thermally distinguishable compartments, *J. Biol. Chem.* **258**:10253–10262.
Olsnes, S., and Sandvig, K., 1985, Entry of polypeptide toxins into animal cell, in *Endocytosis* (I. Pastan and M. C. Willingham, eds.), pp. 195–237, Plenum Press, New York.

Opresko, L. K., and Karpf, R. A., 1987, Specific proteolysis regulates fusion between endocytic compartments in Xenopus oocytes, *Cell* **51**:557–568.

Ose, L., Ose, T., Reinertsen, R., and Berg, T., 1980, Fluid endocytosis in isolated rat parenchymal and nonparenchymal liver cells, *Exp. Cell Res.* **126**:109–119.

Ottosen, P. D., Courtoy, P. J., and Farquhar, M. G., 1980, Pathways followed by membrane recovered from the surface of plasma cells and myeloma cells, *J. Exp. Med.* **152**:1–19.

Pease, R. J., Smith, G. D., and Peters, T. J., 1987, Characterization of insulin degradation by rat-liver low-density vesicles, *Eur. J. Biochem.* **164**:251–257.

Perez, J. H., Branch, W. J., Smith, B., Mullock, B. M., and Luzio, J. P., 1988, Investigation of endosomal compartments involved in endocytosis and transcytosis of polymeric immunoglobulin A by subcellular fractionation of perfused rat liver, *Biochem. J.* **251**:763–770.

Quintart, J., Courtoy, P. J., Limet, J. N., and Baudhuin, P., 1983, Galactose-specific endocytosis in rat liver: biochemical and morphological characterization of a low-density compartment isolated from hepatocytes, *Eur. J. Biochem.* **131**:105–112.

Quintart, J., Courtoy, P. J., and Baudhuin, P., 1984, Receptor-mediated endocytosis in rat liver: purification and enzymic characterization of low-density organelles involved in uptake of galactose-exposing proteins, *J. Cell Biol.* **98**:877–884.

Quintart, J., Baudhuin, P., and Courtoy, P. J., 1989, Marker enzymes of rat-liver vesicles involved in transcellular transport, *Eur. J. Biochem.* **184**:567–574.

Regoeczi, E., Chindemi, P. A., Debanne, M. T., and Charlwood, P. A., 1982, Partial resialylation of human asialotransferrin type 3 in the rat, *Proc. Natl. Acad. Sci. USA* **79**:2226–2230.

Rijnboutt, S., Kal, A. J., Geuze, H. J., Aerts, H., and Strous, G. J., 1991, Mannose 6-phosphate-independent targeting of cathepsin D to lysosomes in HepG2 cells, *J. Biol. Chem.* **266**:23586–23592.

Robert, A., Carpentier, J. L., Van Obberghen, E., Canivet, B., Gorden, P., and Orci, L., 1985, The endosomal compartment of rat hepatocytes: its characterization in the course of ^{125}I-insulin internalization, *Exp. Cell Res.* **159**:113–126.

Roederer, M., Bowser, R., and Murphy, R. F., 1987, Kinetics and temperature dependence of exposure of endocytosed material to proteolytic enzymes and low pH: evidence for a maturation model for the formation of lysosomes, *J. Cell Physiol.* **131**:200–209.

Scharschmidt, B. F., Lake, J. R., Renner, E. L., Licko, V., and Van Dyke, R. W., 1986, Fluid-phase endocytosis by cultured rat hepatocytes and perfused rat liver: implications for plasma membrane turnover and vesicular trafficking of fluid-phase markers. *Proc. Natl. Acad. Sci. USA* **83**:9488–9492.

Schaudies, R. P., Gorman, R. M., Savage, C. R., Jr, and Poretz, R. D., 1987, Proteolytic processing of EGF within endosomes. *Biochim. Biophys. Res. Comm.* **143**:710–715.

Schiff, J. M., Fisher, M. M., Jones, A. L., and Underdown, B. J., 1986, Human IgA as a heterovalent ligand: switching from the asialoglycoprotein receptor to secretory component during transport across the rat hepatocyte, *J. Cell Biol.* **102**:920–931.

Schmid, S. L., Fuchs, R., Mâle, Ph., and Mellman, I., 1988, Two distinct subpopulations of endosomes involved in membrane recycling and transport to lysosomes, *Cell* **52**:73–83.

Shepherd, V. L., Lee, Y. C., Schlesinger, P. H., and Stahl, P. D., 1981, L-fucose-terminated glycoconjugates are recognized by pinocytosis receptors on macrophages, *Proc. Natl. Acad. Sci.* **78**:1019–1022.

Sibille, J. C., Octave, J. S., Schneider, Y-J., Trouet, A., and Crichton, R., 1986, Subcellular localization of transferrin protein and iron in the perfused rat liver: effect of Triton WR1339, digitonin and temperature. *Eur. J. Biochem.* **155**:47–55.

Slot, J. W., Geuze, H. J., Gigengack, S., Lienhard, G. E., and James, D. E., 1991, Immunolocalization of the insulin regulatable glucose transporter in brown adipose tissue of the rat, *J. Cell Biol.* **113**:123–135.

Smith, G. D. and Peeters, T. J., 1982, The localization in rat liver of alkaline phosphodiesterase to a discrete organelle implicated in ligand internalization, *Biochim. Biophys. Acta* **716:**24–30.

Snider, M. D. and Rogers, O. C., 1985, Intracellular movement of cell surface receptors after endocytosis: resialylation of asialotransferrin receptor in human erythroleukemia cells, *J. Cell Biol.* **100:**826–834.

Solari, R., Schaerer, E., Tallichet, C., Braiterman, L. T., Hubbard, A. L., and Kraekenbuhl, J. P., 1989, Cellular location of the cleavage event of the polymeric immunoglobulin receptor and fate of its anchoring domain in the rat hepatocyte, *Biochem. J.* **257:**759–768.

Steinman, R. M., Silver, J. M., and Cohn, Z. A., 1974, Pinocytosis in fibroblasts: quantitative studies *in vitro, J. Cell Biol.* **63:**949–969.

Steer, C. J., Bisher, M., Blumenthal, R., and Steven, A. C., 1984, Detection of membrane cholesterol by filipin in isolated rat liver coated vesicles is dependent upon removal of the clathrin coat, *J. Cell Biol.* **99:**315–319.

Stockert, R. J., Haimes, H. B., Morell, A. G., Novikoff, P. M., Novikoff, A. B., Quintana, N., and Sternlieb, I., 1980, Endocytosis of asialoglycoprotein-enzyme conjugates by hepatocytes, *Lab. Invest.* **43:**556–563.

Stoorvogel, W., Geuze, M. J., Griffith, J. M., Schwartz, A. L., and Strous, G. J., 1989, Relations between the intracellular pathways of the receptors for transferrin, asialoglycoprotein and mannose-6-phosphate in human hepatoma cells, *J. Cell Biol.* **108:**2137–2148.

Stoorvogel, W., Strous, G. J., Geuze, H. J., Oorschot, V., and Schwartz, A. L., 1991, Late endosomes derive from early endosomes by maturation, *Cell* **65:**417–427.

Sztul, E. S., Howell, K. E., and Palade, G. E., 1983, Intracellular and transcellular transport of secretory component and albumin in rat hepatocytes, *J. Cell Biol.* **97:**1582–1591.

Sztul, E., Kaplin, A., Saucan, L., and Palade, G. E., 1991, Protein traffic between distinct plasma membrane domains: isolation and characterization of vesicular carriers involved in transcytosis, *Cell* **64:**81–89.

Takahashi, T., Nakada, H., Okumara, T., Sawamura, T., and Tashiro, Y., 1985, Phosphorylation of the rat hepatocyte asialoglycoprotein receptor, *Biochem. Biophys. Res. Comm.* **126:**1054–1060.

Thinès-Sempoux, D., 1972, A comparison between the lysosomal and the plasma membrane, in *Lysosomes in biology and pathology*, vol. 3, (J. T. Dingle, ed.), Elsevier, Amsterdam, pp. 278–299.

Tolleshaug, H. and Berg, T., 1981, The effect of leupeptin on intracellular digestion of asialofetuin in rat hepatocytes, *Exp. Cell Res.* **134:**207–217.

Tolleshaug, H., Berg, T., and Holte, K., 1982, Effects of local anesthetics and related compounds on the endocytosis and catabolism of asialoglycoproteins in isolated hepatocytes, *Biochim. Biophys. Acta.* **714:**114–121.

Tooze, J., and Hollinshead, M., 1991, Tubular early endosomal networks in AtT20 and other cells, *J. Cell Biol.* **115:**635–653.

Tran, D., Carpentier, J. L., Sawano, F., Gorden, P., and Orci, L., 1987, Ligands internalized through coated or noncoated invaginations follow a common intracellular pathway, *Proc. Natl. Acad. Sci. USA* **84:**7957–7961.

Van der Sluijs, P., Braakman, I., Meijer, D. K. F., and Groothuis, G. M. M., 1988, Heterogeneous acinar localization of the asialoglycoprotein internalization system in rat hepatocytes, *Hepatology* **8:**1521–1529.

Van der Sluijs, P., Hull, M., Zahraoui, A., Tavitian, A., Goud, B., and Mellman, I., 1991, The small GTP-binding protein rab 4 is associated with early endosomes, *Proc. Natl. Acad. Sci. USA* **88:**6313–6317.

Van Deurs, B., Sandvig, K., Petersen, O. W., Olsnes, S., Simons, K., and Griffiths, G., 1988, Estimation of the amount of internalized ricin that reaches the trans-Golgi network, *J. Cell Biol.* **106:**253–267.

Van Deurs, B., Petersen, O. W., Olsnes, S., and Sandvig, K., 1989, The ways of endocytosis, *Int. Rev. Cytol.* **117:**131–177.

Van Dyke, R. W., Hornick, C. A., Belcher, J., Scharschmidt, B. F., and Havel, R. J., 1985, Identification and characterization of the ATP-dependent proton transport by rat-liver multivesicular bodies, *J. Biol. Chem.* **260:**11021–11026.

Vergés, M., Evans, W. H., and Enrich, C., 1992, Calmodulin-binding proteins in rat liver endosomes. Identification of tau proteins in the endosome fractions, in *Mechanisms of hepatic endocytosis* (Windler E. and Greten H., eds.) Zuckschwerdt, Munich, pp. 192–198.

Von Figura, K. and Hasilik, A., 1986, Lysosomal enzymes and their receptors, *Ann. Rev. Biochem.* **55:**167–193.

Wada, I., Lai, W. E., Posner, B. I., and Bergeron, J. J. M., 1992, Association of the tyrosine-phosphorylated EGF receptor with a 55-kD tyrosine-phosphorylated protein at the cell surface and in endosomes, *J. Cell Biol.* **116:**321–330.

Wall, D. A. and Hubbard, A. L., 1985, Receptor-mediated endocytosis of asialoglycoproteins by rat-liver hepatocytes: biochemical characterization of the endosomal compartments. *J. Cell Biol.* **101:**2104–2112.

Wall, D. A., Wilson, G., and Hubbard, A. L., 1980, The galactose-specific recognition system of mammalian liver: the route of ligand internalization in rat hepatocytes, *Cell* **21:**79–93.

Weibel, E. R., Staübli, W., Gnägi, H. R., and Hess, F. A., 1969, Correlated morphometric and biochemical studies on the liver cell. I. Morphometric model, stereologic methods and normal morphometric data for rat liver, *J. Cell Biol.* **42:**68–91.

Weigel, P. H. and Oka, J. A., 1983, The large intracellular pool of asialoglycoprotein receptors functions during the endocytosis of asialoglycoproteins by isolated rat hepatocytes, *J. Biol. Chem.* **258:**5095–5102.

Wibo, M., Thinès-Sempoux, D., Amar-Costesec, A., Beaufay, H., and Godelaine, D., 1981, Analytical study of microsomes and isolated subcellular membranes from rat liver. VIII. Subfractionation of preparations enriched with plasma membranes, outer mitochondrial membranes, or Golgi complex membranes, *J. Cell Biol.* **89:**456–474.

Wisse, E., De Zanger, R. B., Jacobs, R., and McCuskey, R. S., 1983, Scanning electron microscope observations on the structure of portal veins, sinusoids, and central veins in rat liver, *Scanning Electron Microscopy* **3:**1441–1452.

Zijderhand-Bleekemolen, J. E., Schwartz, A. L., Slot, J. W., Strous, G. J., and Geuze, H. J., 1987, Ligand- and weak base–induced redistribution of asialoglycoprotein receptors in hepatoma cells, *J. Cell Biol.* **104:**1647–1654.

Chapter 3

Reconstitution of Endosomal Transport and Proteolysis

J. S. Blum, R. Diaz, L. S. Mayorga, and P. D. Stahl

1. INTRODUCTION

Receptor-mediated endocytosis is initiated by the accumulation of receptor-ligand complexes in clathrin-coated pits on the plasma membrane. The clustering of receptors in these domains is followed by membrane invagination and the subsequent formation of coated vesicles. Ligand binding induces the clustering of some receptors—for example, epidermal growth factor (EGF) receptor—in coated pits (Beginout *et al.*, 1985). Alternatively, many other receptors—low density lipoprotein receptor (LDL), transferrin receptor—are internalized constitutively in the absence of ligand (Anderson *et al.*, 1977; Harding *et al.*, 1983). Numerous studies suggest the signals for clustering and internalization are encoded in the cytoplasmic tail of endocytosed receptors (Davis *et al.*, 1987; Lobel *et al.*, 1989). Proteins associated with clathrin in the vesicle coat, such as adaptins, may interact directly with receptor tails to regulate these early steps in endocytosis (Pearse, 1988; Glickman *et al.*, 1989). Shortly after their formation,

J. S. Blum, R. Diaz, and P. D. Stahl Department of Cell Biology and Physiology, Washington University School of Medicine, St. Louis, Missouri 63110. **L. S. Mayorga** Facultad de Cincias Medican, Instituto de Histologia y Embriologia, 550 Mendoze, Argentina. *Current address for J. S. B.:* Virginia Mason Research Center, Seattle, Washington 98101.

Subcellular Biochemistry, Volume 19: Endocytic Components: Identification and Characterization, edited by J. J. M. Bergeron and J. R. Harris. Plenum Press, New York, 1993.

coated vesicles containing receptor-ligand complexes lose their clathrin lattice in an ATP-dependent reaction (Rothman and Schmid, 1986). The resulting smooth vesicles (early endosomes) may retain some peripheral proteins and membrane components that function in intracellular transport and sorting reactions.

Endosomes are a heterogeneous vesicular compartment where the sorting of receptors and ligands begins. Internalized ligands traverse endosomal compartments of various morphological characteristics during transport to their final destination. Studies have shown that ligands for the asialoglycoprotein (ASGP) receptor (Geuze *et al.*, 1983) and for the mannose receptor (Harding *et al.*, 1985) are first found in small smooth vesicles followed by larger endocytic vesicles with tubular extensions. Formation of large diameter (approximately 600–900 nm) endosomes containing ligand most likely requires vesicle fusion. The increased diameter of these vesicles carrying ligand cannot be caused only by the swelling of small-diameter vesicles (approximately 40–100 nm) that appear after clathrin uncoating. Ligands are later localized in multivesicular structures (multivesicular bodies) prior to their transport to lysosomes. The sequential localization of ligand in morphologically different endosomal vesicles may be caused by a maturation process whereby newly formed endosomes are transient structures that undergo mophological changes along the cycle. Alternatively, these different endosomes may be stable structures with specialized functions and with transport of ligand between them requiring shuttle vesicles. Early and late endosomal compartments have been distinguished kinetically by monitoring ligand transport (Schmid *et al.*, 1988). These compartments differ in protein composition and are distinct from the plasma membrane. Changes in the morphological and biochemical features of endosomes may come about as proteins and membranes are sorted away or added during the course of endocytosis. In each case endosomal vesicles must undergo membrane fission and fusion events to maintain the flux of molecules through this compartment.

Endosomes play an active role in the sorting and targeting of both intracellular and internalized proteins. The proton pump on endosomal membranes maintains an acidic vesicle lumen, which in turn favors the dissociation of various ligands from their receptors (mannose receptor, asialoglycoprotein receptor, mannose-6-phosphate receptor). After dissociation, ligands and receptors are segregated into different intracellular compartments. This sorting appears to take place in an endosomal compartment. A tubulovesicular endosomal organelle, referred to as the compartment for uncoupling of receptor and ligand (CURL) has been described (Geuze *et al.*, 1983). This compartment was characterized by the accumulation of receptors in tubular regions while ligand remained in the lumen of the vesicular areas. From these observations, it was inferred that receptors could be sorted into vesicles that would pinch off from the tubular regions and recycle back to the plasma membrane while ligand was delivered to lysosomes. Recycling endosomal vesicles also intersect the trans-Golgi reticulum to a limited extent. A fraction of the endocytosed transferrin receptors and mannose-6-phosphate receptors are slowly shunted to the trans-Golgi and subsequently these receptors recycle to the plasma membrane (Snider and Rogers, 1985; Duncan and Kornfeld, 1988). Endosomes also play a role in the sorting of proteins that originate from the Golgi apparatus. As pointed out later in this review, lysosomal proteases are found in a functional form in early endosomes (Diment and Stahl, 1985). The complexity of endosomal transport suggests that specific signals must regulate the fate of a molecule transiting through this organelle.

The existence of an extensive endosomal network involved in both the transport and sorting of proteins has raised much interest in understanding the mechanisms and regulatory components required for these processes. In this review we focus on two aspects of our work that relate to fundamental questions of endosome function. First, what are the properties of endosomal proteases and are endosomes a biologically important processing compartment within cells? Second, is it possible to reconstitute endocytosis in broken cell preparations and to use this system to identify regulatory events in endosome transport and sorting?

2. PROTEOLYSIS IN ENDOSOMES

2.1. Characteristics of Endosomal Proteolysis

Endosomal vesicles serve as an important proteolytic processing compartment within cells. Hormones, growth factors, toxins, and foreign antigens are cleaved by endosomal proteases. Changes in the biological activity of macromolecules following processing in endosomes have been observed. In the case of parathyroid hormone, endosomal proteolysis results in the generation of active peptides derived from this hormone (Diment *et al.*, 1989). Proteolytic products generated in endosomes may be rapidly shuttled back to the plasma membrane or sorted to intracellular compartments such as lysosomes, Golgi, or (via transmembrane transport) the cytoplasm. The endosomal recycling of processed foreign antigens back to the cell surface appears to be essential for immune recognition and response to these antigens (Unanue and Allen, 1987). The cleavage of proteins following endocytosis has been observed in a number of cells including macrophages, fibroblasts, hepatocytes, B-cells, and oocytes.

Proteases were initially identified in endosomal vesicles during studies of receptor-ligand transport using macrophages. The mannose receptor is expressed on the surface of macrophages, and ligands bound to this receptor are rapidly delivered into the endocytic cycle. The fate of two ligands, β-glucuronidase and mannosylated bovine serum albumin (mannose-BSA), varies following their transport into endosomes (Stahl *et al.*, 1980; Wileman *et al.*, 1984; Diment and

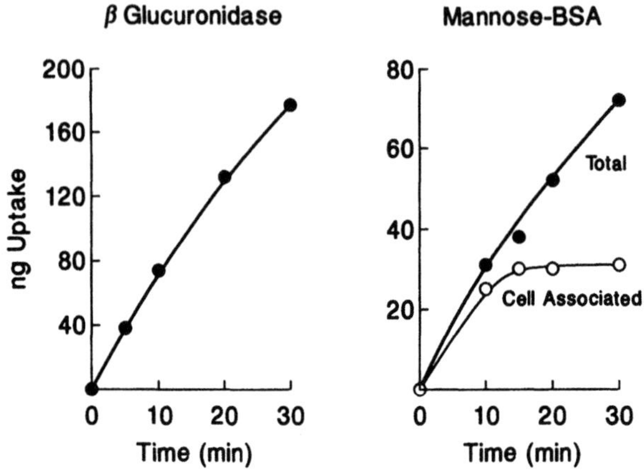

FIGURE 1. Differential proteolysis of ligands following receptor-mediated endocytosis in macrophages. Macrophages were incubated at 37°C with radiolabeled ligands specific for the mannose receptor. Cells accumulate the ligand, β-glucuronidase in a continuous fashion over time (left, ●). Mannose-BSA is also endocytosed via the mannose receptor; uptake of this ligand by cells plateaus after a few minutes, however (right, ○). Proteolytic fragments of mannose-BSA are detected in the medium surrounding cells. The sum of the cell-associated mannose-BSA and proteolyzed ligand present in the medium is indicated (right, ●).

Stahl, 1985). In comparing the incorporation of each ligand into macrophages, differences in the amount of cell-associated ligand are observed (Figure 1). β-Glucuronidase binds the mannose receptor, and intracellular accumulation of this ligand by cells is linear with time. In contrast, the accumulation of mannose-BSA by macrophages appears to plateau after only 5–10 min. Cells incubated at 37°C with mannose-BSA rapidly release proteolytic fragments of this ligand into the surrounding medium. Summation of the cell-associated ligand with the released ligand indicates that the net accumulation of each ligand into cells via this receptor occurs by a continuous, linear process.

These results demonstrated that mannose-BSA is rapidly transported to an intracellular compartment containing proteases. Subcellular fractionation studies showed mannose-BSA and β-glucuronidase are each transported from the plasma membrane into endosomes after 5–10 min (Wileman *et al.*, 1985). Mannose-BSA is very susceptible to proteolysis, and trichloroacetic acid soluble fragments of this ligand are detected in cells 6 min after internalization (Diment and Stahl, 1985). β-Glucuronidase, a lysosomal enzyme, is resistant to endosomal proteolysis. Approximately 20–30 min after the uptake, β-glucuronidase is transported from endosomes to lysosomes as shown by subcellular fractionation using Percoll density gradients (Wileman *et al.*, 1985). Proteolysis of mannose-BSA occurs in light-density endosomal vesicles and significant transport of this ligand to dense lysosomes is not typically observed *in vivo*. Treatment of macrophages with the protease inhibitor pepstatin A results in a diminution of endosomal protease activity (Diment and Stahl, 1985). In cells exposed to pepstatin A, proteolysis of mannose-BSA is reduced. Fractionation of membrane organelles from pepstatin-treated cells indicated that mannose-BSA is first delivered into endosomes. With endosomal proteolysis inhibited, however, this ligand is subsequently transported from endosomes to lysosomes with kinetics similar to β-glucuronidase.

Semipurified endosomal vesicles containing ligand can be used to study proteolysis *in vitro*. Ligand proteolysis is observed in light-density endosomes incubated in iso-osmotic buffered medium. Proteolysis is dependent on the colocalization of ligand and protease within the same endosomal vesicle (Diment and Stahl, 1985; Mayorga *et al.*, 1989a). The addition of purified lysosomes to endosomal vesicles during *in vitro* incubations does not stimulate proteolysis. The cleavage of mannose-BSA in endosomes is stimulated by ATP, which drives endosome acidification. Weak bases or ionophores added to vesicles inhibit the proteolysis of mannose-BSA. In the absence of ATP, degradation of mannose-BSA is observed by incubating vesicles at acid pH. These results demonstrate that proteases with an acidic pH optima are present in endosomes. Experiments using different ligands indicate that proteolysis can also be observed at neutral pH using endosomal vesicles. Ricin A chain, isolated from the plant toxin *Ricinus communis,* is selectively incorporated into macrophages via the mannose receptor (Simmons *et al.*, 1986; Fiani and Stahl, 1988). Cleavage of ricin A chain in endosomes occurs when vesicles are incubated with nigericin, an ionophore that dissipates the internal proton gradient (Fiani and Stahl, 1988; Blum *et al.*, 1989). Insulin cleavage was observed when endosomal vesicles were incubated in neutral as well as acidic buffers (Pease *et al.*, 1985; Hamel *et al.*, 1988). Differences in ligand susceptibility or the presence of several proteases in endosomes may account for these results. Studies with specific protease inhibitors indicate that both pepstatin- and leupeptin-sensitive proteases are present in isolated endosomes from macrophages (Diment and Stahl, 1985; Blum *et al.*, 1989). Whether the protease composition of endosomes varies in different cell types is at present unknown (Table I). The intralumenal pH of endosomes drops from neutral to acidic as these vesicles cycle through the cell (Tycko and Maxfield, 1982). Thus, the presence of proteases with different pH optima within endosomes may regulate the sequence and specificity of proteolytic cleavage in these vesicles.

2.2. Identification of Specific Proteases Found in Endosomes

Several proteases have been identified in endosomal vesicles using different substrates and specific inhibitors. Only one endosomal protease has been extensively characterized using both biochemical and morphological techniques. The

Table I
Proteases Localized in Endosomal Vesicles

Protease	Cell type	Substrate
Cathepsin D	Macrophages	Mannose-BSA; parathyroid hormone; ricin A chain
	Oocytes	Vitellogenin
	B-lymphocytes	
Cathepsin B	3T3 fibroblasts	N-carbobenzyloxy-ala-arg-arg-4-methoxy-β-naphthylamine
	B-lymphocytes	
Insulin-specific protease	Hepatocytes	Insulin
Carboxypeptidase B	3T3 fibroblasts	Epidermal growth factor
Leucyl-β-naphthyl-amidase	HeLa	1-leucine-β-naphthylamine-HCl

aspartyl protease cathepsin D was localized in endosomes isolated from rabbit alveolar macrophages (Diment and Stahl, 1985; Diment *et al.*, 1988). Endosomal proteolysis of mannose-BSA was greatly reduced in the presence of pepstatin A, an inhibitor of cathepsin D. By radiolabeling the contents of endosomes, the 46-kD form of cathepsin D was detected using either affinity chromatography or specific antibodies.

Biosynthetic studies demonstrated cathepsin D is synthesized as a 53 kD inactive precursor protein and converted to the active 46 kD protease following transport through Golgi compartments (Hasilik and Neufeld, 1980). The newly synthesized cathepsin D is tightly associated with membranes, and a significant proportion of the endosomal protease is still bound to membranes (Diment *et al.*, 1988). In lysosomes, cathepsin D is present as a soluble enzyme. These results suggest that the protease is first delivered to endosomes and then to lysosomes. Immunocytochemistry has been utilized to confirm the localization of cathepsin D in endosomes. Early endosomes, marked after 2 or 5 min of endocytosis with gold-labeled mannose-BSA, contained measurable amounts of cathepsin D (Rodman *et al.*, 1990). Quantitation revealed the bulk of the cellular cathepsin D was present in lysosomes, yet a significant proportion was present in endosomes and small peripheral vesicles. Similar small vesicles have also been identified in B-lymphoblastoid cells and were found to stain positive for cathepsin D and B (Guagliardi *et al.*, 1990). These vesicles may serve to transport proteases from the Golgi to endosomes or from endosome to endosome. Cathepsin D was not detected on the plasma membrane of cells using biochemical and morphological techniques (Rodman *et al.*, 1990). In rat kidney cells cathepsin D was colocalized with the mannose-6-phosphate receptor in a reticular-vesicular prelysosomal compartment (Griffiths *et al.*, 1988). A number of lysosomal proteins were also found in this compartment. The transport of internalized ligands to this prelysosome was blocked at 20°C suggesting it differs from early endosomes. Such prelysosomal vesicles may serve as a specialized sorting compartment in cells. Proteases destined for secretion or for early endosomes may be segregated in these vesicles away from proteins transported directly to lysosomes.

Several other proteases present in endosomes have been identified and partially characterized (Table I). The cysteine protease cathepsin B was detected in endosomes using a sensitive fluorometric assay (Roederer *et al.*, 1987). In fibroblasts as early as 2 min after endocytosis, a peptide substrate for cathepsin B is cleaved, releasing products that can be measured fluorometrically. Endocytosis of the substrate was required for proteolysis, and methylamine, which blocks endosome acidification, partially prevented substrate hydrolysis. Cathepsin B was localized in endosomes of B-lymphoblastoid cells using electron microscopy (Guagliardi *et al.*, 1990). Immunoglobulin molecules internalized from the plasma membrane were used to mark endocytic vesicles in these cells. Neither cathepsin B nor cathepsin D was detected on the plasma membrane of B-cells. In contrast, an insulin-specific protease derived from the plasma membrane may be responsible for cleaving this hormone in rat-liver endosomes (Hamel *et al.*, 1988). Proteolysis of insulin occurs in early endosomes 1–2 min after internalization. Several of the proteolytic fragments of insulin detected in endosomes are similar to those generated by the insulin-specific protease. A carboxypeptidase B-like enzyme was observed to cleave the carboxy-terminal arginine from epidermal growth factor in early endosomes. (Schaudies *et al.*, 1987). Leucyl-β-naphthyl amidase (leucine aminopeptidase) activity has also been detected in endosomes of HeLa cells (Ajoika and Kaplan, 1987).

2.3. Biological Importance of Endosomal Proteases

The biological importance of endosomal proteolysis is suggested by the cleavage of specific proteins and peptides within this intracellular compartment. Conditions within endosomes, such as the fluctuating pH and limited number of proteases, may make this a selective processing environment in comparison to lysosomes.

A number of hormones are cleaved upon endocytosis, including parathyroid hormone (Diment *et al.*, 1989), insulin (Pease *et al.*, 1985; Hamel *et al.*, 1988), and epidermal growth factor (Schaudies *et al.*, 1987; Wiley *et al.*, 1985). Activation of parathyroid hormone to its bioactive peptide 1-34 is mediated in endosomes by cathepsin D (Diment *et al.*, 1989). The plant toxin ricin A chain is cleaved by both cathepsin D and cathepsin B within endosomes (Fiani and Stahl, 1988; Blum *et al.*, 1989). Inhibition of endosomal proteolysis alters ricin A chain toxicity to cells. Vitellogenin is taken up by oocytes and proteolyzed in an endosomal compartment prior to its transit and storage in yolk platelets (Opresko and Karpf, 1987). Although mature lysosomal enzymes are resistant to endosomal proteases, precursor forms of these hydrolases are processed in endosomes. Macrophages secrete precursors of many lysosomal enzymes. These enzymes can be taken into cells by receptor-mediated processes and proteolytically cleaved to their mature form in endosomes (Gabel and Foster, 1987).

Endosomal proteolysis may be especially relevant to the regulatory and recognition properties of immune cells. Foreign protein antigens are internalized by cells and proteolytically processed (Unanue and Allen, 1987). Some of the peptides generated bind intracellular class II histocompatibility antigens. Complexes of peptide and class II molecules are then presented on the cell surface for recognition by T-cells. Several studies suggest endosomal proteases are important in antigen processing and presentation. Cells defective in endosomal acidification have a reduced capacity to process foreign antigens (McCoy *et al.*, 1989). In addition, class II histocompatibility antigens are transported through endosomes containing proteases prior to appearing on the cell surface (Blum and Cresswell, 1988; Guagliardi *et al.*, 1990). Within these vesicles the invariant chain, a protein associated with class II molecules, is released following its proteolysis. Cleavage and release of the invariant chain may be required to permit foreign peptide binding to class II molecules. Other immunologically relevant proteins also appear to be proteolyzed in endosomes. Antigen-antibody complexes are bound and internalized via Fc receptors present on immune cells. In macrophages, susceptible antigens are rapidly proteolyzed with endosomes (Mayorga *et al.*, 1989a) while resistant immunoglobulin molecules are typically transported to lysosomes for catabolism (Mellman and Plutner, 1984). Some monoclonal antibodies are more sensitive to proteolysis, and their cleavage in endosomes has been observed (Rudd *et al.*, 1988). Several aspects of the immune response are regulated via the binding of antibodies or antigen-antibody complexes to cells (Virgin *et al.*, 1985). The removal and catabolism of these complexes may therefore serve as an important control step *in vivo*.

The identification of active proteases in endosomes provides new insight into the complex function of these vesicles within cells. A scheme for the sorting and delivery of proteases and ligands into endosomes is presented in Figure 2. Protein ligands are delivered from the cell surface into vesicles which then fuse to form early endosomes. Proteases are transported into these vesicles and ligand cleavage occurs. Endosomal proteases such as cathepsin D are derived from intracellular sources such as the Golgi apparatus. Some proteases, however, such as those responsible for insulin cleavage, may be sorted into endosomes from the plasma membrane. Future studies must address the molecular mechanisms responsible for targeting and retaining proteases in endosomes. Equally important are the mechanisms for sorting ligands and the proteolytic products of endosomal processing reactions. Critical to this model are membrane fusion and fission events that occur between intracellular organelles. Reconstitution of the fusion

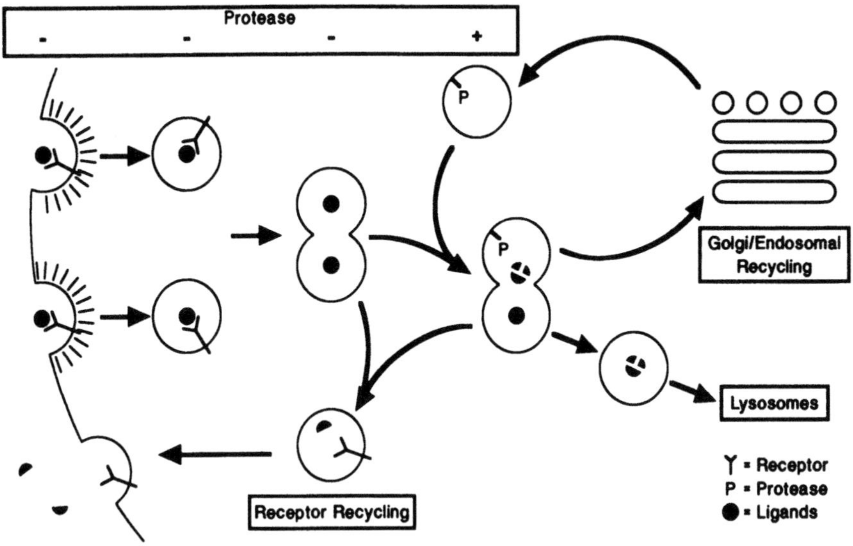

FIGURE 2. Model for the transport of ligands and proteases into endosomal vesicles.

events required for protease delivery into endosomes may be the first step in understanding the regulation of endosomal proteolysis (Mayorga *et al.*, 1989a).

3. RECONSTITUTION OF VESICLE FUSION DURING ENDOCYTOSIS

The formation, function, and fate of endocytic vesicles and their contents is brought about by vesicle fusion events at various steps along the endocytic pathway (Figure 3). It is likely that vesicle fusion events control the overall rate and direction of the endocytic process. Putative intracellular membrane fusion events include (1) the pinching off of coated vesicles from the plasma membrane, (2) the fusion of endosomes with each other as suggested by morphological studies, (3) the formation of receptor-rich vesicles from endosomes (e.g., CURL) during receptor recycling, and (4) fusion of these vesicles with the plasma membrane. As endosomes "mature," (5) invaginations of their membranes may yield multivesicular bodies. The transport of ligand from endosomes to lysosomes may require (6) endosome-lysosome fusion. Finally, vesicle fusion may be required during transport between the trans-Golgi reticulum and endosomal compartment (7).

Vesicle transport during endocytosis in whole cells has been widely studied,

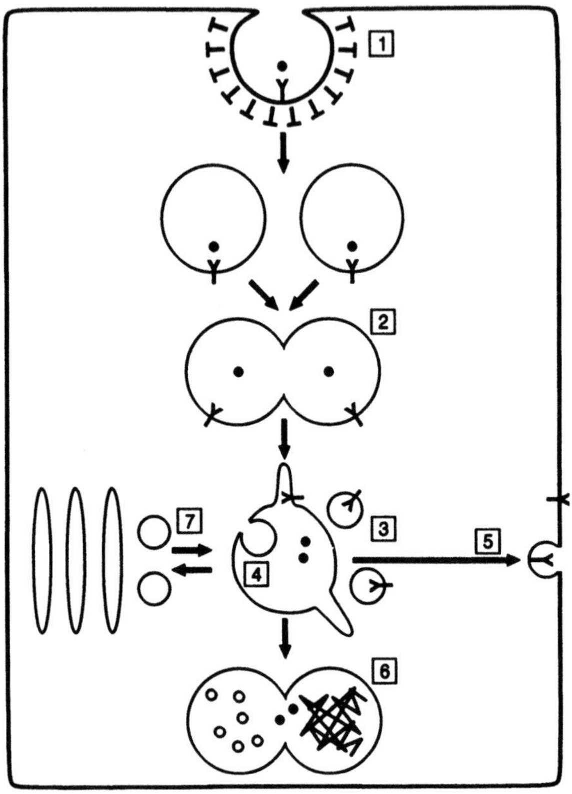

FIGURE 3. Putative vesicle fusion events during endocytosis. These include (1) coated vesicle formation, (2) endosome-endosome fusion, (3) budding of receptor-rich vesicles from endosomes, (4) multivesicular formation, (5) fusion of recycling vesicles to the plasma membrane, (6) endosome-lysosome fusion, and (7) transport between trans Golgi and endosomes.

but little information has emerged on the mechanism of membrane recognition and fusion. A major obstacle has been the difficulty in manipulating the cytosolic environment surrounding intracellular vesicles. The study of vesicle transport in the exocytic pathway is hampered by similar constraints. One approach to overcoming this problem has been to reconstitute vesicle transport events in broken-cell preparations. With an *in vitro* experimental strategy, both cytosol and transport vesicles are more accessible to experimental manipulation. The transport of proteins, both through the Golgi apparatus and from the endoplasmic reticulum to Golgi, has been successfully reconstituted in this fashion and is becoming amenable to study by biochemical means (Balch *et al.*, 1984; Beckers *et al.*, 1987).

Experimentally, endosome fusion is reconstituted using various markers internalized into endosomes of different cell populations. Once the cells are homogenized, reconstitution studies can be carried out by mixing endocytic vesicle fractions containing different probes and monitoring the interaction of the probes upon fusion of the vesicles to each other.

The first reconstitution of vesicle fusion was reported by Oates and Touster (1980). In these studies, the conditions required for phagolysosome fusion in *Acanthamoeba* were characterized. Vesicles containing two different biological or synthetic markers in separate phagosomes were incubated together, and mixing of the markers in the same compartment constituted a measure of vesicle fusion. These experiments provided evidence that phagolysosome fusion could take place *in vitro*.

Alstiel and Branton (1983) described an assay using a population of purified coated vesicles loaded with nonfluorescent carboxyfluorescein acetate, a membrane-permeable probe. When these vesicles were allowed to fuse to lysosomes, after being artificially stripped of their clathrin coat, lysosomal enzyme hydrolysis of the probe's acetate group produced carboxyfluorescein. The fluorescence of this reaction product served as a measure of fusion. Fusion could only be induced in the presence of Ca^{2+} at high concentrations and did not require cytosol or membrane protein. The specificity of this reaction could not be studied because of the heterogeneous source of coated vesicles and the nonselective method used for loading the probe.

3.1. Endosome-Endosome Fusion

Several studies have reported reconstituting fusion between endosomes. Davey *et al.* (1985) have described an assay for fusion between two populations of endocytic vesicles from Chinese hamster ovary (CHO) cells. Two viruses, Fowl Plague virus (FPV) and Semliki Forest virus (SFV), were internalized into endosomes in separate populations of CHO cells. Upon fusion of endosomal vesicles obtained from homogenates of these cells, neuraminidase activity on the surface of FPV could hydrolyze radiolabeled sialic acid residues of SFV present in the same endosomal compartment. Endosome fusion required ATP, but no cytosol requirement was observed. The fusion signal was very low when compared to background activity, making the significance of these results difficult to evaluate. Moreover, viruses that use membrane fusion as a mechanism of insertion may alter endosome fusion when they are present in the extravesicular buffer. The fusion mixtures used in this study contained large numbers of free viral particles. Despite its limitations, this was the first assay that reconstituted fusion among endosomes *in vitro*.

Our laboratory has designed both biochemical and morphological techniques to study endosome-endosome fusion (Diaz *et al.*, 1988, 1989a). The biochemical assay is based on two probes that are recognized by the macrophage mannose receptor and internalized into endosomes by receptor-mediated endocytosis: mannose-derivitized IgG specific for the hapten dinitrophenol (anti-DNP IgG) and dinitrophenol derivitized β-glucuronidase. These probes have a high affinity for each other and are not degraded after relatively long periods of internalization. Vesicle preparations from macrophages that have internalized both fusion probes separately are used in these studies. Upon fusion of endosomes, immune complexes form that can be quantitated by measuring β-glucuronidase activity. The amount of immune complex formed within closed vesicles constitutes a measure of vesicle fusion. Alternatively, if endosomes are marked with colloidal gold particles of different sizes, endosome fusion can be assessed by monitoring colocalization of different-size colloidal gold within the same vesicle. Both approaches have proved useful in vesicle-fusion studies.

In vitro fusion of macrophage endosomes containing ligands internalized via the mannose receptor required ATP, cytosol, KCl, and both membrane and cytoplasmic proteins (Table II). ATP was commonly added in the presence of an ATP regenerating system. Nonhydrolyzable analogues of ATP were not effective in supporting fusion. Other nucleotide triphosphates (GTP) were only partially effective in the absence of added ATP. Trypsinization of cytosol or membrane-associated proteins inhibits the fusion reaction. Fusion was not inhibited by proton ionophores or amines, which suggests that endosome acidification is not necessary for the fusion step. Fusion was temperature-sensitive, being fully inhibited at temperatures below 18°C. Below this temperature, transfer of endocytosed ligand to lysosomes is inhibited *in vivo*. The block of endocytic transport may be partially caused by the inability of early endosomes to fuse to each other as suggested by the *in vitro* studies. The alkylating agent N-ethylmaleimide (NEM) inhibited the fusion reaction. This agent has also been shown to inhibit vesicle-mediated transport in the secretory pathway (Glick and Rothman, 1987).

Endosome fusion was only detected between endosomal preparations from cells that had internalized fusion probes for short periods of time (5 min at 37°C). As ligand was chased to lysosomes or late endocytic compartments, there was a concomitant loss of fusogenicity. These early endosomes had a distinctive light density when fractionated by equilibrium Percoll density gradients that commonly distinguish endosomes from more dense lysosomes (Diaz *et al.*, 1989a). Endosomal fission or budding, thought to take place as part of receptor recycling, could not be assessed since the assay distinguishes only vesicle fusion events (in effect, mixing of luminal contents).

Endosomes first form clusters when incubated under fusogenic conditions (Diaz *et al.*, 1989a). These clusters were very labile (formation and/or disassembly was inhibited by low temperatures and dilution) which may indicate a low affinity interaction. However, endosomal clustering was observed when preparations were chemically fixed while being incubated in media that supports fusion.

Table II
Requirements for Endosome Fusion

Condition	Fusion
Complete[a]	+++
Nucleotides	
ATP-RS[b]	+++
ATP (1mM)	++
GTP (1mM)	+
PNP-AMP[c] (1mM)	−
ATPase inhibitors	
Azide (1mM)	+++
Vanadate (1mM)	+++
NEM (1mM)	−
Ions	
+KCl (50mM)	+++
+Sucrose	−
Ionophores	
$+NH_4Cl$ (10mM)	+++
+Nigericin (10μM)	+++
Temperature	
37°C	+++
18°C	−
Trypsin treatment of	
Cytosol	−
Vesicles	−

[a]Vesicles containing probes + cytosol (1.5mg protein/ml) + KCl (50mM) + 250 mM sucrose + ATP-regenerating system.
[b]ATP regenerating system.
[c]Adenyl-imidodiphosphate.

Endosome aggregates were observed by electron microscopy using the morphological assay with colloidal gold particles as ligands (Figure 4A). In this instance colloidal gold of varying size could be used to show clustering and multiple fusion of endosomes. Aggregation required cytosol and ATP; hence, it is likely that a cytosol component(s) plays a role in the specific recognition of endosomes. The limiting membranes of the vesicles found in these aggregates appear to undergo multiple membrane fusion reactions. The resulting large-diameter vesicles maintained the same density as endosomes when analyzed by Percoll gradient sedimentation. These large-diameter vesicles were no longer capable of fusion using this assay. Fusion-dependent inactivation of vesicle fusion could be a result of (1) fusion per se, (2) the increase in vesicle size that occurs as a consequence of vesicle fusion, or (3) some maturation step occurring under fusion conditions.

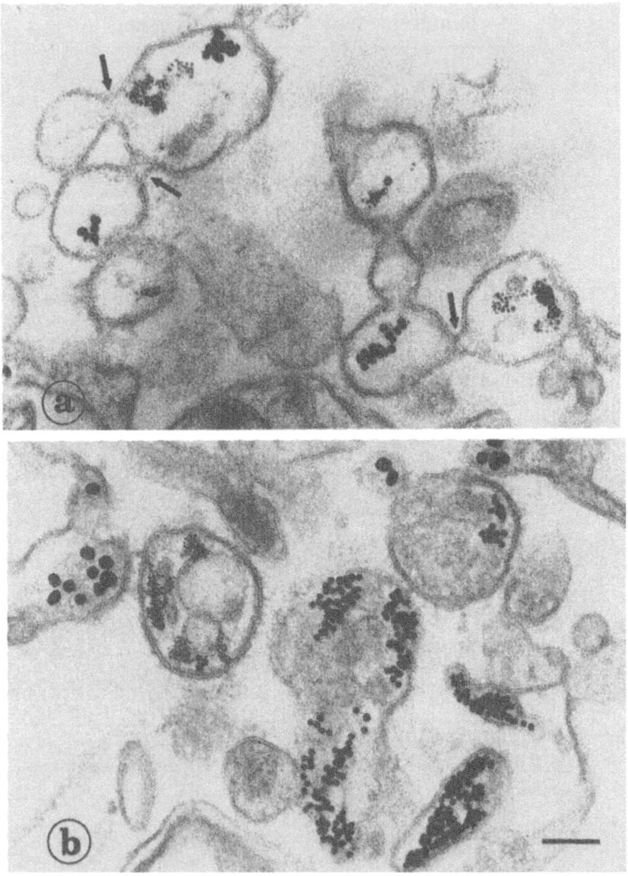

FIGURE 4. Clustering and fusion of endosomes under normal conditions or after GTPγS stimulation. Subcellular fractions containing 5-, 10-, or 20 nm gold-labeled ligands loaded in 5 min endosomes were incubated *in vitro* for 60 min at 37°C under fusion conditions containing (a) 1.5 mg of cytosolic proteins/ml or (b) 0.1 mg of cytosolic proteins/ml and 20 μM GTPγS. After fusion, the samples were fixed in suspension, pelleted, and analyzed for transmission electron microscopy. Arrows in (b) point to tubular connections. Bar = 100 nM.

Our *in vitro* endosome fusion assays do not reconstitute the transfer of endosomal contents to lysosomes. Fusion of early endosomes *in vitro* did not result in the appearance of assay probes in heavy density vesicles similar to lysosomes. It is possible that our reconstitution studies do not provide the appropriate conditions for either endosome maturation, which alone may be required for lysosome biogenesis, or endosome-lysosome fusion. Luzio and co-workers, however, have recently shown the transfer of asialofetuin from a late endocytic compartment to lysosomes in crude liver homogenates using Nycodenz gradients (Mullock *et al.*, 1989). Ligand transfer required cytosol, ATP. and the presence of lysosomes.

Several other *in vitro* assays have also been developed to study endosome-endosome fusion. The basic requirements for endosome fusion are remarkably similar in each of these diverse assay systems. Braell (1987) has used two probes, avidin-β-galactosidase and biotin-IgG, which have a high affinity for each other and gain access to endosomes by fluid-phase uptake. Fusion of vesicles containing each probe results in the formation of a complex that can be immunoprecipitated and then quantitated by β-galactosidase activity. Woodman and Warren (1988) have taken a similar approach using the transferrin receptor. IgG bound to transferrin but with available sites for additional transferrin binding was internalized into endosomes of A431 cells via the transferrin receptor. A preparation of these endosomes was incubated with a second vesicle population containing radiolabeled transferrin, which can interact with the free binding sites of the antibody. Immunoprecipitation of radiolabeled transferrin was a measure of fusion. Gruenberg and Howell (1986) made use of an antibody to vesicular stomatitis virus (VSV) G protein to immunoprecipitate, from a cell homogenate, endosomes with G protein inserted in their membranes. These vesicles fused with lactoperoxidase-loaded endosomes, allowing lactoperoxidase to radioiodinate the VSV G protein. Using electron-dense markers, the characteristics of fused endosomes could be observed morphologically. These studies suggest that only early tubulovesicular structures were active in fusion (Gruenberg *et al.*, 1989). When membrane-associated VSV G protein was chased into larger spherical vesicles, the latter were not fusogenic with early endosomal elements or with each other. This nonfusogenic compartment was observed to inhibit transfer of ligand to lysosomes when microtubule disrupting agents were present.

Vesicle fusion mediating transport from endosomes to the Golgi apparatus has been reconstituted in semiintact cells (Goda and Pfeffer, 1988). Mannose-6-phosphate receptors lacking sialyated oligosaccharides were localized in endosomal vesicles. Fusion of these endosomes with elements of the trans-Golgi resulted in the addition of sialic acid residues to the receptor. This fusion process was shown to require cytosol and ATP. Whether an intermediate or "shuttle" vesicle is required for transport from endosomes to the trans-Golgi is not known. Carrier vesicles mediate the transfer of newly synthesized proteins from the endoplasmic reticulum to the Golgi apparatus and between Golgi stacks in the secretory pathway.

3.2. Endosome–Plasma Membrane Fusion

The fusion of endosomes to each other may not constitute a transport step per se in the endocytic cycle since ligand is not necessarily transferred from one cellular compartment to another. Vectorial transport is achieved only if two distinct membrane-bound compartments are mixed during the fusion process; such fusion-mediated transport has been well-characterized in the secretory pathway. Carrier vesicles that bud from one compartment and fuse with the compartment of destination are responsible for the transport of newly synthesized proteins from the ER to the Golgi apparatus and between Golgi stacks (Pfeffer and Rothman, 1987). The first transport step in endocytosis is the transfer of ligands bound to receptors on the cell surface to an intracellular compartment. These two compartments may differ in several aspects (e.g., pH, presence of proteases) and transport between them probably requires membrane-fusion events.

Only small portions of the cell surface (in effect, clathrin-coated pits) are active in the internalization of receptor-ligand complexes. Two reports have addressed this step in the endocytic cycle. Coated vesicle formation has been observed to occur in semiintact cells (Smythe *et al.*, 1989). This was measured by the loss of accessibility of antitransferrin antibodies to transferrin in coated pits as they bud off from the plasma membrane. ATP and cytosol appeared to be required for this step. We have observed that vesicles derived from the plasma membrane, formed after homogenization by the inside-out sealing of receptor-ligand–containing regions of the cell surface, are able to fuse with early endosomes (Mayorga *et al.*, 1988). The same biochemical approach as that designed for endosome-endosome fusion was used in these studies. In these experiments, cell-surface Fc receptors were saturated with aggregated anti-DNP IgG by incubation at 4°C prior to homogenization. Aggregation of the antibody was achieved by adding an anti–mouse IgG polyclonal rabbit antibody. Early endosomes were loaded with the "fusion partner," dinitrophenol-derivitized β-glucuronidase by a 5 min uptake at 37°C. Formation of immune complexes within vesicles served as a measurement of vesicle fusion. This biochemical assay showed that plasma membrane–derived vesicles were able to fuse with early endosomes. Vesicle fusion in this case was less efficient than endosome-endosome fusion. As an alternative assay, colloidal gold of two different sizes were coated with the ligands and used as morphological markers of endosomes and plasma membrane–derived vesicles. Prior to homogenization, colloidal gold coated with aggregated antibody was localized in coated invaginations of the plasma membrane. After homogenization, a population of sealed vesicles containing the ligand was observed. These vesicles were often decorated with the typical coat observed on clathrin-coated vesicles. The density of these vesicles, assessed by fractionation in Percoll gradients, showed that they were lighter than endosomes. This confirmed that plasma membrane–derived vesicles were selectively loaded since these are typically less dense than endosomes. When preparations containing plasma membrane–derived vesicles were incubated, under fusogenic conditions with a homogenate containing early endosomes loaded with colloidal gold of a different size, the appearance of vesicles containing both sizes of gold particles was observed.

The requirements for both endosome-endosome- and endosome–plasma membrane–derived vesicle fusion were very similar, suggesting that they are mediated by the same mechanism. In both instances, vesicle fusion required ATP and cytosol, and it was inhibited by NEM. Membrane factors were also required since fusion was inhibited by treatment of the vesicles with trypsin. Moreover, plasma membrane–derived vesicles appear to fuse with early endosomes. Late endosomes, loaded with DNP-β-glucuronidase by a 2 min uptake followed by 15 min chase, were not able to fuse with early endosomes or plasma membrane–derived vesicles. Fusion with endosomes was also observed using plasma membrane–derived vesicles loaded with mannosylated antidinitrophenol IgG, suggesting that interaction with endosomes is not a peculiar property of Fc receptor–containing, plasma membrane–derived vesicles. Coated regions of the plasma membrane, involved in the internalization of receptor-ligand complexes, must therefore contain the necessary information to fuse with early endosomes. This fusion event may mediate the transport of ligands from the cell surface to an endocytic compartment. These results suggest that fusion is not only involved in the formation of large vesicles from small early endosomes, but also plays a role in the transport of ligands and receptors along the endocytic pathway.

3.3. Proteolysis

The presence of proteolytic enzymes in endosomal vesicles may have important implications for the processing of extracellular molecules delivered into this compartment. Specifically, proteases may be involved in the processing of antigens internalized by antigen-presenting cells such as macrophages. However, not all proteins are equally susceptible to endosomal proteases. Our fusion probes, DNP-β-glucuronidase and anti–DNP IgG remain fully active long after internalization. A protein that has proved to be very useful to study endosomal proteolysis is bovine serum albumin (BSA). Proteolytic fragments of mannose-BSA start to appear in the medium a few minutes after endocytosis by macrophages (Diment and Stahl, 1985). Endosomal proteolysis was also observed using an immune complex formed with radiolabeled DNP-BSA and anti–DNP IgG (Mayorga *et al.*, 1989a). This complex is bound by the Fc receptor of macrophages. To study hydrolysis *in vitro,* endosomal vesicles were incubated in iso-osmotic sucrose buffered to pH 4.5, or alternatively, at pH 7.0 with the addition of an ATP-regenerating system. Proteolysis of the radiolabeled DNP-BSA was observed in endosomes isolated 5–10 min after ligand internalization. Plasma membrane–derived vesicles and very early endosomes are unable to degrade ligands upon *in vitro* incubation (Mayorga *et al.*, 1989a). This presumably is because of the lack of proteases in very early endocytic vesicles.

As outlined above, we have observed that plasma membrane–derived vesicles are able to fuse with endosomes (Mayorga *et al.*, 1988). Plasma membrane–

endosome fusion should mix ligand-containing vesicles with protease-positive vesicles, and hence, trigger the proteolytic processing of ligands. Fusion of very early endosomes containing immune complexes of radiolabeled DNP-BSA and anti–DNP IgG, resulted in the onset of degradation. Vesicle fusion as measured by ligand proteolysis was ATP and cytosol dependent, as well as inhibitible by NEM treatment and treatment of the membranes with trypsin. Fusion-dependent proteolysis was strongly affected by dilution of the incubation reaction. This indicates that the stimulation of hydrolysis is mediated by the interaction of two separate compartments. Fusion-dependent hydrolysis was observed under conditions that are able to support fusion with endosomes; however, fusion with other intracellular compartments could not be ruled out. Fusion-dependent proteolysis was observed using membrane fractions from which lysosomes had been removed. Fusion of light-density protease-containing vesicles coming from, for example, the Golgi apparatus may, however, occur during this *in vitro* assay. Golgi-derived vesicles may be the primary origin of the endosomal enzymes, and fusion among early endocytic compartments may spread these enzymes throughout the endocytic compartment.

4. REGULATION OF VESICLE-MEDIATED TRANSPORT DURING ENDOCYTOSIS

Endosome fusion requires a mechanism of specific recognition and membrane fusion. The two processes may be closely linked. Possible candidates that may play a role in vesicle recognition are endocytic receptors. These membrane proteins may help distinguish endosomes from other compartments. However, the lack of characterization of endosomal membrane proteins does not exclude other proteins as possible target molecules. Since cytosol appears to be required for endosome clustering, a soluble factor may be required for endosome recognition or, alternatively, clustering may be activated or stabilized by a factor present in cytosol. Further studies are needed to address these possibilities.

A limited number of mechanisms may have evolved to induce vesicle fusion. Whether two vesicles fuse depends on their proximity and their capacity to recognize each other as suitable fusion partners. Vesicles can be brought together by random diffusion or by directed translocation via microtubule tracks. Such tracks have been shown to mediate endosome translocation (De Brabander *et al.*, 1988). The juxtaposition of endosomal membranes could be sufficient to induce their fusion without a priming event. Alternatively, an endosome recognition step may activate the fusion process. Several observations indicate that vesicle fusion steps in different transport pathways share similar characteristics and are affected by many of the same agents. Specifically, it appears that an N-ethylmaleimide sensitive factor(s) and at least one GTP-binding protein play a role in both exocytic and endocytic vesicle transport.

4.1. NEM-Sensitive Factor

Reconstitution studies of vesicle transport and fusion between endosomes have shown, in all cases tested, that N-ethylmaleimide (NEM) is a potent inhibitor of the process (Braell, 1987; Woodman and Warren, 1988; Diaz *et al.*, 1988). Our laboratory has shown that separate treatment with NEM of either cytosol or vesicle fractions alone only partially inhibits endosome fusion. This suggested that the NEM-sensitive factor distributed itself between soluble- and membrane-associated forms. In fact, fusion could be restored to NEM-inactivated vesicles and cytosol by adding fresh fractions of cytosol. Such an assay has been used to show that a NEM-sensitive factor purified from CHO cells and active in Golgi transport, was able to restore endosome fusion (Diaz *et al.*, 1989b). This factor is a tetramer with 76,000 d subunits, which shares significant homology with the product of the yeast *SEC18* gene (Wilson *et al.*, 1989). *SEC18* yeast mutants are defective in transport along the exocytic pathway. This NEM-sensitive factor (NSF) can mediate the fusion of plasma membrane–derived vesicles to endosomes as well as fusion-dependent proteolysis in endosomes. Thus, elements of the "catalytic" machinery that mediate fusion events may be shared by various transport pathways. If this were the case, vesicle recognition would be determined by other components to prevent futile membrane fusion events between vesicles active in unrelated intracellular pathways. Alternatively, catalytic units may have several membrane recognition sites, allowing them to exert their fusogenic activity on different membrane fusion events and still provide specificity. Thus, different types of NSF could be active in different transport pathways.

4.2. Role of GTP-Binding Proteins in the Fusion Process

Several independent lines of evidence indicate that GTP-binding proteins play an active role in the fusion-mediated transport of macromolecules along both the endocytic and exocytic pathways. First, at least two GTP-binding proteins have been identified that are required for normal secretion in yeast. The *YPT1* gene product is a 23,000-d protein required for transport from the endoplasmic reticulum to the Golgi apparatus (Segev *et al.*, 1988). The *SEC4* gene encodes a 23,500-d GTP-binding protein that associates with secretory vesicles and plasma membrane. The product of the *SEC4* gene is required for a late step of the secretory process (Salminen and Novick, 1987). Second, nonhydrolyzable analogs of GTP inhibit several transport steps when tested in cell-free systems or in permeabilized cell preparations. Transport between ER and Golgi, in both mammalian (Beckers and Balch, 1989) and yeast systems (Baker *et al.*, 1988; Ruohola *et al.*, 1988) is inhibited by GTPγS. Transport between Golgi stacks (Melancon *et al.*, 1987) and recycling from endosomes to trans-Golgi (Goda and Pfeffer, 1988) is also inhibited by GTPγS. Finally, a large number of studies have shown that nonhydrolyzable GTP analogues are able to trigger the exocytosis of secretory granules in permeabilized cells (Burgoyne, 1987).

Nonhydrolyzable GTP analogues also affect fusion between early endocytic compartments. Plasma membrane–endosome fusion, endosome-endosome fusion as well as fusion-dependent proteolysis are inhibited by GTPγS. This inhibitory effect is observed only in the presence of high cytosol concentrations (Mayorga *et al.*, 1989b). A similar cytosol dependency has been observed for the inhibitory effect of GTPγS in Golgi transport. Interestingly, when GTPγS is assessed at concentrations of cytosol that do not support full fusion activity, a prominent activation of endosome-endosome fusion is observed (Mayorga *et al.*, 1989c). In the presence of the GTP analogue, the cytosol concentration required to obtain maximal activity decreases severalfold. This striking GTPγS-mediated activation of fusion at low cytosol concentrations and inhibition at high cytosol concentrations is shown in Figure 5. Moreover, if endosomes are preincubated in cytosol and GTPγS, washed, and then incubated in a fusion reaction in the absence of cytosol, fusion is still observed (Mayorga *et al.*, 1989c). These results suggest that a GTP-binding protein is involved in the attachment of cytosolic

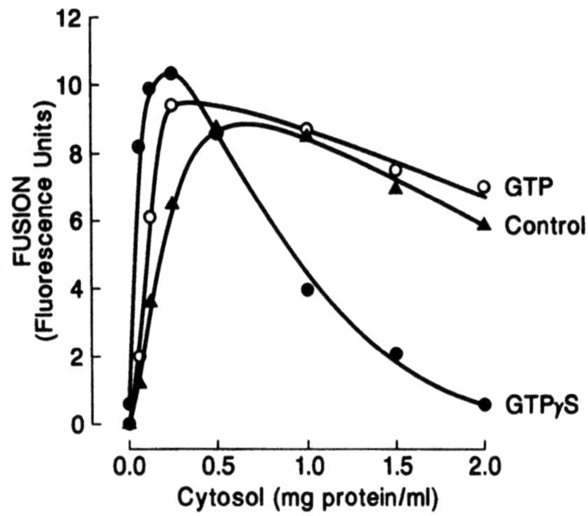

FIGURE 5. Effect of GTPγS on *in vitro* endosome fusion at different cytosol concentrations. Endosomal fractions containing probes were incubated under fusion conditions with different concentrations of gel-filtered cytosol in the presence (●) or absence (▲) of 20 μM GTPγS. After 60 min incubations at 37°C, the samples were lysed and immunoprecipitated, and the amount of complex formed was quantitated by measuring β-glucuronidase activity with a fluorescence assay.

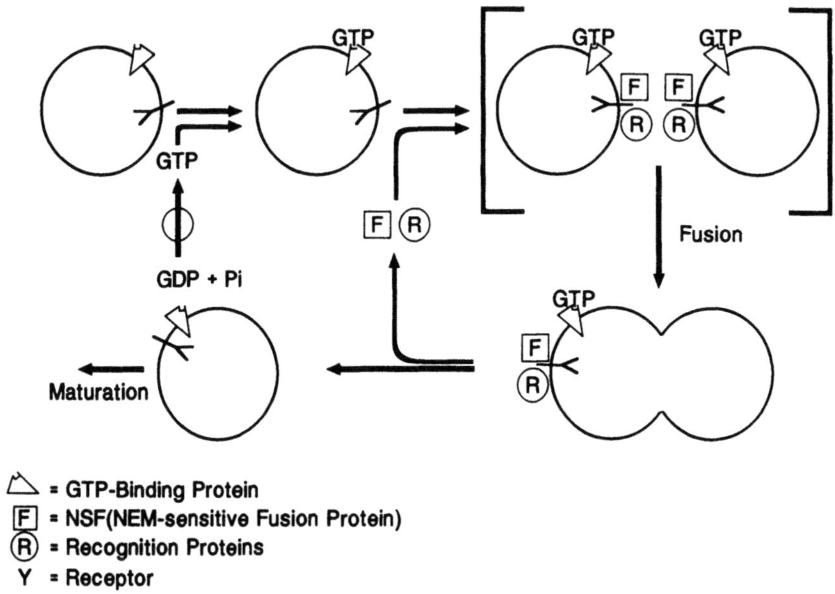

FIGURE 6. Speculative model for endosome fusion.

factors required for fusion. The binding of these factors to the endosome membranes would be irreversible in the presence of a nonhydrolyzable GTP analog and fusion would become cytosol independent after the preincubation.

As mentioned above, endosome clusters have been observed during the fusion reaction. These aggregates have a very peculiar structure when the fusion reaction is carried out under GTPγS stimulation. Under this condition, most of the endosomes within the aggregate are connected by intervesicular channels forming a three-dimensional network (Figure 4B). This network connects round, tubular, annular, and multivesicular endosomes, suggesting that all these structures are fusogenic in the *in vitro* assay.

Thus, it seems clear that GTP-binding proteins have an active role in the mechanism of vesicle fusion in both the endocytic and exocytic pathways. However, several questions about the actual role of these proteins still need to be answered. First, it is not known whether the inhibitory and activating effect of GTPγS is mediated by the same proteins. Second, it is not clear whether these proteins are a necessary component of the fusion reaction or whether they have a regulatory role in the process. Finally, the relationship between the different GTP-binding proteins identified along the exocytic and endocytic pathways is completely unknown. In the future it will be necessary to characterize the precise function of the GTP-binding proteins that have been purified (e.g., *YPT1* and

SEC4 gene products) active in secretion and to purify those GTP-binding proteins whose effect has been characterized in different *in vitro* transport assays. A number of small GTP-binding proteins have been characterized that have unknown cellular functions. This family of proteins includes the ras proto-oncogene (Burgoyne, 1989). Their widespread distribution in different mammalian tissue makes them potential candidates for mediating organelle traffic and vesicle fusion.

Although a precise model to explain vesicle fusion during endocytosis has not yet emerged, it is clear that certain factors are required for the process to take place. One speculative model may invoke the presence of a GTP-binding protein on the endosomal membrane which, in the presence of GTP, induces various soluble factors required for fusion to bind membranes (Figure 6). These factors may include NSF and additional proteins which appear to mediate its binding to membranes (Weidman *et al.*, 1989). Soluble recognition proteins may mediate vesicle aggregation. Receptors may play a role in the process by serving as anchors for some of these soluble factors. Once all required factors are assembled on the membrane, vesicle fusion occurs. The soluble factors may then be released, permitting additional rounds of membrane fusion. This release may be regulated by the hydrolysis of GTP. In this manner the GTP-binding protein could initiate another round of vesicle fusion.

5. REFERENCES

Alstiel, L. and Branton, D., 1983, Fusion of coated vesicles with lysosomes: measurement with a fluorescent assay, *Cell* **32**:921–929.

Ajioka, R. S. and Kaplan, J., 1987, Characterization of endocytic compartments using the horseradish peroxidase–diaminobenizidine density shift technique, *J. Cell Biol.* **104**:77–85.

Anderson, R. G. W., Brown, M. S., and Goldstein, J. L., 1977, Role of the coated endocytic vesicle in the uptake of receptor-bound low-density lipoprotein in human fibroblasts, *Cell* **10**:351–364.

Baker, D., Hicke, L., Rexach, M., Schleyer, M., and Schekman, R., 1988, Reconstitution of *SEC* gene product-dependent intercompartmental protein transport, *Cell* **54**:335–344.

Balch, W. E., Dunphy, W. G., Braell, W. A., and Rothman, J. E., 1984, Reconstitution of the transport of protein between successive compartments of the Golgi measured by the coupled incorporation of N-acetylglucosamine, *Cell* **39**:405–416.

Beckers, C. J. M. and Balch, W. E., 1989, Calcium and GTP: essential components in vesicular trafficking between the endoplasmic reticulum and Golgi apparatus, *J. Cell. Biol.* **108**:1245–1256.

Beckers, C. J. M., Keller, D. S., and Balch, W. E., 1987, Semi-intact cells permeable to macromolecules: Use in reconstitution of protein transport from the endoplasmic reticulum to the Golgi complex, *Cell* **50**:523–534.

Beguinot, L., Hanover, J. A., Ito, S., Richert, N. D., Willingham, M. C., and Pastan, I., 1985, Phorbol esters induce transient internalization without degradation of unoccupied epidermal growth-factor receptors, *Proc. Natl. Acad. Sci. USA* **82**:2774–2778.

Blum, J. S. and Cresswell, P., 1988, Role for intracellular proteases in the processing and transport of class II HLA antigens, *Proc. Natl. Acad. Sci. USA* **85**:3975–3979.

Blum, J. S., Fiani, M. L., and Stahl, P. D., 1989, Characterization of neutral and acidic proteases in endosomal vesicles, *J. Cell Biol.* **109:**188a.

Braell, W. A., 1987, Fusion between endocytic vesicles in a cell-free system, *Proc. Natl. Acad. Sci. U.S.A.* **84:**1137–1141.

Burgoyne, R. D., 1987, G proteins: control of exocytosis, *Nature* **328:**112–113.

Burgoyne, R. D., 1989, Small GTP-binding proteins, *TIBS* **14:**393–396.

Davey, J., Hurtley, S. M., and Warren, G., 1985, Reconstitution of an endocytic event in a cell-free system, *Cell* **43:**643–652.

Davis, C. G., van Driel, I. R., Russell, D. W., Brown, M. S., and Goldstein, 1987, The low-density lipoprotein receptor. Identification of amino acids in cytoplasmic domain required for rapid endocytosis, *J. Biol. Chem.* **262:**4075–4082.

De Brabander, M., Nuydens, R., Geerts, H., and Hopkins, C. R., 1988, Dynamic behavior of the transferrin receptor followed in living epidermoid carcinoma (A431) cells with navoid microscopy, *Cell Motil. Cytoskel.* **9:**30–47.

Diaz, R., Mayorga, L., and Stahl, P., 1988, *In vitro* fusion of endosomes following receptor-mediated endocytosis, *J. Biol. Chem.* **263:**6093–6100.

Diaz, R., Mayorga, L. S., Mayorga, L. E., and Stahl, P. D., 1989a, *In vitro* clustering and multiple fusion among macrophage endosomes, *J. Biol. Chem.* **264:**13171–13180.

Diaz, R., Mayorga, L. S., Weidman, P. J., Rothman, J. E., and Stahl, P. D., 1989b, Vesicle fusion following receptor-mediated endocytosis requires a protein active in Golgi transport, *Nature* **339:**398–400.

Diment, S., Leech, M. S., and Stahl, P. D., 1988, Cathepsin D is membrane-associated in macrophage endosomes, *J. Biol. Chem.* **263:**6901–6907.

Diment, S., Martin, K. J., and Stahl, P. D., 1989, Cleavage of parathyroid hormone in macrophage endosomes illustrates a novel pathway for intracellular processing of proteins, *J. Biol. Chem.* **264:**13403–13406.

Diment, S. and Stahl, P., 1985, Macrophage endosomes contain proteases which degrade endocytosed protein ligands, *J. Biol. Chem.* **260:**15311–15317.

Duncan, J. and Kornfeld, S., 1988, Intracellular movement of two mannose-6-phosphate receptors: return to the Golgi apparatus, *J. Cell Biol.* **106:**617–628.

Fiani, M. L. and Stahl, P. D., 1988, Intracellular processing of ricin and ricin A by macrophages, *J. Cell Biol.* **107:**114a.

Gabel, C. A. and Foster, S. A., 1987, Postendocytic maturation of acid hydrolases: evidence of prelysosomal processing, *J. Cell Biol.* **105:**1561–1570.

Geuze, H. J., Slot, J. W., Strous, G. J. A. M., Lodish, H. F., and Schwartz, A. L., 1983, Intracellular site of asialoglycoprotein receptor-ligand uncoupling. Double-label immunoelectron microscopy during receptor-mediated endocytosis, *Cell* **32:**277–287.

Glick, B. S. and Rothman, J. E., 1987, Possible role for fatty acyl-coenzyme A in intracellular protein transport, *Nature* **326:**309–312.

Glickman, J. N., Conibear, E., and Pearse, B. M. F., 1989, Specificity of binding of clathrin adaptors to signals on the mannose-6-phosphate/insulin-like growth factor II receptor, *EMBO J.* **8:**1041–1047.

Goda, Y. and Pfeffer, S. R., 1988, Selective recycling of the mannose-6-phosphate/IGF-II receptor to the trans-Golgi network *in vitro, Cell* **55:**309–320.

Griffiths, G., Hoflack, B., Simons, K., Mellman, I., and Kornfeld, S., 1988, The mannose-6-phosphate receptor and the biogenesis of lysosomes, *Cell* **52:**329–341.

Gruenberg, J. E. and Howell, K. E., 1986, Reconstitution of vesicle fusions occurring in endocytosis with a cell-free system, *EMBO J.* **5:**3091–3101.

Gruenberg, J., Griffiths, G., and Howell, K. E., 1989, Characterization of the early endosome and putative endocytic carrier vesicles *in vivo* and with an assay of vesicle function *in vitro, J. Cell Biol.* **108:**1301–1316.

Guagliardi, L. E., Koppelman, B., Blum, J. S., Mark, M. S., Cresswell, P., and Brodsky, F. M., 1990, Colocalization of molecules involved in antigen processing and presentation in an early endocytic compartment, *Nature* **343:**133–139.

Hamel, F. G., Posner, B. I., Bergeron, J. J. M., Frank, B. H., and Duckworth, W. C., 1988, Isolation of insulin degradation products from endosomes derived from intact rat liver, *J. Biol. Chem.* **263:**6703–6708.

Harding, C., Heuser, J., and Stahl, P., 1983, Receptor-mediated endocytosis of transferrin and recycling of the transferrin receptor in rat reticulocytes, *J. Cell Biol.* **97:**329–339.

Harding, C., Levy, M. A., and Stahl, P., 1985, Morphological analysis of ligand uptake and processing: the role of multivesicular endosomes and CURL in receptor-ligand processing, *Eur. J. Cell Biol.* **36:**230–238.

Hasilik, A. and Neufeld, E. F., 1980, Biosynthesis of lysosomal enzymes in fibroblasts, *J. Biol. Chem.* **255:**4937–4945.

Lobel, P., Fujimoto, K., Ye, R. D., Griffiths, G., and Kornfeld, S., 1989, Mutations in the cycoplasmic domain of the 275 kd mannose-6-phosphate receptor differentially alter lysosomal enzyme sorting and endocytosis, *Cell* **57:**787–796.

Mayorga, L. S., Diaz, R., and Stahl, P. D., 1988, Plasma membrane–derived vesicles containing receptor-ligand complexes are fusogenic with early endosomes in a cell-free system, *J. Biol. Chem.* **263:**17213–17216.

Mayorga, L. S., Diaz, R., and Stahl, P. D., 1989a, Reconstitution of endosomal proteolysis in a cell-free system, *J. Biol. Chem.* **264:**5392–5399.

Mayorga, L. S., Diaz, R., and Stahl, P. D., 1989b, Regulatory role for GTP-binding proteins in endocytosis, *Science* **244:**1475–1477.

Mayorga, L. S., Diaz, R., Colombo, M. I., and Stahl, P. D., 1989c, GTPγS stimulation of endosome fusion suggests a role for a GTP-binding protein in the priming or vesicles before fusion, *Cell Regulation* **1:**113–124.

McCoy, K. L., Miller, J., Jenkins, M., Ronchese, F., Germain, R. N., and Schwartz, R. H., 1989, Diminished antigen processing by endosomal acidification mutant antigen-presenting cells, *J. Immunol.* **143:**29–38.

Melancon, P., Glick, B. S., Malhotra, V., Weidman, P. J., Serafini, T., Gleason, M. L., Orci, L., and Rothman, J. E., 1987, Involvement of GTP-binding G proteins in transport through the Golgi stack, *Cell* **51:**1053–1062.

Mellman, I. and Plutner, H., 1984, Internalization and degradation of macrophage Fc receptors bound to polyvalent immune complexes, *J. Cell Biol.* **98:**1170–1177.

Mullock, B. M., Branch, W. J., van Schaick, M., Gilbert, L. K., and Luzio, J. P., 1989, Reconstitution of endosome-lysosome interaction in a cell-free system, *J. Cell Biol.* **108:**2093–2099.

Oates, P. J. and Touster, O., 1980, *In vitro* fusion of *Acanthamoeba* phagolysosomes. III. Evidence that cyclic nucleotides and vacuole subpopulations respectively control the rate and the extent of vacuole fusion in *Acanthamoeba* homogenates, *J. Cell Biol.* **85:**804–810.

Opresko, L. K. and Karpf, R. A., 1987, Specific proteolysis regulates fusion between endocytic compartments in Xenopus oocytes, *Cell* **51:**557–568.

Pearse, B. M. F., 1988, Receptors compete for adaptors found in plasma membrane coated pits, *EMBO J.* **11:**3331–3336.

Pease, R. J., Smith, G. D., and Peters, T. J., 1985, Degradation of endocytosed insulin in rat liver is mediated by low-density vesicles, *Biochem. J.* **228:**137–146.

Pfeffer, S. R. and Rothman, J. E., 1987, Biosynthetic protein transport and sorting by the endoplasmic reticulum and Golgi, *Ann. Rev. Biochem.* **56:**829–852.

Rodman, J. S., Levy, M. A., Diment, S., and Stahl, P. D., 1990, Immunolocalization of endosomal cathepsin D in rabbit alveolar macrophages, *J. Leuk. Biol.* **48:**116–122.

Roederer, M., Bowser, R., and Murphy, R. F., 1987, Kinetics and temperature dependence of exposure of endocytosed material to proteolytic enzymes and low pH; evidence for a maturation model for the formation of lysosomes, *J. Cell Physiol.* **131**:200–209.

Rothman, J. E. and Schmid, S. L., 1986, Enzymatic recycling of clathrin from coated vesicles, *Cell* **46**:5–9.

Rudd, E., Kindber, G. M., Blomhoff, H. K., Godal, T., and Berg, T., 1988, Degradation of a monoclonal anti-μ chain antibody in a human surface IgM-positive B cell line starts in prelysosomal vesicles, *J. Immunol.* **141**:2951–2958.

Ruohola, H., Kabcenell, A. K., and Ferro-Novick, S., 1988, Reconstitution of protein transport from the endoplasmic reticulum to the Golgi complex in yeast: the acceptor Golgi compartment is defective in the *sec23* mutant, *J. Cell Biol.* **107**:1465–1476.

Salminen, A. and Novick, P. J., 1987, A *ras*-like protein is required for a post-Golgi event in yeast secretion, *Cell* **49**:527–538.

Schaudies, R. P., Gorman, R. M., Savage, C. R., and Poretz, R. D., 1987, Proteolytic processing of epidermal growth factor within endosomes, *Biochem. Biophys. Res. Commun.* **143**:710–715.

Schmid, S. L., Fuchs, R., Mâle, P., and Mellman, I., 1988, Two distinct subpopulations of endosomes involved in membrane recycling and transport to lysosomes, *Cell* **52**:73–83.

Segev, N., Mulholland, J., and Botstein, D., 1988, The yeast GTP-binding YPT1 protein and a mammalian counterpart are associated with the secretion machinery, *Cell* **52**:915–924.

Simmons, B. M., Stahl, P. D., and Russell, J. H., 1986, Mannose receptor-mediated uptake of ricin toxin and ricin A-chain by macrophages, *J. Biol. Chem.* **261**:7912–7920.

Smythe, E., Pypaert, M., Lucocq, J., and Warren, G., 1989, Formation of coated vesicles from coated pits in broken A431 cells, *J. Cell Biol.* **108**:843–853.

Snider, M. D. and Rogers, O. C., 1985, Membrane traffic in animal cells: cellular glycoproteins return to the site of Golgi mannosidase I, *J. Cell Biol.* **100**:826–834.

Stahl, P., Schlesinger, P. H., Sigardson, E., Rodman, J. S., and Lee, Y. C., 1980, Receptor-mediated pinocytosis of mannose glycoconjugates by macrophages: characterization and evidence for receptor recycling, *Cell* **19**:207–215.

Tycko, B. and Maxfield, F. R., 1982, Rapid acidification of endocytic vesicles containing α_2-macroglobulin, *Cell* **28**:643–651.

Unanue, E. R. and Allen, P. M., 1987, The basis for the immunoregulatory role of macrophages and other accessory cells, *Science* **236**:551–557.

Virgin, H. W., Wittenberg, G. F., and Unanue, E. R., 1985, Immune complex affects in murine macrophages, *J. Immunol.* **135**:3735–3743.

Weidman, P. J., Melancon, P., Block, M. R., and Rothman, J. E., 1989, Binding of an N-ethylmaleimide-sensitive fusion protein to Golgi membranes requires both a soluble protein(s) and an integral membrane receptor, *J. Cell Biol.* **108**:1589–1596.

Wileman, T., Boshans, R. L., Schlesinger, P., and Stahl, P., 1984, Monensin inhibits recycling of macrophage mannose-glycoprotein receptors and ligand delivery to lysosomes, *Biochem. J.* **220**:665–675.

Wileman, T., Boshans, R., and Stahl, P., 1985, Uptake and transport of mannosylated ligands by alveolar macrophages, *J. Biol. Chem.* **260**:7387–7393.

Wiley, H. S., Van Nostrand, W., McKinley, D. N., and Cunningham, D. D., 1985, Intracellular processing of epidermal growth factor and its effect on ligand-receptor interactions, *J. Biol. Chem.* **260**:5290–5295.

Wilson, D. W., Wilcox, C. A., Flynn, G. C., Chen, E., Kuang, W. J., Henzel, W. J., Block, M. R., Ullrich, A., and Rothman, J. E., 1989, A fusion-promoting protein needed for the transport from the endoplasmic reticulum and within the Golgi apparatus in both animal cells and yeast, *Nature* **339**:355–359.

Woodman, P. G. and Warren, G., 1988, Fusion between vesicles from the pathway of receptor-mediated endocytosis in a cell-free system, *Eur. J. Biochem.* **173**:101–108.

Chapter 4

Quantitative Fluorescence Techniques for the Characterization of Endocytosis in Intact Cells

Nita H. Salzman and Frederick R. Maxfield

1. INTRODUCTION

Fluorescence techniques have been used extensively in the biological sciences, and have been shown to be powerful tools for studying cell biology. Fluorescence has been used mainly for qualitative study in fixed preparations, as in immunofluorescence which has proved useful in the identification and localization of organelles and cellular proteins. Recently, a number of techniques have been developed for quantitative studies in both fixed and living cells.*

*List of abbreviations used in this chapter. α_2m, alpha$_2$-macroglobulin; ASGP, asialoglycoprotein; AFA, antifluorescein antibody; CHO, Chinese hamster ovary; CCD, charge-coupled device; CURL, compartment of uncoupling receptor and ligand; DiI 3,3′-dioctadecylindocarbocyanine iodide; DiO,; 3,3′-dioctadecyloxacarba-cyanine perchlorate; EGF, epidermal growth factor; FITC, fluorescein isothiocyanate; ISIT, intensifier silicon intensified target; LDL, low-density lipoprotein; LY, lucifer yellow; NBD (N-4-nitrobenzo-2-oxa-1,3-diazole)-aminocaproate; SIT, silicon-intensified target; Tf, transferrin; TR, transferrin receptor; TCA, trichloroacetic acid; TRITC, tetramethylrhodamine isothiocyanate; TRVb-1, a CHO variant cell line into which the human transferrin receptor has been transfected (McGraw *et al.* 1987).

Nita H. Salzman Department of Physiology, College of Physicians and Surgeons, Columbia University, New York, New York 10032, and Department of Pharmacology, New York University School of Medicine, New York, New York 10016. **Frederick R. Maxfield** Department of Pathology, College of Physicians and Surgeons, Columbia University, New York, New York 10032.
Subcellular Biochemistry, Volume 19: Endocytic Components: Identification and Characterization, edited by J. J. M. Bergeron and J. R. Harris. Plenum Press, New York, 1993.

Endocytosis is particularly amenable to quantitative analysis using fluorescence techniques. Endocytosis is the process by which extracellular macromolecules are taken into the cell. These macromolecules may be taken up nonspecifically by fluid-phase pinocytosis, or specifically by receptor-mediated endocytosis.

The initial step of receptor-mediated endocytosis involves the binding of a ligand to its receptor on the plasma membrane of the cell (Goldstein *et al.*, 1985). The receptor-ligand complexes diffuse within the plane of the membrane (Schlessinger *et al.*, 1978; Maxfield *et al.*, 1981) and cluster over clathrin-coated pits (Pearse and Bretscher, 1981; Keen, 1985). Coated pits invaginate and pinch off, forming early endosomes. Although this is the common route of ligand entry, some macromolecules enter cells via noncoated pits (Goldberg *et al.*, 1988; Sandvig *et al.*, 1988). Early endosomes can contain a combination of different receptors. Tf has been colocalized with α_2m (Yamashiro *et al.*, 1984), ASGP (Stoorvogel *et al.*, 1987), and EGF (Dickson *et al.*, 1983), demonstrating that receptors are not sorted from each other at the level of the coated pit, although the coated pit does selectively internalize receptors over other membrane proteins (Pearse and Bretscher, 1981).

At this point on the endocytic pathway, the receptor and ligand may take one of several routes. (1) The receptor recycles while the ligand is degraded as in the case of α_2m (Maxfield *et al.*, 1981), LDL (Goldstein *et al.*, 1985), ASGP (Bridges *et al.*, 1982). (2) The receptor and ligand recycle, as for Tf (Klausner *et al.*, 1983; Dautry-Varsat *et al.*, 1983). (3) The receptor and ligand are degraded, an example of which is EGF and its receptor (Renfrew and Hubbard, 1988). (4) The receptor and/or ligand are transported from one surface of a polarized epithelium to the other, as in Ig transport from the basolateral side of mammary epithelia to the apical surface, to be secreted into milk (Mostov *et al.*, 1984; Solari and Kraehenbuhl, 1984). (5) Ligands may be delivered from the endosome into the cytoplasm, as seen with toxins (Olsnes and Sandvig, 1985), some viruses (White *et al.*, 1983), and possibly hormones (Smith and Jarett, 1987). Figure 1 illustrates the endocytic pathways in CHO cells, a cell line that has been used in many studies of endocytosis employing microscopy as well as biochemical and genetic methods.

Fluorescence techniques have made it possible to study many aspects of endocytosis in living cells since the process of endocytosis provides an efficient means of introducing fluorophores into the cell. Specific and nonspecific proteins and macromolecules may be labeled with fluorescent markers and taken up by cells via receptor-mediated endocytosis or fluid-phase pinocytosis. Recently, fluorescent lipid probes have been synthesized that label the endocytic pathway, allowing observation of lipid membrane traffic (Koval and Pagano, 1989).

An early means of fluorescently labeling lysosomes and other acidic organelles was with the weak base acridine orange (Robbins and Marcus, 1963).

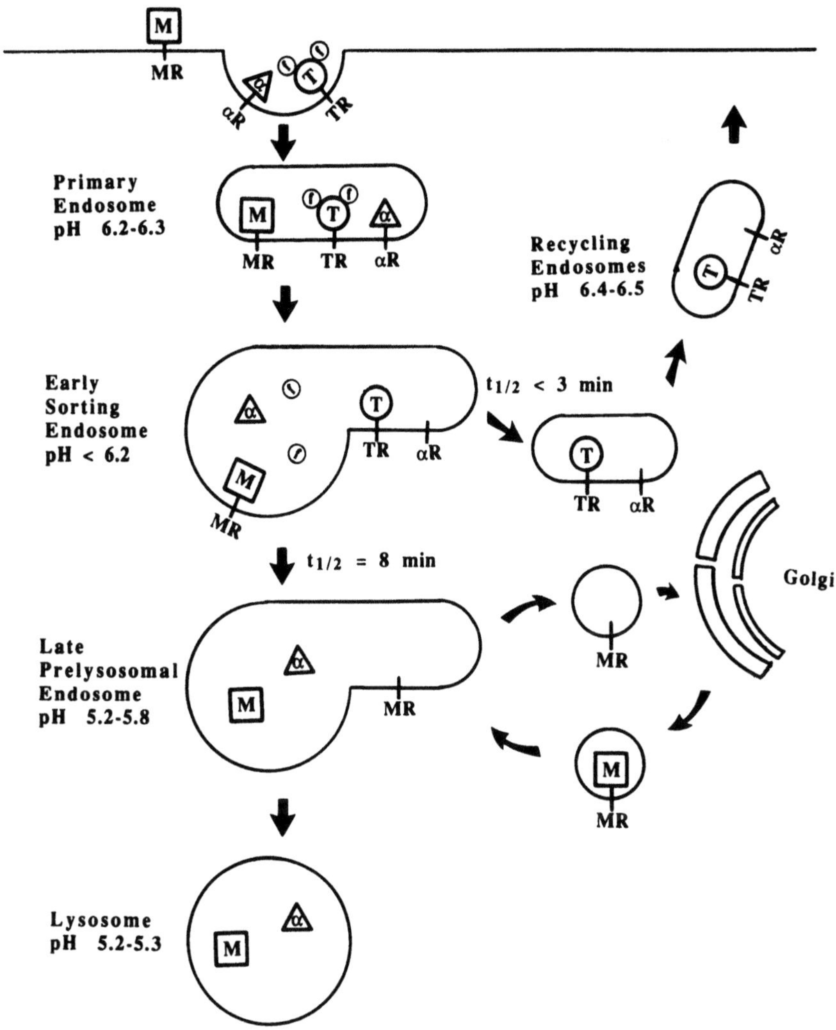

FIGURE 1. Schematic model for endocytosis and receptor recycling in CHO cells. Various ligands bind to receptors that collect in clathrin-coated pits. These include transferrin (T), α_2-macroglobulin (α), and lysosomal enzymes with mannose-6-phosphate (M). The coated vesicles pinch off to form primary endosomes that fuse with the early sorting endosome. Acidification of these endosomes causes release of iron from transferrin, and the apo-transferrin remains receptor bound. Acidification also causes release of α_2-macroglobulin from its receptor. Recycling receptors (and transferrin) are removed from the sorting endosome by budding off in recycling endosomes. In CHO cells these recycling endosomes accumulate near the Golgi complex. Ligands such as α_2-macroglobulin or LDL accumulate in the sorting endosome through several rounds of incoming vesicle fusion. With a half-time of about 8 min, the sorting endosome loses the capacity to fuse with primary endosomes, and it matures into a late endosome. Dissociation of lysosomal enzymes is delayed compared to other ligands, presumably because of a requirement for lower pH to cause dissociation. The late endosome would also be a site where newly synthesized lysosomal enzymes are delivered.

In its neutral form it diffuses through membranes, but in acidic compartments it becomes protonated and trapped within the acidic organelle. As the dye accumulates, acidic organelles are easily observed by fluorescence microscopy. At high concentrations the fluorescence emission of acridine orange shifts from green to red (Robbins and Marcus, 1963; Hart and Young, 1975). This method provides a simple way to visualize acidic compartments (Anderson and Pathak, 1985). Acridine orange labeling is not specific for the endocytic pathway because other organelles are acidic, and the dye also binds to other cellular structures including nuclei. One way of specifically labeling the endocytic pathway is by fluid-phase pinocytosis, using membrane-impermeant fluorescent molecules such as Lucifer Yellow (LY), FITC-dextran, or Texas Red-lactalbumin (for review see Swanson, 1989a). Fluorescent dyes can also be incorporated into specific peptide and hormone ligands, and labeled ligands have been used to investigate the paths of specific ligands or receptor-ligand complexes such as LDL, ASGP, EGF, insulin, and Tf (for review see Maxfield, 1989b).

There are constraints on the use of fluorescence in the study of cells with very low receptor numbers. Sufficient quantities of fluorescent ligand must be endocytosed for the fluorescence to be visible or measurable above the cellular autofluorescence. Generally, a cell must incorporate approximately 10^5 fluorophores to obtain a useful signal by image-intensification fluorescence microscopy, especially when the fluorophores are distributed widely throughout the cell (Maxfield, 1989b). Lower numbers of ligands can be detected if the molecules are intensely fluorescent or if the fluorophores are concentrated in a small region within the cell. For example, the LDL molecule can incorporate as many as 40 to 50 molecules of the fluorophore DiI. This quantity of a highly fluorescent label allows the visualization of individual LDL-receptor complexes (Barak and Webb, 1981).

Other fluorophores commonly used include FITC, TRITC, NBD, LY, and Texas Red. The choice of fluorophore must be carefully considered to maximize its fluorescence and minimize interference by cellular autofluorescence. Usually, the fluorescent signal must exceed the cellular autofluorescence by a considerable amount to obtain interpretable quantitative results. Fluorophores that excite and emit at longer wavelengths are more favorable since cellular autofluorescence is decreased at longer wavelengths (Aubin, 1979).

Along with cellular autofluorescence, other environmental effects should be considered. FITC fluorescence, for example, is sensitive to the pH of its environment between pH 5 and pH 8. FITC fluorescence decreases at low pH, and measurements of FITC fluorescence in living cells may be confounded by this characteristic. This property has been exploited quite effectively to measure the pH of endocytic compartments and lysosomes as discussed in greater detail in a later section.

Fluorophores may be affected by the experimental techniques used. FITC and NBD are subject to significant photobleaching when exposed to intense excitation light and should be used with low light levels and image intensification microscopy when viewing live cells. A dye that is less sensitive to photobleaching such as TRITC may be a better choice for lengthy experiments in living cells. When fixed specimens are viewed, agents such as phenylenediamine (Johnson *et al.*, 1982) or n-propyl gallate (Giloh and Sedat, 1982) that slow photobleaching can largely eliminate this problem. In all cases, fluorescent probes must be carefully evaluated before attempting quantitative experiments.

2. FLUORESCENT PROBES OF FLUID-PHASE PINOCYTOSIS

Commonly used pinocytic probes include LY and FITC or TRITC labeled dextrans. Ideally fluid-phase markers are endocytosed nonspecifically (that is, without binding to cellular components) and fill all compartments along the endocytic pathway (Swanson, 1989a). The markers are retained in lysosomes and are useful for intense labeling and study of lysosomes (Swanson, 1989a). They have also been employed to measure the kinetics of pinocytosis, both influx and efflux, and membrane recycling (Swanson *et al.*, 1985; Swanson, 1989b). Fluorescent fluid-phase markers have been used to label different populations of lysosomes and document exchange of contents between lysosomes (Ferris *et al.*, 1987). Combinations of fluid-phase markers have been used similarly in experiments in polarized epithelia, exploring the convergence of apical and basolateral endocytic pathways using confocal microscopy (Parton *et al.*, 1989).

Although fluid-phase markers label lysosomes well, it is uncertain whether they share the same early endocytic compartments with ligand-receptor complexes internalized by receptor-mediated endocytosis. The endosomes involved in receptor-mediated endocytosis have a fluid phase and will take in bulk-flow markers, it is nevertheless possible that the markers are pinocytosed by additional mechanisms such as endocytosis via noncoated pits (Swanson, 1989a). Therefore, one cannot necessarily extrapolate data using fluid-phase markers to apply to receptor-mediated endocytosis.

Care must be taken to ensure that pinocytosed markers behave as purely fluid-phase molecules (Swanson, 1989a). Adherence to membrane components on the surface or within organelles can alter the dynamics of the probe's influx and complicate the interpretation of experimental results. The nonideal behavior of these probes must be kept in mind whenever they are used.

3. PROBES OF RECEPTOR-MEDIATED ENDOCYTOSIS

Derivatizing native peptides and hormones with fluorophores is a powerful tool for studying endocytosis, allowing the collection of specific information about receptor/ligand behavior and pathways taken by different ligands. Most commonly, FITC and TRITC are used and covalently linked to ligand proteins. Many peptide or protein ligands have been derivatized including α_2m (Maxfield *et al.*, 1978), Tf (Enns *et al.*, 1983; Yamashiro *et al.*, 1984), insulin (Schlessinger *et al.*, 1978), EGF (Schlessinger *et al.*, 1978), LDL (Pitas *et al.*, 1981; Barak and Webb, 1981), and ASGP (Tycko *et al.*, 1983).

The most important precautions to be taken are to minimize altering the natural properties of the ligand in binding affinity, internalization, and biological function, while still linking sufficient numbers of fluorophores to obtain measurable levels of fluorescence (Maxfield, 1989b). Some common problems include underderivatizing the protein, yielding an insufficiently fluorescent probe; and overderivatizing the protein, resulting in a denatured protein which has altered binding properties (Maxfield, 1989b). The best method for demonstrating appropriate binding is to demonstrate that binding of the fluorescent ligand is completed by excess unlabeled ligand (Maxfield *et al.*, 1981). It is also useful to compare binding and other behaviors to that of radiolabeled ligand as an additional demonstration that the fluorescent ligand is behaving similarly (Maxfield, 1989b).

Specific applications of fluorescent probes for the study of endocytosis are discussed later in this article. There are many different quantitative fluorescence techniques available, with individual strengths and weaknesses for the quantitation of dynamic processes in intact or living cells.

4. FLUOROMETRY

Spectrofluorometry is a technique that has been widely used in the physical and natural sciences. In cell biology, the technique is particularly useful for measuring large numbers of cells in a cuvette. It is a practical and easy way to measure properties of living cells in suspension. For studies of adherent cells, a coverslip holder that fits into the top of a fluorometer cuvette can be used (Okhuma and Poole, 1978). This makes it possible to measure fluorescence intensities on a monolayer of cells adherent to a glass coverslip. This technique has been used more recently (Salzman and Maxfield, 1989) for studies of endosome fusion. These studies are discussed in a later section.

Fluorometry has been used in elegant studies by Sklar *et al.*, measuring ligand-receptor interactions in human neutrophils (for review see Sklar, 1987). They used fluorometry and flow cytometry to measure the kinetics of ligand-receptor binding, dissociation, internalization, and cellular responses (Sklar *et al.*, 1982, 1984, 1985). Their methods took advantage of the particular fluorescent properties of fluorescein. Internalized FITC-labeled ligands were distinguished from external ligands by quenching the external FITC fluorescence by lowering the extracellular pH to 4 or by adding an antifluorescein antibody (AFA) which quenches FITC fluorescence (Sklar *et al.*, 1982a,b). They measured the binding of fluorescein-labeled-N-formyl peptide using spectroscopic fluorescence analysis of ligand binding and compared it to cellular responses, including membrane depolarization, superoxide anion generation, and elastase secretion. All of these responses were noted to begin within 10 sec of ligand binding. Elastase secretion is completed within 30 sec and requires less than 20% receptor occupancy. van Renswoude *et al.* (1982) used a similar technique to distinguish internalized from external FITC-Tf while studying the endocytic pathway of Tf and Fe delivery to cells.

5. FLOW CYTOMETRY

Flow cytometry has been used similarly for the study of endocytosis. It is applicable to the use of living cells in suspension and can be used for fluorescence intensity measurements on individual cells. The technique is useful in excluding certain artifacts that may create problems when studying large numbers of cells in a cuvette. These include debris, dead cells, and unbound fluorescent ligand. Autofluorescence can be a significant problem with this technique as with most fluorescence methods. A method for autofluorescence correction with flow cytometry has been developed by Roederer and Murphy (1986). For each cell they measure the FITC fluorescence and either the light scattering (as a measure of cell size) or the fluorescence at a wavelength that the fluorophore itself does not emit. The fluorescence at this second wavelength can be used to estimate the autofluorescence at the FITC excitation and emission wavelengths. Murphy and co-workers have used flow cytometry to study acidification along the endocytic pathways of EGF, insulin, Tf, and FITC-dextran (Murphy *et al.*, 1984; Roederer and Murphy, 1986; Sipe and Murphy, 1987).

6. MICROSCOPE SPECTROFLUOROMETRY

Microscope spectrofluorometry has found diverse uses in cell biology. The equipment usually consists of a fluorescence microscope equipped with interchangeable excitation filters, xenon or mercury light sources, and a photometer mounted in position to collect and measure emitted light (Maxfield, 1989a; Wang and Taylor, 1989; Wampler and Kutz, 1989). This technique is especially useful for observing cells that are adherent and spread flat on the coverslip surface. Instead of forcing cells (e.g., fibroblasts) to remain in suspension as is required for flow cytometry, the microscope method allows the measurement of the fluorescence in single cells attached to a solid substrate. The fluorescence intensities of individual cells or groups of cells can be measured. Debris and dead cells can be excluded from measurements by visual inspection of the measurement field.

Microscope-fluorometry instrumentation can be used along with image-intensification microscopy and digital image processing. Image-intensification microscopy allows the observation and measurement of cells at extremely low light levels, allowing the quantification of extremely low levels of fluorescence. There are several devices used for imaging, including SIT and ISIT video cameras, cameras coupled to image intensifiers, and cooled solid-state cameras (CCD).

SIT cameras are useful for moderately low light input. ISIT cameras add an additional stage of intensification to the SIT camera to provide additional sensitivity. These cameras produce noisy images at low light levels, but signal averaging with an image processor can produce acceptable images even with extremely weak fluorescent signals. As an alternative to the ISIT camera, an image intensifier can be optically coupled to a video camera (Spring and Lowy, 1989).

Cooled CCD cameras provide direct digital readout of images (Aikens *et al.*, 1989). The cooled CCDs combine high resolution, low noise, and broad dynamic range, but they have been limited by relatively slow readout of the images and the requirement to store all data in digital form, which can require large amounts of storage when continuously recording from live cells. This has limited their use in observing real-time signals in living cells, but these problems are being overcome as reasonably priced computers capable of rapidly storing vast amounts of digital data become available. As the technology advances, further application of these devices in the fluorescence microscopy of living cells should be seen.

Digital image processing has increased the power of image intensification microscopy by allowing computer-assisted processing of images. Cells can be digitized from videotaped images or directly from the video camera. Digital imaging allows the manipulation of images, background and autofluorescence subtractions, and fluorescence intensity measurements of individual endocytic compartments (Yamashiro *et al.*, 1984; Dunn *et al.*, 1989; for review see Maxfield, 1989a). Image-intensification microscopy and digital image processing have been used extensively for studies of endocytosis, including the measurement of endosome pH (Tycko *et al.*, 1983), and the identification and quantitation of endocytic vesicle fusion (Dunn *et al.*, 1989). Most recently, pathways taken by pinocytosed fluorescent molecules have been followed using the confocal microscope (Parton *et al.*, 1989) adding a powerful tool for investigating questions about endocytosis that were very difficult to study with conventional, full-field microscopy techniques.

Each of these techniques has its particular merits. Fluorometry is useful for the measurement of average fluorescent values of large numbers of cells in suspension or adherent to a coverslip. It does not allow for the exclusion of dead cells, cell debris, or other fluorescent debris. Flow cytometry is able to avoid the measurements of dead cells and debris, but it is only useful for cells in suspension. Microscope spectrofluorometry requires that the cells be adherent to a glass coverslip, and allows the selection of cells by the operator, enabling the exclusion of dead cells and debris. Large fields of several cells may be measured, or individual cells may be observed. The sample numbers are necessarily reduced by this method, but many artifacts are excluded. Video image-intensification microscopy, especially when coupled with digital image processing, can be used to study individual compartments within cells, allowing the investigation of more sophisticated questions regarding cell function.

7. pH MEASUREMENTS OF ENDOCYTIC COMPARTMENTS

One of the successful uses of microspectrofluorometry has been the measurement of pH of intracellular compartments. A general characteristic of endosomes is that they have been found to contain an electrogenic ATP-dependent proton pump that renders the interior of the endosome acidic (Yamashiro and Maxfield, 1988; Al-Awqati, 1986; Mellman *et al.*, 1986). The low pH of endocytic organelles has been shown to be essential for many of the functional consequences of endocytosis, such as the fusion of viruses with cellular membranes (White *et al.*, 1980), penetration of diphtheria toxin A chain and some other toxins into the cytoplasm (Olsnes and Sandvig, 1985), changes in receptor conformation (DiPaola and Maxfield, 1984), release of ligands from their receptors (Harford *et al.*, 1983, Bridges *et al.*, 1982), the release of Fe from Tf (Klausner *et al.*, 1983; Dautry-Varsat *et al.*, 1983), activation of lysosomal enzymes (Barrett, 1972), and ligand degradation. Clearly, release of ligands from their receptors is a prerequisite for delivery of ligands such as LDL to lysosomes while the receptors are efficiently recycled. In the case of Tf, both the ligand and receptor are returned to the cell surface. It has been shown that apo-Tf remains bound to receptors at mildly acidic pH values, but it is rapidly released at pH 7.4, the situation that would occur when recycling receptors are returned to the cell surface.

The measurement of pH was advanced by the use of the pH sensitive fluorescent dye FITC conjugated to macromolecules (Ohkuma and Poole, 1978; Tycko and Maxfield, 1982). FITC can be coupled to proteins that follow different endocytic pathways, so the pH probe can be used to selectively label lysosomes of specific types of endosomes. As the pH increases from five to seven, the FITC fluorescence intensity increases significantly at 490 nm while changing less at 450 nm (Figure 2). Ratios of fluorescence intensities in cells at these two wavelengths can be used to find the pH in the compartments containing FITC. This characteristic behavior of FITC underlies the pH determination by most of the methods discussed here. The use of a ratio technique renders pH measurements independent of the quantity of dye in the cells or the number of cells being

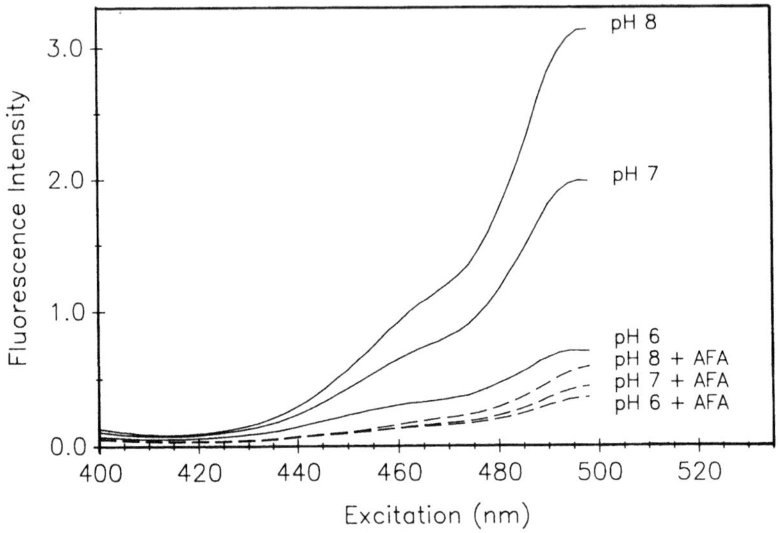

FIGURE 2. pH sensitivity of fluorescein. FITC-α_2m (5 ug/ml) was added to buffers of pH = 6, 7, and 8, with (—) and without (–) AFA (2 mg/ml). Fluorescence was measured using a fluorometer (excitation scanned from 400 to 500 nm, emission = 520 nm). In the absence of AFA, the excitation spectrum is strongly pH dependent. AFA quenches fluorescein fluorescence and makes it nearly independent of pH.

measured. Fluorescein fluorescence is relatively insensitive to other factors such as dye concentration or ions other than H^+ (Heiple and Taylor, 1980). For accurate measurements, however, it is essential to obtain calibration curves with the FITC probe in the same compartments that measurements will be made from. This is accomplished by varying the external pH and collapsing transmembrane pH gradients with ionophores, weak bases, and in some cases weak acids. Ratio measurements of the endocytosed probe at each pH are used to generate the calibration curve.

The methods used to measure the pH of individual endosomes by image intensification microscopy and digital image analysis have been described in detail elsewhere (Maxfield, 1989a; Tycko and Maxfield, 1982; Yamashiro *et al.*, 1984). Cells are incubated with a FITC-labeled ligand for a period of time that will allow entry into the endocytic compartment being investigated. This will be only a few minutes for endosomes and periods up to hours for lysosomes. Images of living cells are obtained with 450-nm and 490-nm fluorescence excitation. It is usually necessary to correct for background arising from autofluorescence and out-of-focus endosomes. This is accomplished by determining the median fluorescence in a local area surrounding each endosome and subtracting this intensity out as a background (Maxfield 1989a). The intensity of each endosome in the 490-nm image is taken as the sum of all contiguous picture elements (pixels) above an arbitrary threshold. The 450-nm intensities are determined for exactly the same pixels in each endosome. The 450/490 intensity ratio is then calculated for each endosome, and the pH of the endosome is obtained by comparison with the calibration curve. The pH of individual endosomes can be determined with a standard deviation of about 0.2 to 0.3 pH units over the range from 5.0 to 6.5 (Yamashiro and Maxfield, 1987b).

Using FITC-labeled dextrans, Ohkuma and Poole (1978) showed that lysosomes in macrophages have a pH between 4.5 and 5.0. Acidification of prelysosomal endosomes has been demonstrated using FITC-labeled ligands in a variety of cell types including Balc/c 3T3 (Tycko and Maxfield, 1982), HepG2 (Tycko *et al.*, 1983), CHO (Yamashiro *et al.*, 1984; Yamashiro and Maxfield, 1987a), Swiss 3T3 fibroblasts (Roederer and Murphy, 1986), and lymphocytes (Murphy *et al.*, 1984). The ligands used for these studies included α_2m (Tycko and Maxfield, 1982; Yamashiro *et al.*, 1984; Yamashiro and Maxfield, 1987a,b), ASGP (Tycko *et al.*, 1983), and insulin (Murphy *et al.*, 1984). The pH of sorting endosomes and late endosomes has been found to be between 5.0 and 6.0 in several cell types.

The pH of the earliest-formed endocytic compartments proved to be difficult to measure using fluorescence microscopy techniques. Primary endosomes are small compartments containing only a few fluorescent ligand-receptor complexes, making the fluorescence signal difficult to distinguish from cellular autofluorescence, and therefore difficult to quantify. Nevertheless, pH measurements of this compartment could be made using a null point method (Yamashiro and Maxfield, 1987a,b). This method involves exposing cells successively to membrane-impermeant and permeant pH buffers. Cells are incubated for short times (about 2 min) with FITC-labeled ligand and rinsed. The medium is exchanged with a lower-pH test buffer (e.g., pH 6.4) that is membrane impermeant. This will affect fluorescein that remains outside the cell, but it does not affect intracellular fluorescein fluorescence or autofluorescence. All transmembrane pH gradients are then collapsed by the addition of methylamine chloride and ammonium acetate without changing the buffer pH. If the FITC-labeled ligand is in a lower pH environment than that of the test buffer, the fluorescein intensity will increase when the pH gradients are collapsed. This process is repeated with test buffers of different pH. The pH of the test buffer at which no intensity change occurs is considered to be the average pH of the compartment containing the FITC-ligand.

The pH of early endosomes in CHO cells, as determined by the null point method, is 6.3. This was found by using a 3 min incubation with FITC-Dextran. The pH values obtained by fluorescence microscopy are generally in good agreement with values obtained by flow cytometry. Both methods show that endosomes on the lysosomal pathway become progressively more acidic (Yamashiro *et al.*, 1984; Yamashiro and Maxfield, 1987b; Sipe and Murphy, 1987).

8. FLUORESCENCE STUDIES OF LIGAND-RECEPTOR SORTING

The sorting of ligands and receptors along endocytic pathways can be observed directly by fluorescence microscopy. This is illustrated in Figure 3, using DiO-LDL to label the pathway leading to lysosomes and TRITC-Tf to label the recycling pathway. Recycling Tf-receptor complexes follow the same pathway back to the cell surface as LDL receptors and other recycling receptors (McGraw and Maxfield, unpublished). Chinese hamster ovary (CHO) cells were incubated with DiO-LDL and TRITC-Tf for 2 min and then fixed immediately or after chase periods of 2 or 4 min. At the earliest time (Figure 4 a,b), most endosomes contain both LDL and Tf as would be expected since they are internalized together in clathrin-coated pits. If the cells are incubated for an additional 2–4 min in the absence of fluorescent ligands, little Tf is found in the LDL-containing endosomes. During these brief chase times, the Tf has begun to move to the center of the cell (Figure 4 d,f). Electron-microscopic studies have shown that in CHO cells this central fluorescence arises from a collection of tubular recycling endosomes that accumulate near the Golgi complex (Yamashiro *et al.*, 1984). In addition to Tf and its receptor, recycling LDL receptors can be immunolocalized in the same recycling endosomes (McGraw and Maxfield, unpublished).

In addition to using the technique of monitoring separation of ligands, one can also see the fusion of endosomes directly by fluorescence microscopy (Pastan and Willingham, 1985). Figure 5 demonstrates the fusion of newly formed endosomes with endosomes previously labeled with LDL. Cells were incubated with DiO-LDL for 2 min and then incubated for an additional 2 or 8 min to allow internalized LDL to move along the endocytic pathway. TRITC-Tf was then added to the cells for 2 min, and the fluorescence was observed. With a 2-min interval between the two incubations, many of the LDL-containing endosomes fused with subsequently formed endosomes containing Tf (Figure 5 a,b). However, with an 8-min interval only a small fraction of LDL-containing endosomes became labeled with TRITC-Tf. This indicates that the ability to fuse with newly formed endosomes decays as endosomes age.

Using digital image processing techniques, Dunn *et al.* (1989) have quantified the relative amounts of fluorescent ligand in individual endosomes by measuring their fluorescence intensity. The methods used to measure the intensity of single endosomes have been described in detail elsewhere (Dunn and Maxfield, 1990). Over time, the increase in brightness of individual endosomes was measured. In LDL-containing endosomes, the endosome brightness increased nearly linearly for 10 min and increased approximately thirtyfold. This

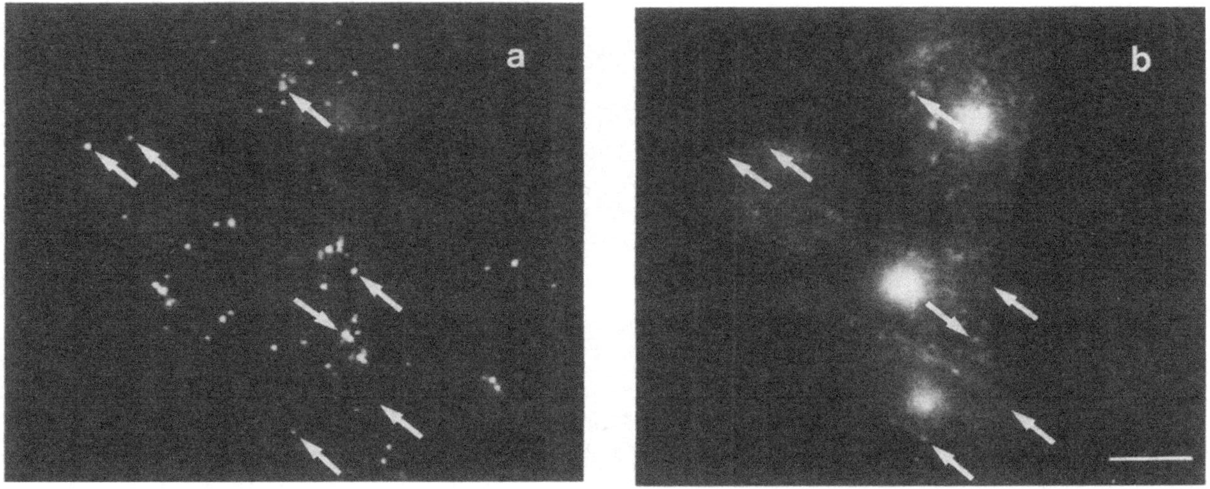

FIGURE 3. Access of TRITC-Tf to an LDL-accumulating sorting endosome. TRVb-1 cells were incubated in 5 ug/ml diO-LDL and 20 ug/ml TRITC-Tf for 10 min. DiO-LDL fluorescence is shown in (a), and TRITC-Tf fluorescence is shown in (b). Arrows indicate examples of endosomes containing both ligands. Note that TRITC-Tf occurs in endosomes containing a range of quantities of DiO-LDL, including those with the largest accumulation of DiO-LDL. Reproduced from Dunn *et al.* (1989). Bar = 10 μm.

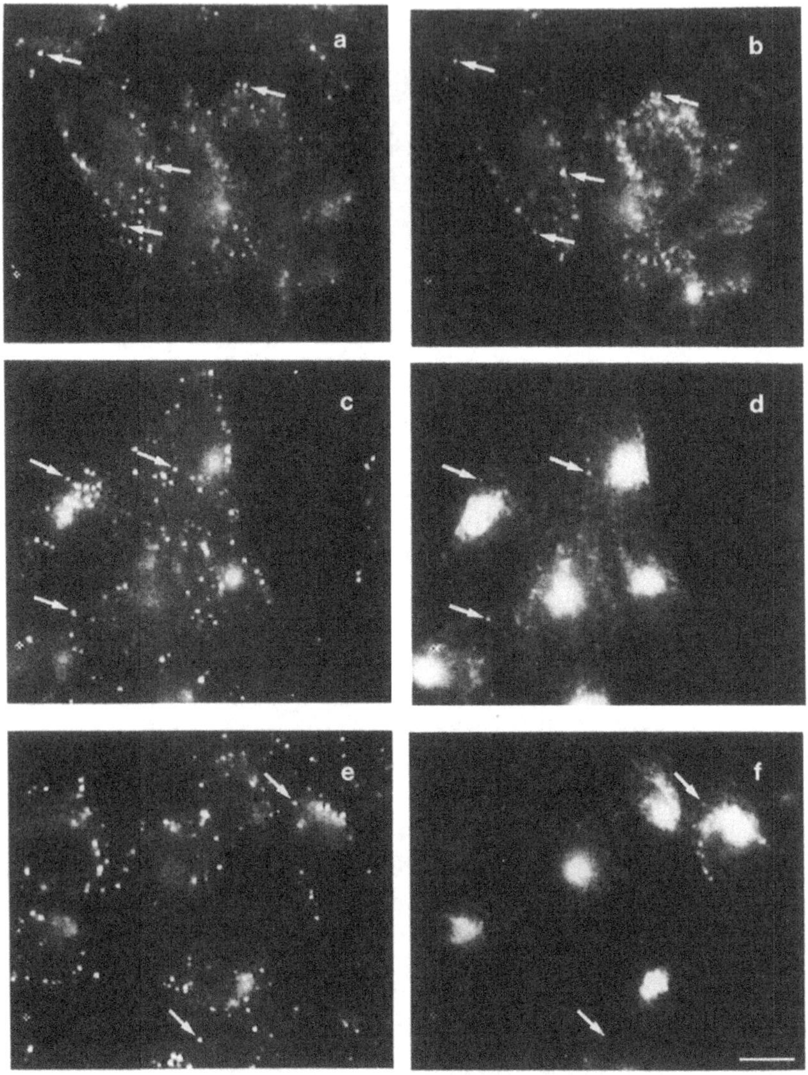

FIGURE 4. Movement of TRITC-Tf through the sorting endosome. TRVb-1 cells were incubated in 5 ug/ml DiO-LDL and 20 ug/ml TRITC-Tf for 2 min and either fixed immediately (a, b), rinsed and incubated for another 2 min in the absence of labels (c, d), or rinsed and incubated for another 4 min in the absence of labels (e, f). a, c, and e were obtained using fluorescein optics to display DiO-LDL. b, d, and f were obtained using rhodamine optics to display TRITC-Tf. Arrows indicate examples of endosomes containing both ligands. Using the arrows as landmarks, one can see that with time Tf appears to depart from the LDL-containing endosomes as it accumulates in the para-Golgi region. Reproduced from Dunn *et al.* (1989). Bar = 10 μm.

indicates that the LDL-containing endosomes underwent at least 30 fusions over a 10-min period. After approximately 10 min the fluorescence intensity per endosome remained nearly constant. When a similar experiment was done with Tf, endosomes containing Tf increased in brightness fourfold over 2 min and then leveled off. This type of ligand accumulation is consistent with a model for endocytosis in which both LDL and Tf enter the sorting endosome by fusion of primary endosomes with the sorting endosome. Most of the LDL is retained in the sorting endosome, but Tf is rapidly removed and delivered to the postsorting recycling compartment. This is illustrated schematically in Figure 6.

An independent fluorescence assay for endosome fusion has been developed that measures the accessibility of previously formed endosomes to fusion with newly formed endosomes (Salzman and Maxfield, 1988, 1989). As described below, this assay permits a quantitative analysis of endosome fusion using either microscope spectrofluorometry or fluorescence spectroscopy on large numbers of cells in a cuvette. The procedure involves the sequential use of two probes: an FITC-labeled ligand and an antifluorescein antibody. When the antibody binds to FITC, it quenches the fluorescein fluorescence. Initially the FITC-ligand is endocytosed by cells, and after a short interval AFA is chased into the cells. If the AFA is delivered to a compartment containing FITC-ligand it will bind to the FITC and quench its fluorescence; otherwise there will be no change. AFA can only be delivered to compartments containing FITC-ligand in this sequential pulse experiment by fusion of newly formed endosomes containing AFA with previously formed vesicles containing FITC-ligand.

Initially the question addressed was whether such sequential fusion of endosomes occurred (Salzman and Maxfield, 1988). The experiments performed to answer this question used FITC-Tf as the ligand. Measurable amounts of fusion did occur along the Tf pathway in CHO cells, and the fusion accessibility of the FITC-Tf decreased with increasing interval times. The quenching of FITC-Tf by AFA, and therefore fusion of successive compartments, occurred as or before Tf reached the recycling endosomes near the Golgi complex.

A second method using the same two-probe system exploited a different attribute of AFA-FITC binding (Salzman and Maxfield, 1989). FITC, as described earlier, is a pH-sensitive probe for pH between 5 and 7. When FITC is saturated with AFA, the fluorescein fluorescence is rendered pH independent (see Figure 2). This characteristic can be used for the measurement of the fraction of FITC saturated by AFA (fractional saturation) and subsequently the quantitation of endosome fusion. The pH dependence of FITC fluorescence depends on the fractional saturation of FITC by AFA. The ratio of FITC fluorescence intensity at two different pH values will vary as a function of saturation by AFA (Figure 7). Using a standard curve, the fractional saturation of FITC by AFA can be determined and directly related to the extent of endosome fusion.

This assay was used to characterize endosome fusion using FITC-Tf as a

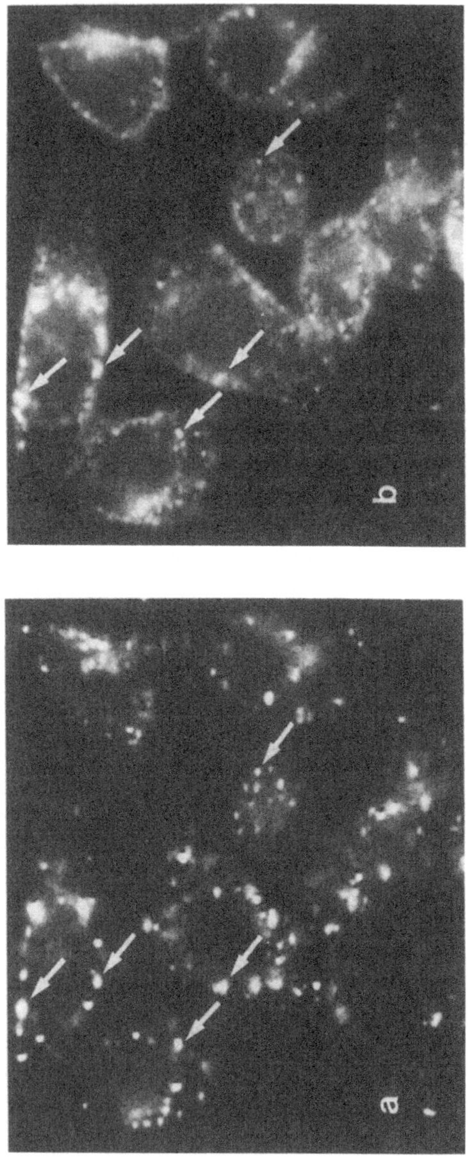

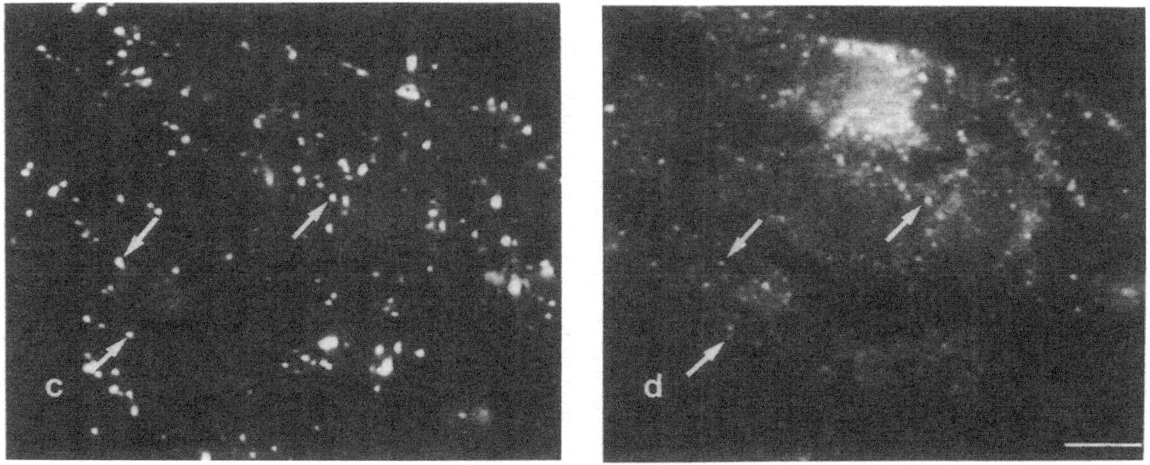

FIGURE 5. Time-dependent loss of fusion competence by the sorting endosome. TRVb-1 cells were incubated in 5 ug/ml DiO-LDL for 2 min, rinsed, and incubated in the absence of label for either 2 min (a, b) or 8 min (c, d), rinsed, and incubated in 20 ug/ml TRITC-Tf for 2 min. As in Figure 4, DiO-LDL fluorescence is displayed in panels (a) and (c). TRITC-Tf fluorescence is displayed in (b) and (d). Arrows indicate examples of endosomes containing both ligands. Using the arrows as landmarks, one can see that although most of the DiO-LDL-containing endosomes can be labeled with Tf endocytosed 2 min later, very few are accessible to Tf endocytosed 8 min later. Reproduced from Dunn *et al.* (1989). Bar = 10 μm.

probe of the receptor-recycling pathway, and FITC-α_2m as a probe of the lysosomal pathway. Appreciable amounts of fusion were found to occur along both pathways. Along the lysosomal pathway, FITC-α_2m accessibility to subsequently pulsed AFA decreased as the interval between the internalization of the probes increased. The decay of fusion accessibility along the lysosomal pathway could be fit to a first order exponential, consistent with the stochastic maturation of a single type of endosomal compartment with a $t_{1/2}$ for maturation of 8 min. This, in conjunction with the fluorescence double-label studies described above, allows the interpretation that this fusion-accessible compartment is the sorting endosome (see Figure 1). The kinetics of loss of fusion accessibility is interpreted as resulting from the maturation of the sorting endosome from a fusion-accessible compartment to a fusion-inaccessible compartment. That is, the $t_{1/2}$ is a measure of the lifetime of the sorting endosome. The interpretation that the sorting endosome has a finite lifetime is consistent with the observation that LDL accumulates for about 10 min in a sorting endosome (see Figure 5, Dunn *et al.*, 1989). Several *in vitro* studies of early-endosome fusion have also demonstrated a window for endosome-endosome fusion of 5–15 minutes, and the ability to fuse decays as ligands move into later compartments (Diaz *et al.*, 1988; Braell, 1987; Gruenberg and Howell, 1987).

Experiments performed at 18°C, a temperature at which ligands are internalized but not delivered to lysosomes and degraded (Dunn *et al.*, 1980), demonstrated that endosome fusion is occurring (Salzman and Maxfield, 1989). Therefore, the sorting endosome is functioning at this temperature. Fusion accessibility is not prolonged indefinitely at this temperature, showing that decreasing the temperature to 18°C does not prevent the maturation of the sorting endosome into a fusion-inaccessible compartment. This compartment is not a lysosome, since the delivery of ligands to lysosomes is prevented at this temperature. This defines

FIGURE 6. Schematic diagram of the iterative fractionation model of endocytic sorting. In this example, we assume that the sorting endosome fuses continuously with early endosomes, each of which carries ten receptor and ten ligand molecules. After fusion, receptors are free to migrate over the membrane, and ligands are free to disperse throughout the lumen of the sorting endosome. Simultaneously (although not necessarily in the lockstep depicted here) 40% of the receptors and 2% of the ligand contained in the sorting endosome are sorted and exported via each recycling vesicle. As shown in the running budget at the right of the figure, receptor recycling efficiency increases and ligand retention efficiency decreases with each round of sorting. One can also see that although the sorting endosome exports only 40% of its receptors with each sorting step, the sorting endosome is operating at 100% efficiency after six rounds of sorting, since from this point onward the number of receptors exported in each round equals the number imported in that round. For simplicity, we have depicted the iterative fractionation sorting model as if the processes of fusion and sorting are linked, but this need not be the case. If recycling vesicles continue to form after fusion has stopped, recycling receptors may continue to be exported from the sorting endosome and the number delivered to lysosomes further decreased. Reproduced from Dunn *et al.* (1989).

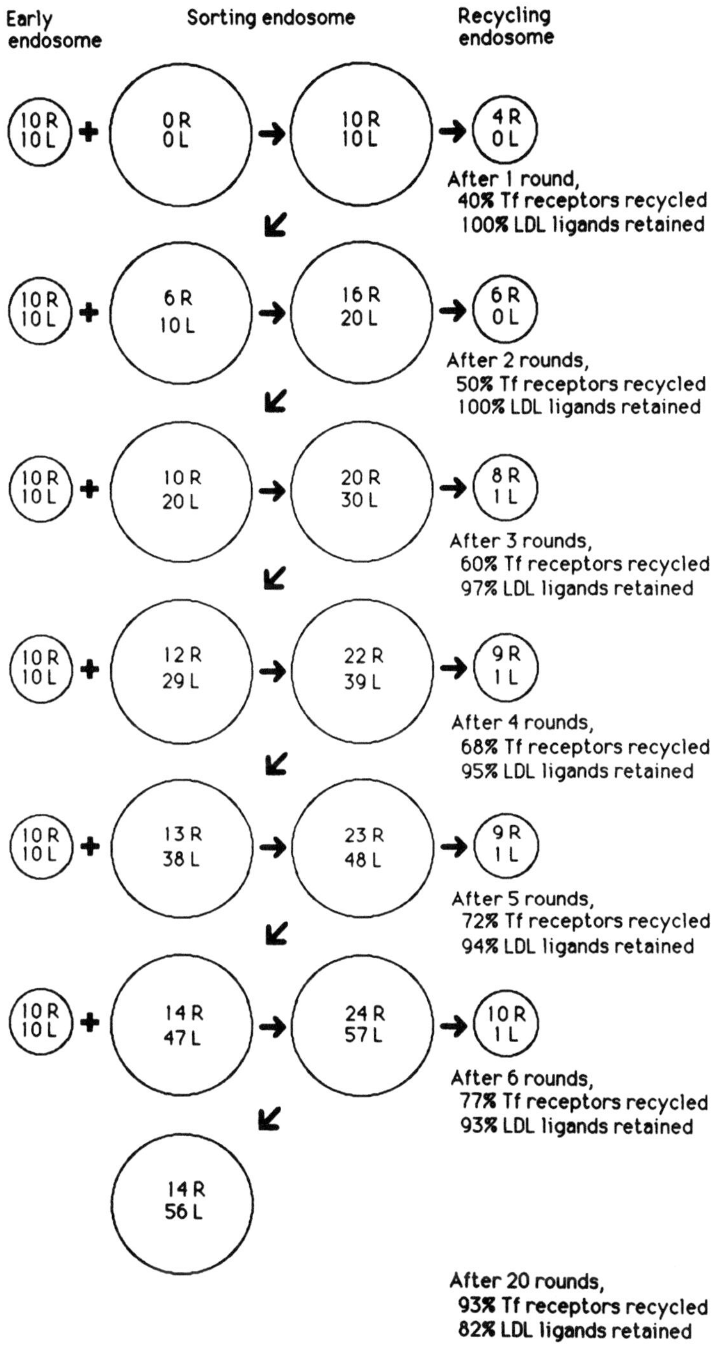

Early endosome
Sorting endosome
Recycling endosome
10 R 10 L
0 R 0 L
10 R 10 L
4 R 0 L
After 1 round,
40% Tf receptors recycled
100% LDL ligands retained
10 R 10 L
6 R 10 L
16 R 20 L
6 R 0 L
After 2 rounds,
50% Tf receptors recycled
100% LDL ligands retained
10 R 10 L
10 R 20 L
20 R 30 L
8 R 1 L
After 3 rounds,
60% Tf receptors recycled
97% LDL ligands retained
10 R 10 L
12 R 29 L
22 R 39 L
9 R 1 L
After 4 rounds,
68% Tf receptors recycled
95% LDL ligands retained
10 R 10 L
13 R 38 L
23 R 48 L
9 R 1 L
After 5 rounds,
72% Tf receptors recycled
94% LDL ligands retained
10 R 10 L
14 R 47 L
24 R 57 L
10 R 1 L
After 6 rounds,
77% Tf receptors recycled
93% LDL ligands retained
14 R 56 L
After 20 rounds,
93% Tf receptors recycled
82% LDL ligands retained

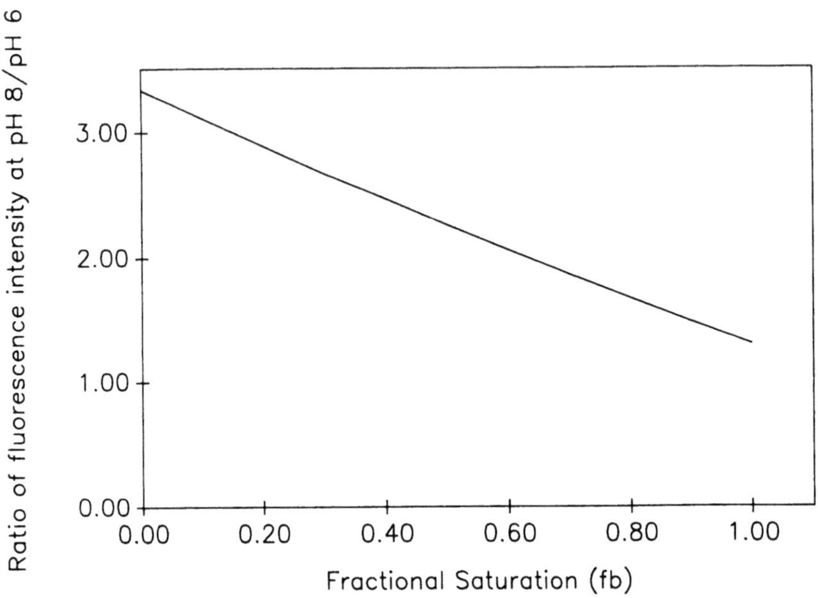

FIGURE 7. The pH dependence of fluorescein fluorescence as a function of increasing AFA saturation. The ratio of fluorescence intensities at pH 8 and pH 6 was calculated using fluorescence intensities at 495 nm. As FITC is progressively saturated by AFA, the pH dependence decreases., Using this curve, the amount of AFA saturation of FITC can be determined from the experimentally measured pH dependence of fluorescein fluorescence. Reproduced from Salzman and Maxfield (1989).

a postsorting late-endosomal compartment on the lysosomal pathway (see Figure 1). Similar compartments or series of compartments have been described in perfused rat liver (Mueller and Hubbard, 1986; Wolkoff *et al.*, 1984).

The recycling pathway showed a significantly different fusion profile (Salzman and Maxfield, 1989). Although fusion along this pathway was found, instead of the single exponential decay of fusion accessibility noted along the lysosomal pathway, the decay along the recycling pathway appears biphasic. The receptors move out of an initial fusion-accessible compartment, the sorting endosome, with a $t_{1/2}$ less than 3 min, and proceed into a second compartment on the recycling pathway that is also fusion accessible. The second compartment is quantitatively less accessible to fusion, but remains accessible indefinitely.

The swift movement of TR out of the initial fusion-accessible endosome is consistent with other biochemical and morphological studies that investigate the separation of lysosomally directed ligands from recycling endocytic components. Stoorvogel *et al.* (1987) have noted that in HepG2 cells, Tf separates from ASGP with a $t_{1/2}$ of approximately 2 min. Yamashiro *et al.* (1984) have found, in CHO cells, that Tf and α_2m are predominantly segregated from each other 5 min after internalization.

The second fusion-accessible compartment, into which Tf moves, is very likely to be the recycling endosomes in the para-Golgi region of CHO cells, since Tf is moved there very swiftly after internalization. This compartment has been morphologically described as a collection of tubules and vesicles (Yamashiro *et al.*, 1984). As mentioned earlier, LDL receptor and Tf have been colocalized to this compartment (McGraw and Maxfield, unpublished data). Recycling fluorescent membrane lipids have also been colocalized with Tf in this recycling compartment (Koval and Pagano, 1989) as described in the next section.

9. ENDOCYTOSIS OF FLUORESCENTLY LABELED LIPIDS

While receptor-ligand pathways have been relatively well characterized, membrane and lipid movement has been difficult to study by conventional cell biological methods. The development of fluorescently labeled lipid analogues has made the investigation of lipid dynamics possible using fluorescence microscopy. The use of image intensification microscopy is particularly suited to the study of a dynamic system like that of bulk membrane flow.

Pagano and his colleagues have described the synthesis and use of fluorescent lipids to study lipid movement and intracellular lipid pathways (Pagano and Sleight, 1985). They have labeled several classes of lipid with the fluorescent label NBD, which replaces one of the lipid's fatty acids. Liposomes containing a nonexchangeable lipid and the fluorescent analogue lipid are incubated with cells at 2°C. The NBD-lipid appears to spontaneously insert into the plasma membrane of the cells. Through the use of fluorescence techniques, lipid incorporation and mobility in the plasma membrane can be assessed. The cells may then be warmed to different temperatures, and the pathways of the NBD-lipid may then be observed by fluorescence microscopy.

Koval and Pagano (1989) have investigated the behavior of fluorescent sphingomyelin analogues. C_6-NBD-SM is delivered to the plasma membrane outer leaflet, and internalization is initiated by warming to 37°C. Within 15 min, much of the intracellular NBD-lipid localizes to a perinuclear region distinct from the Golgi as does Tf (Figure 8). It appears that this lipid is routed along the recycling pathway of endocytosis. When microtubules are disrupted with nocodazole, endocytosis and recycling continues, but the lipid fluorescence becomes localized to vesicles in the periphery of the cell rather than the perinuclear area. The fluorescent lipid remains colocalized with Tf (Koval and Pagano, 1989).

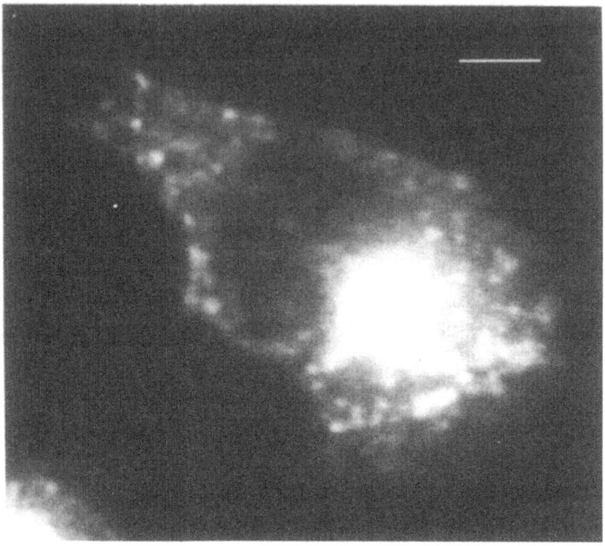

FIGURE 8. Localization of the NBD-SM pathway in TRVb-1 cells. In TRVb-1 cells, NBD-SM appears to follow the same recycling pathway taken by transferrin. TRVb-1 cells were incubated in a 10μm NBD-SM liposome solution in 20mM HEPES buffered F-12 (HF-12, pH 7.4) for 30 min on ice. At the end of the incubation, cells were rinsed and incubated in unlabeled medium for 10 min at 36°C to internalize the label. Cells were rinsed and incubated with medium containing 10% fetal calf serum on ice to return uninternalized NBD-SM from the cell surface to the medium. Cells were rinsed and fixed for observation. The fixed cells were treated with phenylenediamine to stabilize the rapidly bleaching fluorescence and visualized using fluorescein optics. Bar = 20 μm.

Because C_6-NBD-SM labels the receptor recycling pathway, the kinetics of lipid recycling were measured. The $t_{1/2}$ was determined to be approximately 40 min (Koval and Pagano, 1989). The recycling kinetics were unperturbed by monensin or nocodazole. The $t_{1/2}$ for Tf/TR recycling in CHO cells is approximately 20 min. There are a number of possible explanations for this difference in recycling rate. It may be that at certain steps on the recycling pathway recycling receptors are preferentially concentrated relative to the bulk membrane in the compartment. This would be analogous to the concentration of receptors in coated pits on the cell surface. For example, Geuze *et al.* (1987) found that the distribution of recycling receptors in tubular extensions of sorting endosomes (CURL) is not random. The distribution of receptors appears to be concentrated in the tubules, which presumably pinch off to form recycling endosomes. This suggests that there may be an active mechanism for moving receptors into the tubular extensions, or for keeping them from redistributing into the vesicular portion of the sorting endosome.

10. A MODEL FOR EFFICIENT SORTING OF LIGANDS AND RECEPTORS

The fluorescence studies of endocytosis described in the preceding sections, along with biochemical and electron microscopic studies, have led to the development of models for describing sorting of ligands and receptors (Salzman and Maxfield, 1989; Dunn *et al.*, 1989; Linderman and Lauffenburger, 1988). A model of this process must account for the overall efficiency of the sorting. The efficiency of delivery of ligands to lysosomes is variable. For example, 15–30% of α_2m internalized by CHO cells is released back into the medium in a TCA-precipitable form rather than being degraded in lysosomes (Yamashiro *et al.*, 1989). In contrast, the efficiency of receptor return to the cell surface is extraordinarily high, and more than 99% of internalized receptors may be returned back to the surface. It has been estimated that each Tf receptor recycles as many as 300 times (Omary and Trowbridge, 1981), and each LDL receptor recycles 150 times (Goldstein *et al.*, 1985). In addition to the efficient recycling of proteins, Koval and Pagano (1989) found that the C_6-NBD-SM is efficiently returned to the surface, and very little was delivered to lysosomes.

How is this level of sorting efficiency achieved? One possible mechanism would involve selective sorting based on molecular recognition within the sorting endosome. As described above, some receptors do seem to be concentrated over tubular extensions of the sorting endosome (Geuze *et al.*, 1987). Models based on this type of distribution have been proposed by Linderman and Lauffenburger (1988). In one such model ligands diffuse freely and rapidly between tubular and vesicular portions of the sorting endosome, but once receptors move into tubules they are trapped, perhaps by interactions with the cytoskeleton, or a clathrinlike protein. There is no direct evidence, as yet, to demonstrate the existence of a tubular trapping mechanism. This type of model could explain receptor retrieval but cannot easily account for highly efficient lipid recycling.

An alternative type of model for this process depends on the geometry of the sorting endosome. Electron microscopy studies (Geuze *et al.*, 1983; Marsh *et al.*, 1986) have shown that the sorting compartments are roundish vesicles with a number of narrow tubular extensions emanating from them. The surface area of the tubules accounts for approximately 70% of the total surface area of the sorting endosome, while excluding most of the volume that is in the mostly spherical part of the organelle (Marsh *et al.*, 1986). By random distribution of integral membrane proteins throughout the endosome, 70% of the proteins would be in the tubules. If the tubules pinch off and become recycling endosomes, the cells achieve 70% recycling efficiency merely by the geometry of the endosome. This alone would be insufficient to account for the overall efficiency of sorting.

A model described by Dunn *et al.* (1989), the iterative fractionation model, does not require a high-efficiency trapping mechanism for receptor-ligand sorting. This model postulates that a sufficient number of low-efficiency steps are capable of giving very high-efficiency sorting. If a sorting endosome has multiple fusions with primary endosomes and repeatedly buds off recycling endosomes, each with only about 20% of the recycling molecules contained in the sorting endosome, the overall efficiency of recycling quickly reaches very high levels (Figure 6). This type of model was proposed by Rome (1985). The data shown in Figures 3, 4, and 5 (Dunn *et al.*, 1989) show that multiple fusions and simultaneous removal of recycling components do occur in the sorting endosome. Regardless of the mechanism for enriching recycling endosomes with recycling molecules, this type of iterative sorting will greatly improve the efficiency of return to the cell surface. An attractive possibility is that the geometry of the sorting endosome can provide sufficient efficiency for recycling when the fractionation cycle is repeated many times. This would account for the high efficiency of membrane recycling to the plasma membrane, and it would enhance the efficiency of receptor recycling even if additional molecular recognition mechanisms were also used.

The delivery of ligands to lysosomes is less efficient than the return of recycling receptors to the surface. For many ligands, the low pH in the endosomes releases them from their receptors. It is reasonable to assume that many of these ligands will then be distributed according to the relative volumes of various parts of the sorting endosome. The narrow tubules that presumably bud off to form recycling endosomes contain relatively little volume, so most of the soluble ligands will be retained in the sorting endosome. It should be noted that repeated budding off of recycling endosomes will tend to reduce the efficiency of delivery of ligands to lysosomes.

11. CONCLUSION

Quantitative fluorescence techniques have brought about significant advances in the study of endocytosis. Image-intensification fluorescence microscopy has made the observation and measurement of dynamic cell functions possible even at the level of single organelles. Specific fluorescent ligands have been used to measure certain endosomal properties such as acidification, and processes like endosome-endosome fusion. With the use of appropriate probes, other physiologically important endocytic processes can be quantified,including hydrolysis and proteolysis. Fluorescence microscopy, with its high spatial and temporal resolution, is notably useful for the measurement of dynamic processes such as those occurring in living cells. Such understanding could not be gained by conventional electron microscopy methods without laborious counting and sectioning; and even with such techniques, the result might not be comparable to those obtained in living cells. Advances in spatial resolution should soon be achieved with the further development of confocal microscopy which provides a three-dimensional view of organelle location and movement.

These techniques have advanced our understanding of the endocytic process, and have aided in the development of testable models of receptor, ligand, and membrane movement along the endocytic pathway.

12. REFERENCES

Aikens, R. S., Agard, D. A., and Sedat, J. W., 1989, Solid-state imagers for microscopy, *Meth. Cell Biol.* **29**:292–314.

Ajioka, R. S. and Kaplan, J., 1986, Intracellular pools of transferrin receptors result from constitutive internalization of unoccupied receptors, *Proc. Natl. Acad. Sci. USA* **83**:6445–6449.

Al-Awqati, Q., 1986, Proton-translocating ATPases, *Ann. Rev. Cell Biol.* **2**:179–199.

Anderson, R. G. W. and Pathak, R. K., 1985, Vesicles and cisternae in the trans-Golgi apparatus of human fibroblasts are acidic compartments, *Cell* **40**:635–643.

Arndt-Jovin, D. J., Robert-Nicoud, M., Kaufman, S. J., and Jovin, T. M., 1985, Fluorescence digital imaging microscopy in cell biology, *Science* **230**:247–256.

Aubin, J. E., 1979, Autofluorescence of viable cultured mammalian cells, *J. Histochem. Cytochem.* **27**:36–43.

Barak, L. S. and Webb, W. W., 1981, Fluorescent low-density lipoprotein for observation of dynamics of individual receptor complexes on cultured human fibroblasts, *J. Cell Biol.* **90**:595–604.

Barrett, A. J., 1972, Lysosomal Enzymes, in *Lysosomes: A laboratory handbook* (J. T. Dingle, ed.), North-Holland/American Elsevier, Amsterdam/New York, pp. 46–135.

Borden, L. A., Einstein, R., Gabel, C. A., and Maxfield, F. R., 1990, Acidification-dependent dissociation of endocytosed insulin precedes that of endocytosed proteins bearing the mannose-6-phosphate recognition marker, *J. Biol. Chem.* **265**:8497–8504.

Braell, W. A., 1987, Fusion between endocytic vesicles in a cell-free system, *Proc. Natl. Acad. Sci. USA* **84**:1137–1141.

Bridges, K., Harford, J., Ashwell, G., and Klausner, R. D., 1982, Fate of receptor and ligand during endocytosis of asialoglycoproteins by isolated hepatocytes, *Proc. Natl. Acad. Sci. USA* **79**:350–354.

Dautry-Varsat, A., Ciechanover, A., and Lodish, H. F., 1983, pH and the recycling of transferrin during receptor-mediated endocytosis, *Proc. Natl. Acad. Sci. USA* **80**:2258–2262.

Diaz, R., Mayorga, L., and Stahl, P., 1988, *In vitro* fusion of endosomes following receptor-mediated endocytosis, *Biochemistry* **22**:5667–5674.

Dickson, R. B., Hanover, J. A., Willingham, M. C., and Pastan, I., 1983, Prelysosomal divergence of transferrin and epidermal growth factor during receptor-mediated endocytosis, *Biochemistry* **22**:5667–5674.

DiPaola, M. and Maxfield, F. R., 1984, Conformational changes in the receptors for epidermal growth factor and asialoglycoproteins induced by the mildly acidic pH found in endocytic vesicles, *J. Biol. Chem.* **259**:9163–9171.

Duncan, R. and Pratten, M. K., 1977, Membrane economics in endocytic systems, *J. Theor. Biol.* **66**:727–735.

Dunn, K. W. and Maxfield, F. R., 1990, The use of fluorescence microscopy in the study of receptor-mediated endocytosis, in *Modern Cell Biology, Vol. 9: Non-Invasive Techniques in Cell Biology* (J. K. Forkett and S. Grinstein, eds.), Wiley-Liss, New York, pp. 153–176.

Dunn, K. W., McGraw, T. E., and Maxfield, F. R., 1989, Iterative fractionation of recycling receptors from lysosomally destined ligands in an early sorting endosome, *J. Cell. Biol.* **109:**3303–3314.

Dunn, W. A., Hubbard, A. L., and Aronson, N. N., Jr., 1980, Low temperature selectively inhibits fusion between pinocytic vesicles and lysosomes during heterophagy of 1251-asialofetuin by the perfused rat liver, *J. Biol. Chem.* **255:**5971–5978.

Enns, C. A., Larrick, J. W., Suomalainen, H., Schroder, J., and Sussman, H. H., 1983, Comigration and internalization of transferrin and its receptor in K562 cells, *J. Cell Biol.* **97:**579–585.

Ferris, A. L., Brown, J. C., Park, R. D., and Storrie, B., 1987, Chinese hamster ovary cell lysosomes rapidly exchange contents, *J. Cell Biol.* **105:**2703–2712.

Geuze, H. J., Slot, J. W., and Strous, G. J. A. M., 1983, Intracellular site of asialoglycoprotein receptor-ligand uncoupling: double-label immunoelectron microscopy during receptor-mediated endocytosis, *Cell* **32:**277–287.

Geuze, H. J., Slot, J. W., Strous, G. J. A. M., Peppard, J., von Figura, K., Hasilik, A., and Schwartz, A. L., 1984, Intracellular receptor sorting during endocytosis: comparative immunoelectron microscopy of multiple receptors in rat liver, *Cell* **37:**195–204.

Geuze, H. J., Slot, J. W., and Schwartz, A. L., 1987, Membranes of sorting organelles display lateral herogeneity in receptor distribution, *J. Cell Biol.* **104:**1715–1723.

Giloh, H. and Sedat, J. W., 1982, Fluorescence microscopy; reduced photobleaching of rhodamine and fluorescein protein conjugates by n-propyl gallate, *Science* **217:**1252–1255.

Goldberg, R. I., Smith, J. A., and Jarett, L., 1988, Insulin and alpha-2-macroglobulin-methylamine undergo endocytosis by different mechanisms in rat adipocytes. I. comparison of cell surface events, *J. Cell. Physiol.* **133:**203–212.

Goldstein, J. L., Brown, M. S., Anderson, R. G. W., Russell, D. W., and Schneider, W. J., 1985, Receptor-mediated endocytosis: concepts emerging from the LDL receptor system, *Ann. Rev. Cell Biol.* **1:**1–39.

Gruenberg, J. and Howell, K. E., 1987, An internalized transmembrane protein resides in a fusion-competent endosome for less than 5 minutes, *Proc. Natl. Acad. Sci. USA* **84:**5758–5762.

Harford, J., Bridges, K., Ashwell, G., and Klausner, R. D., 1983, Intracellular dissociation of receptor-bound asialoglycoproteins in cultured hepatocytes, *J. Biol. Chem.* **258:**3191–3197.

Hart, P. D. and Young, M. R., 1975, Interference with normal phagosome-lysosome fusion in macrophages using ingested yeast cells and suramin, *Nature (London)* **256:**47–49.

Heiple, J. M. and Taylor, D. L., 1980, Intracellular pH in single motile cells, *J. Cell Biol.* **86:**885–890.

Inoue, S., 1986, *Video Microscopy,* Plenum Press, New York.

Johnson, G. D., Davidson, R. S., McNamee, K. C., Russell, G., Goodwin, D., and Holborow, E. J., 1982, Fading of immunofluorescence during microscopy: a study of the phenomenon and its remedy, *J. Immunol. Meth.* **55:**231–242.

Kam, Z., 1987, Microscopic imaging of cells, *Quart. Rev. Biophys.* **20:**201–259.

Keen, J. H., 1985, The structure of clathrin-coated membranes: assembly and disassembly, in *Endocytosis* (I. Pastan and M. C. Willingham, eds.), pp. 85–126, Plenum Press, New York.

Klausner, R. D., Ashwell, G., van Renswoude, J., Harford, J. B., and Bridges, K. R., 1983, Binding of apotransferrin to K562 cells: explanation of the transferrin cycle, *Proc. Natl. Acad. Sci. USA* **80:**2263–2266.

Koval, M. and Pagano, R. E., 1989, Lipid recycling between the plasma membrane and intracellular compartments: transport and metabolism of fluorescent sphingomyelin analogues in cultured fibroblasts, *J. Cell Biol.* **108:**2169–2181.

Linderman, J. J. and Lauffenburger, D. A., 1988, Analysis of intracellular receptor/ligand sorting in endosomes, *J. Theor. Biol.* **132:**203–245.

Marsh, M., Griffiths, G., Dean, G. E., Mellman, I., and Helenius, A., 1986, Three-dimensional structure of endosomes in BHK-21 cells, *Proc. Natl. Acad. Sci. USA* **83:**2899–2903.

Maxfield, F. R., 1982, Weak bases and inonophores rapidly and reversibly raise the pH of endocytic vesicles in cultured mouse fibroblasts, *J. Cell Biol.* **95**:676–681.

Maxfield, F. R., 1985, Acidification of endocytic vesicles and lysosomes, in *Endocytosis* (I. Pastan and M. C. Willingham, eds.), pp. 235–257, Plenum Press, New York.

Maxfield, F. R., 1989a, Measurements of vacuolar pH and cytoplasmic calcium in living cells using fluorescence microscopy, *Methods in Enzymology* **173**:745–770.

Maxfield, F. R., 1989b, Fluorescent analogs of peptides and hormones, *Meth. Cell Biol.* **29**:13–28.

Maxfield, F. R., Willingham, M. C., Haigler, H. T., Dragsten, P., and Pastan, I. H., 1981, Binding, surface mobility, internalization, and degradation of rhodamine-labeled α_2m-macroglobuli, *Biochemistry* **20:** 5353–5358.

Maxfield, F. R., Schlessinger, J., Shector, Y., Pastani, I., and Willingham, M. C., 1978, Collection of insulin, EGF, and α_2m in the same patches on the surface of cultured fibroblasts and common internalization, *Cell* **14**:805–810.

Maxfield, F. R. and Yamashiro, D. J., 1987, Endosome acidification and the pathways of receptor-mediated endocytosis, in *Immunobiology of Proteins and Peptides* (M. Z. Atassi, ed.), pp. 189–198, Plenum Press, New York.

McGraw, T. E., Greenfield, L., and Maxfield, F. R., 1987, Functional expression of the human transferrin receptor cDNA in Chinese hamster ovary cells deficient in endogenous transferrin receptor, *J. Cell Biol.* **105**:207–214.

Mellman, I., Fuchs, R., and Helenius, A., 1986, Acidification of the endocytic and exocytic pathways, *Ann. Rev. Biochem.* **55:** 663–700.

Mostov, K. E., Freidlander, M., and Blobel, G., 1984, The receptor for transepithelial transport of IgA and IgM contains multiple immunoglobulinlike domains, *Nature* **308**:37–43.

Mueller, S. C. and Hubbard, A. L., 1986, Receptor-mediated endocytosis of asialglycoproteins by rat hepatocytes: receptor-positive and receptor-negative endosomes, *J. Cell Biol.* **102**:932–942.

Murphy, R. F., Bisaccia, E., Cantor, C. R., Berger, C., and Edelson, R. L., 1984, Internalization and acidification of insulin by activated human lymphocytes, *J. Cell. Phys.* **121**:351–356.

Ohkuma, S. and Poole, B., 1978, Fluorescent probe measurement of the intralysosomal pH in living cells and the pertubation of pH by various agents, *Proc. Natl. Acad. Sci. USA* **75**:3327–3331.

Olsnes, S. and Sandvig, K., 1985, Toxins, in *Endocytosis* (I. Pastan and M. C. Willingham, eds.), pp. 195–234, Plenum Press, New York.

Omary, M. B. and Trowbridge, I. S., 1981, Biosynthesis of the human transferrin receptor, *J. Biol. Chem.* **256**:12888–12892.

Pagano, R. E. and Sleight, R. G., 1985, Defining lipid transport pathways in animal cells, *Science* **229**:1051–1057.

Parton, R. G., Prydz, K., Bomsel, M., Simons, K., and Griffiths, G., 1989, Meeting of the apical and basolateral endocytic pathways of the Madin-Darby canine kidney cell in late endosomes, *J. Cell Biol.* **109**:3259–3272.

Pastan, I. and Willingham, M. C., 1985, The pathway of endocytosis, in *Endocytosis* (I. Pastan and M. Willingham, eds.), pp. 1–44, Plenum Press, New York.

Pearse, B. M. F. and Bretscher, M. S., 1981, Membrane recycling by coated vesicles, *Ann. Rev. Biochem.* **50**:85–101.

Pitas, R. E., Innerarity, T. L., Weinstein, J. N., and Mahley, R. W., 1981, Acetoacetylated lipoproteins used to distinguish fibroblasts from macrophages *in vitro* by fluorescence microscopy, *Arteriosclerosis.* **1**:177–185.

Renfrew, C. and Hubbard, A. L., 1988, Characterization of domain specific antibodies and their use in the study of epidermal growth factor receptor (EGFR) degradation, *J. Cell Biol.* **107**:551a abstract.

Robbins, E. and Marcus, P., 1963, Dynamics of acridine orange-cell interaction. I. interrelationships of acridine orange particles and cytoplasmic reddening, *J. Cell Biol.* **18**:237–250.

Roederer, M. and Murphy, R. F., 1986, Cell-by-cell autofluorescence correction for low signal-to-noise systems: application to epidermal growth factor endocytosis by 3T3 fibroblasts, *Cytometry* **7**:558–565.

Rome, L. H., 1985, Curling receptors, *Trends Biochem. Sci.* **10**:151.

Roos, A. and Boron, W. F., 1981, Intracellular pH., *Physiologic Reviews* **61**:297–434.

Salzman, N. H. and Maxfield, F. R., 1989, Fusion accessibility of endocytic compartments along the recycling and lysosomal endocytic pathways in intact cells, *J. Cell Biol.* **109**:2097–2104.

Salzman, N. H. and Maxfield, F. R., 1988, Intracellular fusion of sequentially formed endocytic compartments, *J. Cell Biol.* **106**:1083–1091.

Sandvig, K., Olsnes, S., Petersen, O. W., and van Deurs, B., 1988, Inhibition of endocytosis from coated pits by acidification of the cytosol, *J. Cellular Biochemistry* **36**:73–81.

Schlessinger, J., Shechter, Y., Cuatrecasas, P., Willingham, M. C., and Pastan, I., 1978, Quantitative determination of the lateral diffusion coefficients of the hormone-receptor complexes of insulin and epidermal growth factor on the plasma membrane of cultured fibroblasts, *Proc. Natl. Acad. Sci. USA* **75**:5353–5357.

Sipe, D. M. and Murphy, R. F., 1987, High-resolution kinetics of transferrin acidification in BALB/c3T3 cells: exposure to pH 6 followed by temperature-sensitive alkalinization during recycling, *Proc. Natl. Acad. Sci. USA* **84**:7119–7123.

Sklar, L. A., 1987, Real-time spectroscopic analysis of ligand-receptor dynamics, *Ann. Rev. Biophys. Chem.* **16**:479–506.

Sklar, L. A., Finney, D. A., Oades, Z. G., Jesaitis, A. J., Painter, R. G., and Cochrane, C. G., 1984, The dynamics of ligand-receptor interactions, *J. Biol. Chem.* **259**:5661–5669.

Sklar, L. A., Hyslop, P. A., Oades, Z. G., Omann, G. M., Jesaitis, A. J., Painter, R. G., and Cochrane, C. G., 1985. Signal transduction and ligand-receptor dynamics in the human neutrophil, *J. Biol. Chem.* **260**:11461–11467.

Sklar, L. A., Jesaitis, A. J., Painter, R. G., and Cochrane, C. G., 1982a, Ligand/receptor internalization: a spectroscopic analysis and a comparison of ligand binding, cellular response, and internalization by human neutrophils, *J. Cell. Biochem.* **20**:193–202.

Sklar, L. A., McNeil, V. M., Jesaitis, A. J., Painter, R. G., and Cochrane, C. G., 1982b, A continuous spectroscopic analysis of the kinetics of elastase secretion by neutrophils, *J. Biol. Chem.* **257**:5471–5475.

Smith, R. M. and Jarett, L., 1987, Ultrastructural evidence for the accumulation of insulin in nuclei of intact 3T3-L1 adipocytes by an insulin-receptor mediated process, *Proc. Natl. Acad. Sci. USA* **84**:459–463.

Solari, R. and Kraehenbuhl, J. P., 1984, Biosynthesis of the IgA receptor: a model for the transepithelial sorting of a membrane glycoprotein, *Cell* **36**:61–71.

Spring, K. R. and Lowy, R. J., 1989, Characteristics of low light level television cameras, *Meth. Cell. Biol.* **29**:270–291.

Stoorvogel, W., Geuze, H., and Strous, G. J., 1987, Sorting of endocytosed transferrin and asialoglycoprotein occurs immediately after internalization in HepG2 cells, *J. Cell Biol.* **104**:1261–1268.

Swanson, J. A., 1989a, Fluorescent labeling of endocytic compartments, *Meth. Cell Biol.* **29**:137–152.

Swanson, J. A., 1989b, Phorbol esters stimulate macropinocytosis and solute flow through macrophages, *J. Cell Science* **94**:135–142.

Swanson, J. A., Yirinec, B. D., and Silverstein, S. C., 1985, Phorbol esters and horseradish peroxidase stimulate pinocytosis and redirect the flow of pinocytosed fluid in macrophages, *J. Cell Biol.* **100**:851–859.

Taylor, D. L. and Salmon, E. D., 1989, Basic fluorescence microscopy. *Meth. Cell Biol.* **29**:208–238.

Tycko, B. and Maxfield, F. R., 1982, Rapid acidification of endocytic vesicles containing α_2m-macroglobulin, *Cell* **28**:643–651.

Tycko, B., Keith, C. H., and Maxfield, F. R., 1983, Rapid acidification of endocytic vesicles containing asialoglycoprotein in cells of a human hepatoma line, *J. Cell Biol.* **97**:1762–1776.

van Renswoude, J., Bridges, K. R., Harford, J. B., and Klausner, R. D., 1982, Receptor-mediated endocytosis of transferrin and the uptake of Fe in K562 cells: identification of a nonlysosomal acidic compartment, *Proc. Natl. Acad. Sci. USA* **79**:6186–6190.

Wampler, J. E. and Kutz, K., 1989, Quantitative fluorescence microscopy using photomultiplier tubes and imaging detectors, *Meth. Cell Biol.* **29**:239–269.

Wang, Y. and Taylor, D. L., eds., 1989, *Methods in Cell Biology, vol. 29, Fluorescence Microscopy of Living Cells in Culture, Part A.*, Academic Press, New York.

White, J., Kartenbech, J., and Helenius, A., 1980, Fusion of Semliki Forest virus with the plasma membrane can be induced by low pH., *J. Cell Biol.* **87**:264–272.

White, J., Keilian, M., and Helenius, A., 1983, Membrane fusion proteins of enveloped animal viruses, *Quart. Rev. Biophys.* **16**:151–195.

Wolkoff, A. W., Klausner, R. D., Ashwell, G., and Harford, J., 1984, Intracellular segregation of asialoglycoproteins and their receptor: a prelysosomal event subsequent to dissociation of the ligand-receptor complex, *J. Cell Biol.* **98**:375–381.

Yamashiro, D. J., Borden, L. A., and Maxfield, F. R., 1989, Kinetics of α_2-macroglobulin endocytosis and degradation in mutant and wild-type Chinese hamster ovary cells, *J. Cell. Phys.* **139**:377–382.

Yamashiro, D. J. and Maxfield, F. R., 1987a, Kinetics of endosome acidification in mutant and wild-type Chinese hamster ovary cells, *J. Cell Biol.* **105**:2713–2721.

Yamashiro, D. J. and Maxfield, F. R., 1987b, Acidification of morphologically distinct endosomes in mutant and wild-type Chinese hamster ovary cells, *J. Cell Biol.* **105**:2723–2733.

Yamashiro, D. J. and Maxfield, F. R., 1988, Regulation of endocytic processes by pH, *Trends Pharmacol. Sci.* **9**:190–193.

Yamashiro, D. J., Tycko, B., Fluss, S. R., and Maxfield, F. R., 1984, Segregation of transferrin to a mildly acidic (pH 6.5) para-Golgi compartment in the recycling pathway, *Cell* **37**:789–800.

Chapter 5

Endocytosis and Function of the Hepatic Asialoglycoprotein Receptor

Paul H. Weigel

1. THE HEPATOCYTE ASIALOGLYCOPROTEIN (GALACTOSYL/N-ACETYLGALACTOSAMINYL) RECEPTOR

1.1. General Characteristics

When desialylated, plasma glycoproteins, including some hormones, become ligands for recognition by this galactosyl (Gal) receptor (Ashwell and Harford, 1982; Schwartz, 1984). The interaction of the receptor with most ligands is of high affinity; dissociation constants are in the nM range. The ligand used most predominantly is desialylated α_1-acid glycoprotein (also called orosomucoid), which is a soluble-plasma acute-phase protein. Neoglycoproteins, synthetic polymers, or solid surfaces containing a sufficient density of Gal groups will also be bound by the receptor. The receptor is found only in hepatocytes. Only 10–25% of the total cellular receptors are on the cell surface, while 75–90% are inside the cell. There are about a half-million receptors per cell. The receptor recycles both *in vivo* and *in vitro*. Ligand dissociation occurs below pH 6, and Ca^{2+} is required for binding. Oligosaccharide binding to the receptor requires a relatively

---

Paul H. Weigel Department of Human Biological Chemistry and Genetics, University of Texas Medical Branch, Galveston, Texas 77555-0647.

Subcellular Biochemistry, Volume 19: Endocytic Components: Identification and Characterization, edited by J. J. M. Bergeron and J. R. Harris. Plenum Press, New York, 1993.

high Ca^{2+} concentration; the apparent K_d is approximately 1.5 mM, and about 3 Ca^{2+} are bound/polypeptide chain of the rabbit receptor (Andersen *et al.*, 1982). Very little ligand binding (<1%) occurs in the absence of Ca^{2+} or the presence of chelators such as ethylenediaminetetraacetic acid (EDTA). This fact is often used as the basis for most easily determining the nonspecific binding of radiolabeled ligands.

1.2. Specificity of the Gal Receptor

The receptor is specific for terminal-clustered Gal or N-acetylgalactosaminyl (GalNAc) residues, and it requires two or three sugars close together to give efficient binding. Lee *et al.* (1982) used saccharides of known structure to map the sugar binding site(s) of the receptor (Ashwell and Harford, 1982). More is known about the effect of carbohydrate structure on recognition by the Gal receptor than any of the other carbohydrate receptors. Potentially, five groups in a galactopyranoside can participate in the sugar-receptor binding interaction. These include the aglycone, C2-OH_{eq} or C2-$NHAc_{eq}$, C3-OH_{eq}, C4-OH_{ax} and C6-CH_2-groups. The appropriate group and stereoconformation must be present in at least four of these positions to observe significant binding.

Y. C. Lee and co-workers have extensively studied the effect of oligosaccharide structure on the affinity of the binding of Gal/GalNAc-terminal oligosaccharides to the rabbit receptor. The receptor-ligand affinity can vary over many orders of magnitude (K_ds from 10^{-6} to 10^{-10} M) depending on the structure. The receptor interaction with a wide range of different oligosaccharide structures is best described by a model (Lee *et al.*, 1984) in which three individual monosaccharide binding sites are arranged at the vertices of a triangle with sides of 15, 22, and 25 Å. This proposal represents a major contribution to understanding the molecular design of the carbohydrate receptors in general. These workers also proposed that clusters of receptors on the cell surface can present a lattice of monosaccharide binding sites that will be able to recognize a great diversity of different oligosaccharide structures containing different numbers of branches and terminal sugars, and different spatial organization of those sugars (Hardy *et al.*, 1985). This proposal nicely explains the broad specificity and affinity displayed by native receptors on cells for Gal/GalNAc-terminal glycoconjugates of different structure.

1.3. Polypeptide Composition of the Gal Receptor

The Gal receptor is an integral transmembrane glycoprotein. Based on the primary amino acid sequence, determined directly on the protein or deduced from the complementary DNA (cDNA) sequence (Drickamer, 1987), the rat Gal receptor is composed of polypeptides that contain at least four domains; N-terminal cytoplasmic, transmembrane, extracellular stalk, and C-terminal carbohydrate binding.

The size of the native Gel receptor is unknown. In nonionic detergents the rat receptor is approximately 264 kDa (Andersen *et al.*, 1982). The binding domain for asialo-orosomucoid (ASOR) is approximately 148 kDa in the presence and 105 kDa in the absence of detergent (Steer *et al.*, 1981). In affinity-purified receptor preparations, SDS-PAGE reveals one major band and two minor bands. These three subunits, designated rat hepatic lectins (RHL) 1, 2, and 3 are, respectively, M_r = 41.5, 49, and 54 kDa and are the products of two different genes (Drickamer *et al.*, 1984). RHL 2 and RHL 3 differ only in the type and extent of posttranslational carbohydrate modification and show considerable sequence homology with RHL 1 (Halberg *et al.*, 1987). RHL 2 and RHL 3 have more carbohydrate than RHL 1 and migrate on SDS-PAGE as though they are larger by, respectively, about 8 kDa and 14 kDa. RHL 2/3 is only 18 amino acids (about 1.5 kDa) larger than RHL 1, which cannot account for the size difference of the mature proteins. RHL 3 has polylactosaminoglycan chains and is recognized by the lectin from *Datura stramonium* (Halberg *et al.*, 1987).

The stoichiometry and subunit composition of the native rat receptor is unknown. The mole ratio for RHL 1, 2, and 3 in affinity-purified rat receptor preparations varies, respectively, from 2.5:1:1 to 8:1:1 (Drickamer, 1987; Ashwell and Harford, 1982). The human hepatic Gal receptor is also the product of two genes and consists of two different polypeptides, H1 ($M_r \simeq$ 46,000) and H2 ($M_r \simeq$ 50,000) (Spiess and Lodish, 1985). The smaller polypeptide H1 is severalfold more abundant than H2. Likewise, the rabbit receptor is composed of two polypeptides with M_rs of 40,000 and 48,000 that are also isolated in a molar ratio of, respectively, 2:1 (Steer and Ashwell, 1986; Schwartz, 1984).

1.4. Proposed Structure of the Gal Receptor

Based on chemical-affinity cross-linking studies using N-hydroxysuccinimide derivatives of ASOR, we concluded that active native receptors on intact hepatocytes are heterooligomers (Herzig and Weigel, 1989). These results do not exclude the possibility that homooligomers of RHL 1 and RHL 2/3 may also exist. In the absence of ligand, antiserum specific for RHL 1 coprecipitates RHL 2/3 in Triton X-100 extracts of whole cells (Herzig and Weigel, 1988). The converse is also true. This indicates that heterooligomeric complexes exist. Furthermore, each of these antisera (which were a generous gift from Dr. K. Drickamer) inhibited identically and almost completely [^{125}I]-ASOR binding to hepatocytes, a result not predicted by a completely homooligomeric model for the subunits. Sawyer *et al.* (1988) also reported that the three RHL subunits exist

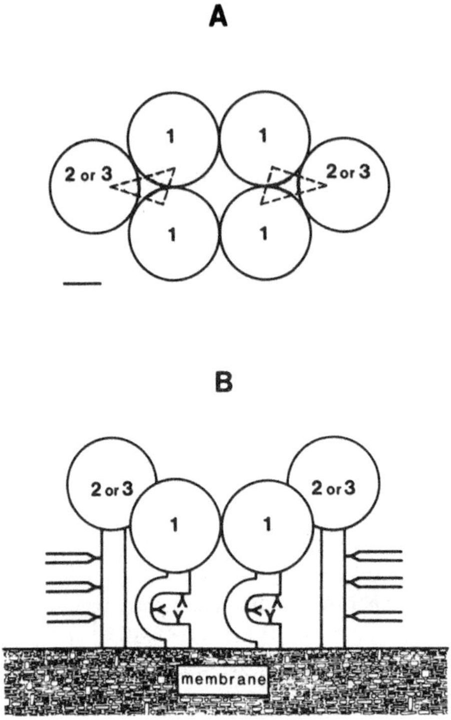

FIGURE 1. A model for the structure of the parenchymal Gal receptor. The proposed heterohexamer is viewed from above the plasma membrane in A and from the side, parallel to the membrane, in B. The circles represent the carboxyl terminal carbohydrate binding domains. The oligosaccharide chains are shown as Ys. RHL 1, RHL 2, and RHL 3 subunits are indicated by the numbers 1, 2, or 3. The shaded triangles denote oligosaccharide binding sites (Lee *et al.*, 1984; Hardy *et al.*, 1985). The vertices of the triangles are individual monasaccharide (Gal or GalNAc) binding sites on each subunit. The bar represents 1.5 nm.

as heterooligomers on the rat hepatocyte surface. Bischoff *et al.* (1988) reported similar results supporting a heterooligomeric model for the human receptor in hepatoma cells.

Understanding the arrangement of the rat native Gal receptor is complicated by the unknown number of subunits per native receptor, the stoichiometry of the three RHL subunits in a native receptor, and the existence of two functionally distinct receptor populations (Weigel, 1987). We recently proposed a model (Herzig and Weigel, 1989) for the structure of the rat Gal receptor (Figure 1) based on our data and the results of others. We suggest that in the rat, native Gal receptors are heterohexamers composed of four RHL 1 subunits and two subunits of either RHL 2 and/or RHL 3. Each subunit would contribute a binding site for one or two sugar residues (Lee and Lee, 1988). Each hexamer would have two potential binding domains for tri- or tetra-antennary oligosaccharide ligands. Each ligand-binding domain would be a heterotrimer composed of two RHL 1 subunits and one RHL 2 or RHL 3 subunit. These domains are arranged at the vertices of a triangle with sides of 15, 22, and 25 Å (Lee *et al.*, 1984). Because of the large carbohydrate chains in their stalk regions, RHL 2 and RHL 3 may have their Gal-binding domains farther away from the plasma membrane than those of RHL 1.

2. MULTIPLE PATHWAYS FOR RECEPTOR-MEDIATED ENDOCYTOSIS

2.1. The Hepatic Gal-Receptor System Functions via Two Pathways

We and others have characterized two pathways for ligand processing and Gal-receptor recycling in rat hepatocytes (Weigel, 1987). Our current working model of the receptor system is summarized in Figure 2. Although this figure is complicated, I will (1) summarize briefly the key evidence for the proposal that two receptor pathways operate in isolated cells and (2) summarize data suggesting that two Gal-receptor pathways operate *in vivo.*

2.1.1. Evidence for Two Receptor Pathways *in Vitro*

The first report that hepatocytes process ASOR in a nonuniform way was by Harford *et al.* (1983a), who found that internalized receptor-ASOR complexes dissociated with biphasic kinetics, suggesting that dissociation occurs in two different compartments and/or by two kinetic pathways. Within months Oka and Weigel (1983a) also reported that dissociation is biphasic. Although the groups used different measuring techniques, both found identical rates of dissociation. About 50% of internalized receptor-ASOR complexes dissociate slowly ($t_{1/2}$ = 50 min) and 50% dissociate 20 times faster ($t_{1/2} \leq 2.5$ min). Since this initial finding, nine characteristics have been found that define these two pathways, which we have designated the State 1 and the State 2 pathway (Table I). These pathways appear to be parallel, not serial. The State 1 pathway is a minor pathway that accounts for about 20% of the ligand-processing ability of hepatocytes. The State 2 pathway is the predominant pathway. However, the numbers of surface and total cellular State 1 and State 2 Gal receptors are approximately equal. About 50% of surface Gal receptors operate via the State 1 pathway and 50% via the State 2 pathway.

The ability to detect and characterize the two pathways has depended on experiments using a single round of surface-bound ligand and on the properties of the hepatocytes isolated by our collagenase-perfusion procedure. After removal from the liver, these cells are purified for 20 min at about 30°C by filtration and differential centrifugation (Oka and Weigel, 1987). Other investigators keep the cells on ice at this stage. Our unpublished studies show that at higher temperature and high cell density, the cells are anoxic (depleted of oxygen) and therefore they become ATP depleted, and State 2 receptor activity is downregulated. Therefore,

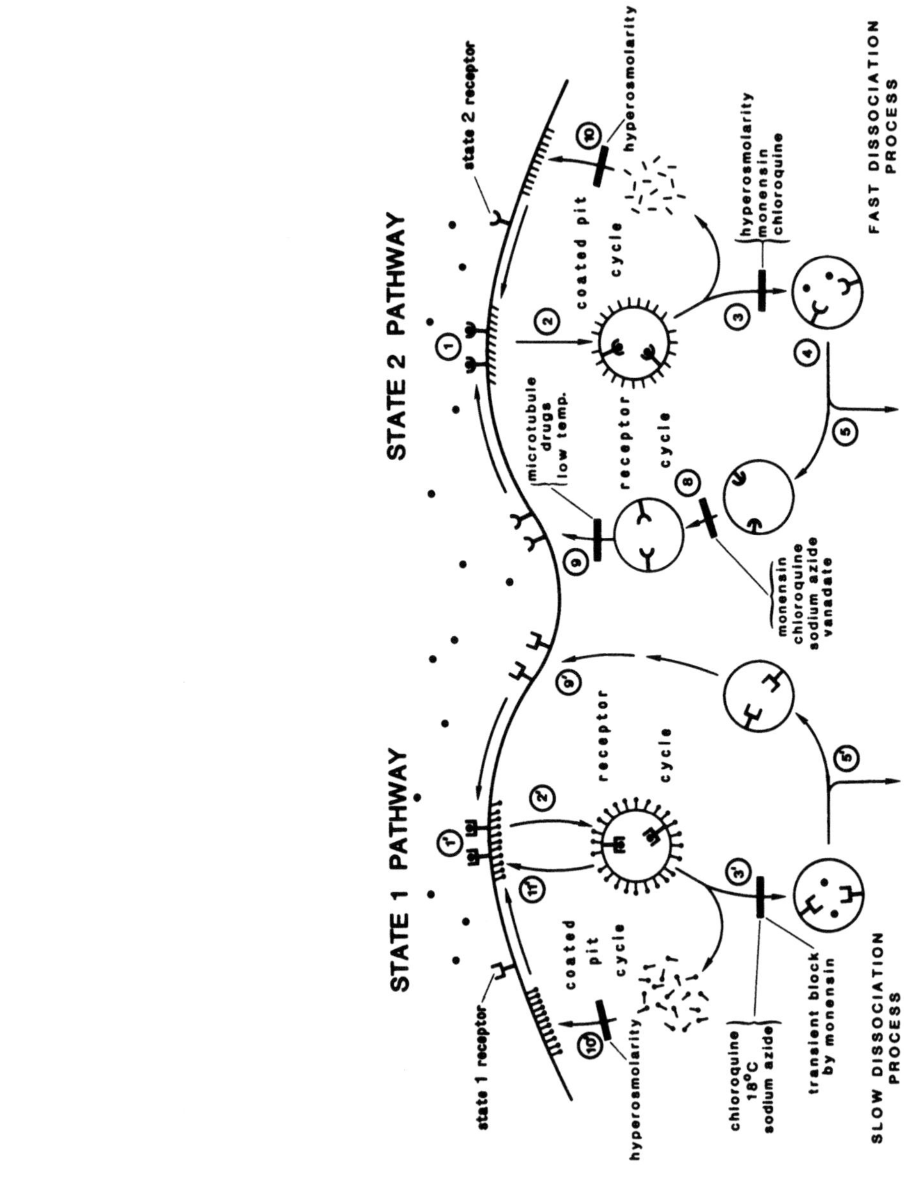
STATE 1 PATHWAY
STATE 2 PATHWAY
state 1 receptor
state 2 receptor
coated pit cycle
receptor cycle
receptor cycle
coated pit cycle
hyperosmolarity
hyperosmolarity
microtubule drugs low temp.
monensin chloroquine sodium azide vanadate
chloroquine 18°C sodium azide
transient block by monensin
hyperosmolarity monensin chloroquine
SLOW DISSOCIATION PROCESS
FAST DISSOCIATION PROCESS

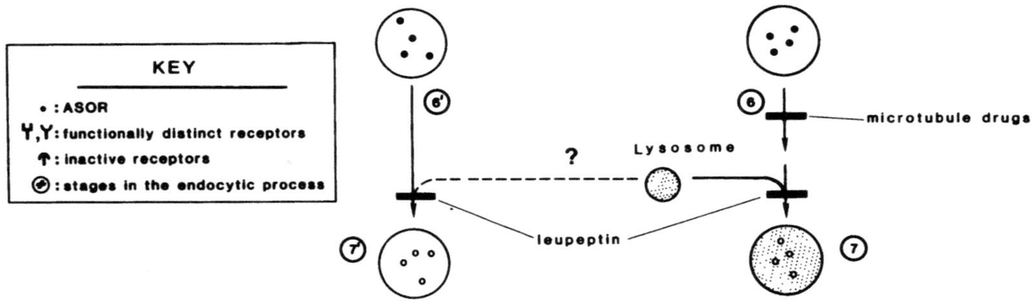

FIGURE 2. The two-pathway model for endocytosis mediated by the hepatic Gal receptor. The State 1 and State 2 Gal receptors and their respective parallel pathways are discussed in the text. The points of action of various inhibitors are indicated by the solid bars. The numbers denote the following stages in these pathways: 1,1′: ligand binding; 2,2′: receptor-ligand internalization; 3,3′: receptor-ligand dissociation; 4: State 2 Gal receptor inactivation; 5,5′: segregation of receptor and ligand; 6,6′: ligand delivery to a degradative compartment (lysosomes in the State 2 pathway, possibly an earlier endosome in the State 1 pathway); 7,7′: ligand degradation; 8: reactivation of State 2 Gal receptors; 9,9′: recycling of receptors to the cell surface; 10,10′: formation of new coated pits; 11′: diacytosis of receptor-ligand complexes in the State 1 pathway.

Table I
Characteristics of the Two Hepatic Gal Receptor Pathways

Feature	The (minor) State 1 pathway	The (major) State 2 pathway
1. Surface receptor activity is modulated by temperature, monesin, ATP depletion, microtubule drugs, chloroquine and vanadate.	No	Yes
2. Receptors constitutively recycle.	No	Yes
3. Recycling receptors undergo a transient inactivation-reactivation cycle.	No?	Yes
4. Diacytosis of internalized receptor-ligand complexes.	Yes	No
5. Dissociation of receptor-ligand complexes.	Slow	Fast
6. Ligand dissociation is blocked by monensin.	No	Yes
7. Ligand dissociation is blocked at 18°C.	Yes	No
8. A lag period occurs before ligand degradation.	No	Yes
9. Colchicine (1 μM) inhibits ligand degradation.	No	Yes

our freshly isolated cells express only the State 1 Gal-receptor population; the State 2 receptors are already inactivated. When these cells are put at 37°C, surface Gal receptor activity increases to a new steady-state level because of reexpression of State 2 Gal receptors (Weigel and Oka, 1983). This fortuitous circumstance enables us to generate two cell populations with different surface-receptor properties. Freshly isolated cells express only active State 1 receptors, whereas cells equilibrated at 37°C express both State 1 and State 2 receptors. This has allowed us to assign various properties and kinetic characteristics to one pathway or the other. Cultured cells expressing essentially only surface State 1 Gal receptors are prepared by incubation at 24°C for 1 to 2 hr (Oka and Weigel, 1987; Clarke *et al.*, 1987). The following summarizes some of the key differences between the two Gal-receptor subpopulations and pathways.

Modulation of Surface Gal Receptor Activity. Cells incubated at 37°C with microtubule drugs, monensin, chloroquine, azide, or sodium vanadate or at low temperatures (24–37°C) without ligand lose about 50% of their surface receptor activity in 1 hr (McAbee and Weigel, 1991; McAbee *et al.*, 1990; Oka and Weigel, 1991; Weigel and Oka, 1983). No further loss occurs regardless of incubation time or concentration. These agents affect the same population of receptors, and combinations of drugs do not cause further loss of receptor activity (McAbee *et al.*, 1990). The modulated receptors are the State 2 Gal receptors; State 1 receptor activity is unaffected. Other investigators have similarly observed only a partial loss of surface Gal-receptor activity with monensin (Fiete *et al.*, 1983; Berg *et al.*, 1983), azide (Tolleshaug *et al.*, 1985; Scarmato *et*

Table II
Effects of Perturbants on Gal Receptor Distribution and Activity

Category	Treatment	Receptor redistribution	Receptor inactivation
Type I	Low temperature (24–37°C), mi-	Yes	No
	crotubule depolymerizing drugs	Yes	No
Type II	Monesin, chloroquine	No	Yes
		No	Yes
Type III	Azide/fluoride, N_2 atmosphere	Yes	Yes
		Yes	Yes

Different perturbants can be classified into three categories depending on their effects on State 2 Gal receptor activity or distribution in isolated hepatocytes, as discussed in the text.

al., 1986), microtubule drugs (Kolset *et al.*, 1979), or weak bases like chloroquine (Tolleshaug and Berg, 1979; Schwartz, 1984). We have been the only laboratory to determine whether these agents affect the same receptor subpopulation (McAbee *et al.*, 1990). Likewise, the cell surface receptors for low-density lipoprotein (Basu *et al.*, 1981), α_2-macroglobulin (Kaplan and Keogh, 1981), mannose (Tietze *et al.*, 1980, 1982), mannose-6-phosphate (Gonzalez-Noriega *et al.*, 1980), and transferrin (Stein *et al.*, 1984) are also only partially modulated by these agents.

The modulating agents mentioned above fall into three categories depending on whether Gal receptors are redistributed in the cell and/or inactivated (Table II). This point is discussed further below.

Ligand Dissociation. This occurs very rapidly ($k > 0.28$/min) in the State 2 Gal-receptor pathway but very slowly ($k = 0.014$/min) in the State 1 pathway (Oka and Weigel, 1983b; Weigel et al., 1986a). Furthermore, ligand dissociation in the major State 2 pathway is completely stopped by monensin (Harford *et al.*, 1983b; Maxfield, 1982), whereas dissociation in the State 1 pathway is delayed for about 40 min but then proceeds to completion (Oka and Weigel, 1987). This surprising result suggests that dissociation of receptor-ASOR complexes in the State 1 pathway may not be mediated by low pH. It is currently thought that low pH (around 5.5) in the endosome mediates dissociation (Yamashiro and Maxfield, 1988), because ligand dissociation *in vitro* occurs below pH 6 and conformational changes in the Gal receptor occur at pH 5.6 (Dipaola and Maxfield, 1984). We have observed no differences in the pH sensitivity for dissociating prebound ASOR at 4°C from State 1 or State 2 Gal receptors.

Degradation. Degradation of one wave of surface-bound [^{125}I]-ASOR is biphasic at 37°C (Clark *et al.*, 1987). In cells with active State 1 and State 2 Gal receptors on their surface, the kinetics of degradation of one wave of surface-bound ligand is biphasic. Each of the two phases of this degradation process is specific to one of the two Gal-receptor pathways. Degradation in the State 1 pathway occurs immediately after internalization. The State 2 pathway has a lag of approximately 15 min prior to ligand degradation. The lag is the time needed to deliver ligand to lysosomes by a process sensitive to microtubule inhibitors (Oka and Weigel, 1983a; Clarke and Weigel, 1989). In cells not initially expressing State 2 surface Gal-receptor activity, the kinetics of ASOR degradation are monophasic. Under conditions of continuous ASOR uptake the low degradation rate in the State 1 pathway is easily swamped by the approximately sevenfold greater rate in the State 2 pathway.

2.1.2. Evidence for Two Receptor Populations/Pathways *in Vivo*

Although no one has specifically looked for the presence of two Gal-receptor populations or pathways in intact mammalian liver *in vivo,* the available data suggest that this is the case. In particular, studies in which perturbing agents or drugs are used consistently show only partial affects on receptor function. Using temperatures < 37°C, Kloppel (1989) demonstrated two intracellular transport pathways for asialo-orosomucoid in perfused rat liver. Gal-receptor activity is only partially lost on hepatocytes isolated from rats treated with a variety of drugs or subjected to partial hepatectomy (Andersson *et al.*, 1988) or jejunoileal bypass (Serbource-Goguel *et al.*, 1985). Chronic phenobarbital (Evarts *et al.*, 1985) or ethanol (Casey *et al.*, 1987) treatment of rats decreases both total immunodetectable Gal receptors and hepatocyte surface receptor activity by about 50%. However, these cells are unimpaired in their ability to diacytose ^{125}I-asialo-orosomucoid (Casey *et al.*, 1990), a property of only the minor State 1 Gal-receptor endocytic pathway (Weigel and Oka, 1984). Hepatocytes isolated from diabetic rats lose 35% of all cellular Gal-receptor activity with no loss of receptor protein (Slama *et al.*, 1988). Lost Gal-receptor activity is not caused by occupation of receptors with endogenous ligand (Durand *et al.*, 1984). These results suggest that during chronic liver dysfunction hepatocytes lose roughly 50% of their functional Gal receptors, either by loss of receptor protein or by accumulation of inactive receptors. The data are explainable by our results with isolated cells. We believe that the altered receptors in these *in vivo* studies are the State 2 Gal receptors described above and that the residual functional receptors are the State 1 receptors. Unfortunately, direct tests of this hypothesis are presently not possible until we have an analytical tool or probe to differentiate between the two functional receptor populations.

2.2. Other Receptor Systems Function via Two Pathways

The inhibitors discussed above (and shown in Figure 2) also induce similar patterns of surface-receptor modulation in other receptor systems. Methyl amine, whose action is similar to that of chloroquine, depletes surface α_2-macroglobulin receptor activity on alveolar macrophages by 50% in the absence of ligand (Kaplan and Keogh, 1981). Chloroquine and NH_4Cl reduce surface mannose-6-phosphate receptor activity (Gonzales-Noriega *et al.*, 1980) as well as mannose receptors (Tietze *et al.*, 1980) to a similar extent. Monensin depletes human fibroblast-surface low-density lipoprotein-receptor activity by half in the absence of added ligand (Basu *et al.*, 1981) and depletes 50% of the surface transferrin-receptor activity of K562 cells (Stein *et al.*, 1984). Thus the overall pattern of surface-receptor activity modulation for the Gal receptor system and other Class II receptors is very similar. We believe that a hallmark characteristic of multiple-receptor pathways is the presence of these drug-sensitive and drug-resistant populations of cell-surface receptor activity. Differential sensitivity of cell-surface receptor activity to various perturbants is likely diagnostic for multiple receptor subpopulations and/or receptor-recycling pathways. These data strongly suggest that these other recycling receptors also exist in discrete subpopulations and function by two pathways. More direct evidence has been reported that both the transferrin- (Stein and Sussman, 1986) and mannose- (Tietze *et al.*, 1982) receptor systems exhibit multiple functional receptor pools and ligand-processing or receptor-recycling pathways.

3. CONSTITUTIVE RECEPTOR RECYCLING AND THE GAL-RECEPTOR INACTIVATION/REACTIVATION CYCLE

3.1. Discovery of a Receptor Inactivation/Reactivation Cycle

Experiments have shown that Gal receptors in the major State 2 pathway undergo constitutive recycling (McAbee and Weigel, 1988; McAbee *et al.*, 1990). They are able to go into and out of the cell even in the absence of ligand. During this constitutive recycling process these receptors are transiently inactivated and then reactivated inside the cell. We first considered this possibility several years ago when we observed that the different agents discussed above, such as metabolic energy poisons, would cause a loss of cell-surface receptor activity at 37°C in the absence of ligand. For example, hepatocytes were incubated at 37°C in the presence of different amounts of sodium azide without ligand for 45 min and then chilled to 4°C and washed. Viable cells were isolated over discontinuous Percoll gradients and radioactive ligand was then used to measure receptor activity, or an affinity-purified polyclonal antireceptor antibody was used to measure receptor protein (McAbee and Weigel, 1987). As the cellular ATP level decreased, there was a loss of approximately 50% of both receptor activity and protein from the cell surface. This suggests that there is constitutive receptor recycling but that only a portion of receptors are involved.

When the azide-treated cells are permeabilized with the nonionic detergent digitonin, we expect to recover the antibody-binding activity and the ligand-binding activity inside the cell. All of the antibody reactive binding is recovered in the permeabilized cells; there is virtually no change even at severe ATP depletion (>95%). However, as cellular ATP is depleted there is a parallel decrease in the total cellular content of active receptors as measured by the ligand-binding assay (McAbee and Weigel, 1987). There is about a 50% loss of total cell Gal-receptor activity over about 30 min. There is a sensitive population of receptors, the State 2 receptor population, and an insensitive population of receptors, the State 1 receptors (McAbee *et al.*, 1990). The receptor activity loss cannot be explained by the degradation and loss of receptor protein because the receptors are still detected with the antibody. Furthermore, the receptor activity loss is reversible in the absence of protein synthesis. If cells are incubated with sodium azide, then washed and kept at 37°C in the presence of cycloheximide, within a few hours the cells recover their total receptor activity, ATP levels, and the normal distribution of receptors between the cell surface and the cell interior. The same results are obtained with a nitrogen atmosphere, although the kinetics are different because of problems with gas exchange. We do not know if State 1 receptors are induced by binding ligand to undergo a similar inactivation/reactivation cycle.

To determine the cellular site of receptor inactivation, we measured the rate of change in ligand-binding capability on the cell surface and inside the cell. When azide is added to the cells, there is an immediate decrease in receptor activity on the cell surface (McAbee and Weigel, 1988). However, no decrease is seen when the cells are permeabilized and the total cellular content (i.e., both surface and intracellular) of active receptors is measured. For the first 2 or 3 min the apparently lost receptor activity from the surface is recovered inside the cell; there is virtually no change in the total cellular receptor activity. Only after about 3 min is there a decrease. The intracellular content of active receptors actually increases transiently before it decreases. This result indicates that receptor inactivation is an intracellular event.

Our interpretation of these experiments is depicted in Figure 3. This model applies only to the State 2 receptor pathway. Receptors in this pathway are able to enter the cell, through coated pits, even in the absence of ligand. Once inside the cell in an active form, they are inactivated by a mechanism that remains to be determined (see below). Reactivation of receptors and externalization, getting them back to the cell surface, requires ATP. When cells are depleted of ATP by using metabolic energy poisons or nitrogen gas, then the receptors cannot be reactivated and reexternalized, and inactive receptors accumulate inside the cell. At the start of the experiment there is a substantial pool of active receptors still inside the cell beyond the block point(s); the ATP-requiring step(s). More inactive receptors can accumulate in the whole cell than the net number lost from the cell surface, because some of these receptors are externalized and then inter-

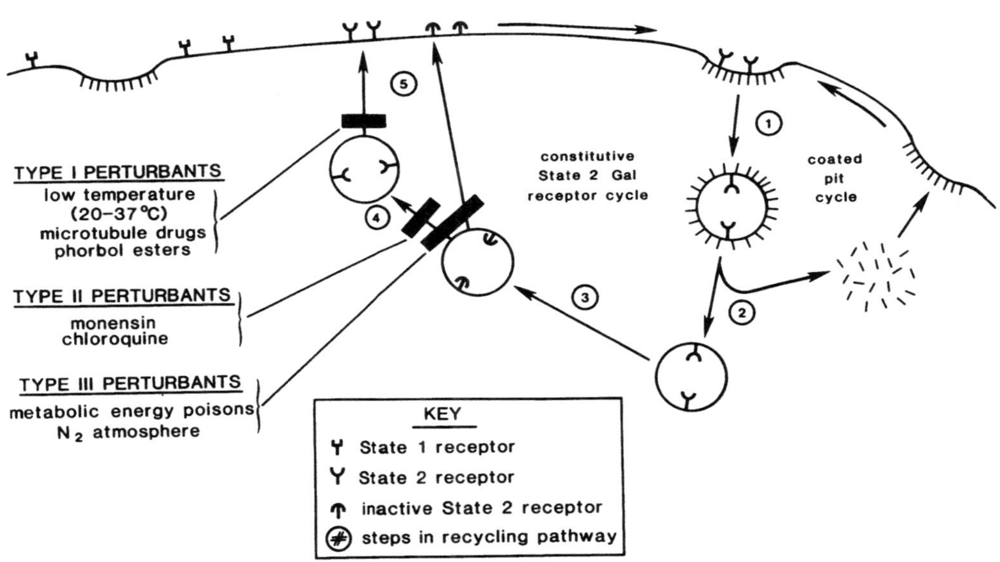

FIGURE 3. Scheme for the effects of Type I, II, and III perturbants on the constitutive recycling pathway of State 2 Gal receptors. State 2 but not State 1 Gal receptors constitutively enter coated pits, which are themselves continually recycling, and follow a multistep intracellular pathway. These steps are (1) internalization into endosomes, (2) uncoating of clathrin from endosomes, (3) receptor inactivation, (4) receptor reactivation, and (5) receptor recycling back to the cell surface. Normally, State 2 Gal receptors are transiently inactivated and then reactivated before reinsertion into the plasma membrane (McAbee and Weigel, 1987, 1988). Type I perturbants block this reinsertion but do not affect the inactivation and reactivation cycle, Type II perturbants block the receptor reactivation but not the reinsertion steps, and Type III perturbants block both the reactivation and recycling steps. The branch point depicted in the model to differentiate between the Type II and III agents need not correspond to a morphologically distinct pathway but could reflect a functional biochemical difference. For example, in the presence of Type II but not Type III agents, inactive Gal receptors may follow the same physical pathway and arrive at the plasma membrane recycled but not reactivated.

nalized and get trapped inside the cell at this block point. Some receptors may traverse through several cycles before being inactivated and trapped. We believe that this cycle is a fundamental discovery that will be a characteristic of recycling pathways and that inactivation cycles will be found in other receptor systems.

There are three possible explanations for the loss of receptor activity during receptor recycling. Receptors may become inaccessible to ligand, occupied by endogenous ligand, or truly inactivated. The first possibility is eliminated because the antibody used to quantitate receptor number is larger than the ligand asialo-orosomucoid, which is used to quantitate receptor activity, and shows no binding decrease in permeable, azide-treated cells. Inactivation of State 2 Gal receptors is also not caused by receptor occupancy by an endogenous ligand. To test this, cells were azide-treated at 37°C in the absence of ligand and then washed at 4°C in the presence or absence of EGTA to expose cryptic Gal receptors (Figure 4). The cells were then assessed for their ability to bind radio-

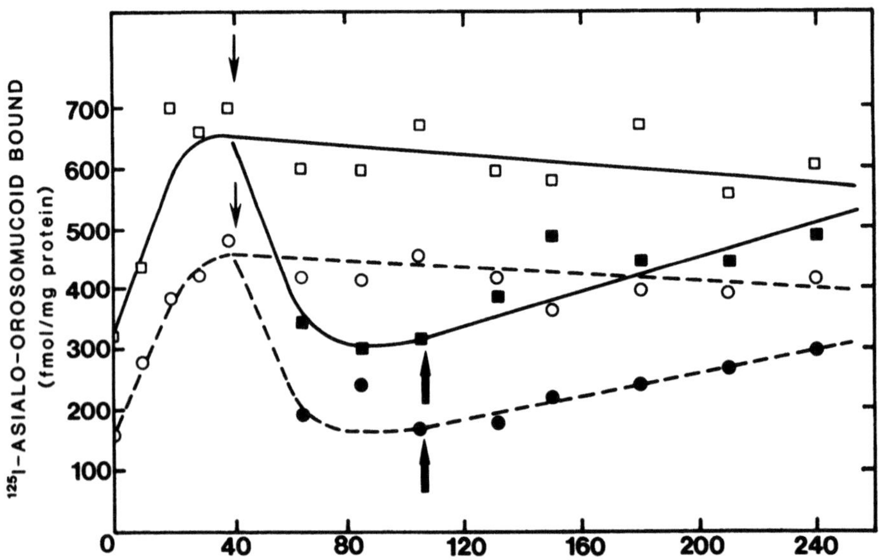

FIGURE 4. Effect of EGTA on the detection of inactive Gal receptors. Freshly isolated hepatocytes (2×10^6 cells/ml) were incubated at 37°C. After 40 min, either sodium azide to a final concentration of 10 mM (●, ■) or PBS (○, □) was added (small arrows) and the 37°C incubation was continued. After 100 min, the cells were collected by centrifugation, resuspended in fresh medium without azide and put at 37°C (large arrows). At the indicated times, samples of the cell suspensions were chilled on ice and 10 mM EGTA (□, ■) or PBS (○, ●) was added. After 10 min, the cells were collected, washed, and incubated in medium containing 1.5 μg/ml [^{125}I]-ASOR for 60 min at 4°C. After washing away unbound ligand, the cells were assessed for bound radioactivity. Points represent the mean of duplicates.

active ligand at 4°C. There was still a loss of cellular Gal-receptor activity whether or not occupied, EGTA-sensitive receptors had been exposed prior to measuring receptor activity. The best explanation for receptor inactivation is that receptor activity is modulated either by a noncovalent interaction with a regulatory protein (i.e., an allosteric modulator) or by covalent modification.

Receptor inactivation may have an important function. It provides an efficient way to segregate internalized receptor from ligand during the sorting process. Ligands internalized by receptor-mediated endocytosis are concentrated in endosomes 10^3–10^4-fold more than their extracellular concentration. Ligand concentrations within endosomes would be $\geq 10^{-5}$M. Consequently, even if the K_d for receptor-ligand binding at the low pH of the endosome increases from about 10^{-9}M to 10^{-5}M, this lower affinity would be offset by the higher ligand concentration. To ensure approximately 99% dissociation of Gal receptor–ligand complexes in the endosome, the K_d would have to increase by about 6 orders of magnitude (to about 10^{-3}M)! There is no evidence to suggest that such dramatic changes occur at pH 5.5. Therefore, the ability to separate active receptors from dissociated ligand, even at low pH, may not be very efficient. Receptor inactivation would, however, make this segregation process efficient. An attractive explanation for segregation is a mechanism that inactivates the receptor, so that the ligand dissociates and then cannot rebind to the receptor. The high concentration of free ligand in the endosomal pathway would, therefore, not hinder the efficient segregation of receptor, to be recycled, from ligand, to be degraded. We believe that transient Gal-receptor inactivation is the mechanism that allows the continuous endocytosis process to be so efficient.

3.2. Uncoupling of State 2 Gal-Receptor Inactivation and Receptor Redistribution

The cellular process by which the State 2 Gal-receptor population can be inactivated is distinct from the process by which these receptors become redistributed in the cell, that is, accumulate inside the cell and not return to the cell surface. We conclude this because the two processes can be uncoupled. They can occur independently depending on the perturbing agent to which the cells are exposed. We examined a variety of treatments and inhibitors and followed Gal-receptor activity and localization by measuring ASOR or antireceptor antibody binding capacity in intact and digitonin-permeabilized rat hepatocytes (McAbee *et al.*, 1991). The various agents fell into three categories (see Table II). Type I agents such as low temperature (24–37°C) or microtubule drugs do not cause receptor inactivation but do lead to receptor redistribution. These treatments cause intracellular accumulation of active State 2 Gal receptors. Type II agents such as monensin or chloroquine cause receptor inactivation but not redistribution, as first reported by Fiete *et al.* (1983). These agents cause accumulation of inactive receptors on the cell surface. However, accumulation of inactive surface-Gal receptors induced by monensin or chloroquine requires receptor internalization, because action of these drugs is blocked by hyperosmolar pretreatment of cells (McAbee *et al.*, 1989). Hyperosmolarity stops Gal-receptor recycling and endocytosis by inhibiting clathrin recycling and coated-pit formation (Oka and Weigel, 1988; Oka *et al.*, 1989). Type III agents, those that deplete cellular ATP, cause both receptor redistribution and inactivation. They cause intracellular accumulation of inactive receptors.

3.3. The Role of Gal-Receptor Phosphorylation

The rat and human hepatic Gal receptors and the chicken hepatic N-acetylglucosaminyl receptor are phosphoproteins (Drickamer, 1987; Schwartz, 1984; Weigel, 1987; Breitfeld *et al.*, 1985). All of the subunit polypeptides of these receptors have one or more Ser/Thr or Tyr residues in their cytoplasmic domains. This is also true for all the carbohydrate receptors whose sequences have been determined. Most of the Class I signal-transducing receptors that operate via either the phosphoinositide or the G-protein/adenylate-cyclase pathway are phosphorylated and dephosphorylated as part of the regulation of the receptor function. Often these receptors are also tyrosine kinases and modify themselves as well as other target intracellular proteins (Blackshear *et al.*, 1988). It has not been determined if the Class II carbohydrate receptor cycle or function is regulated by phosphorylation and dephosphorylation of these receptors. Since exposure to vanadate (Oka and Weigel, 1989) or depletion of cellular ATP (McAbee and Weigel, 1987, 1988) causes Gal-receptor inactivation, it is likely that phosphorylation/dephosphorylation is involved in the regulation of receptor activity. This possibility is presently being examined.

4. POSSIBLE FUNCTIONS OF THE ENDOCYTIC MEMBRANE-BOUND CARBOHYDRATE RECEPTORS

The hepatic parenchymal Gal/GalNAc receptor belongs to a larger family of related mammalian receptors that mediate the binding, endocytosis, and degradation of carbohydrate ligands (Steer and Ashwell, 1986; Drickamer, 1988). An intriguing idea for the function of the carbohydrate receptors, particularly for those located in liver, is that these are the last steps in a housekeeping pathway. That is, cells or tissues at other sites in the body may process a given family of glycoconjugates and convert them to soluble molecules containing the oligosaccharide-recognition marker (like a zip code) needed to send the molecule to the liver where it will be completely degraded. There are good candidate functions for the two mannose-6-phosphate receptors, the glucosyl–advanced glycation end products (AGE) receptor and the liver-hyaluronan receptor. The likely endogenous ligands are known for these receptors. Several functions of the macrophage mannose receptor have also been proposed. The possible function(s) of the other galactosyl and fucosyl receptors is much less clear.

4.1. The Mannose-6-Phosphate (Man-6-P) Receptor(s)

The discovery of two forms of the Man-6-P receptor (present in most mammalian cells) and the identity of the insulinlike growth factor II and the cation-independent Man-6-P receptor (von Figura and Hasilik, 1986; Kornfeld, 1987) means that the previous studies to determine the function of a single receptor need to be reexamined. The cation-independent Man-6-P receptor appears to be primarily involved in endocytosis (Stein *et al.*, 1987b). A possible function for the endocytically active cation-independent Man-6-P receptor could be the removal of extracellular lysosomal enzymes released into the circulation as a result of cell injury, disease, or death. Such endogenous enzymes could clearly be detrimental to normal cells and to the organism because of their degradative activity, even at neutral pH. It is reasonable to believe that cells could have evolved several pathways to deal with such an important yet potentially dangerous set of enzymes. No one has yet proposed an explanation for why there are two apparently unrelated ligand-binding activities, one for a growth factor and one for an oligosaccharide, residing on the same cell-surface receptor. The function(s) of the cation-independent Man-6-P receptor will almost certainly go beyond its role as a carbohydrate receptor.

The vast majority of the cation-dependent Man-6-P receptors are located intracellularly, with only 3% on the cell surface (Stein *et al.*, 1987a). This receptor, therefore, may only be able to bind glycoproteins containing Man-6-P within intracellular luminal compartments with a high Ca^{2+} concentration. Based on what is presently known, this cation-dependent Man-6-P receptor is more likely to be involved in intracellular lysosomal enzyme sorting and packaging than the cation-independent Man-6-P receptor. In cells containing only the 46 kDa cation-dependent Man-6-P receptor, antibodies to this receptor cause loss of intracellular lysosomal enzymes. In cells with both receptors, the same antibody does not induce this loss or secretion (Stein *et al.*, 1987a). These latter investigators concluded that the cation-dependent receptor mediates transport of endogenous newly synthesized lysosomal enzymes and that the cation independent receptor is responsible for internalization of exogenous molecules containing Man-6-P.

4.2. The Reticuloendothelial Mannosyl (Man) Receptor

This receptor is abundant, localized in the reticuloendothelial cells of liver, and active in the cell-surface binding and uptake of injected glycoproteins *in vivo* and in isolated cells. It probably plays a role in the turnover and degradation of

Man-containing glycoproteins. Isolated Kupffer/endothelial cells have about 0.5–1.5 × 10^5 surface receptors/cell and can internalize 13–80 × 10^3 molecules/cell/min (Maynard and Baenziger, 1981); this is a recycling receptor. These features are typical characteristics of high-capacity endocytic Class II recycling receptor systems.

The contribution of endothelial cells to the *in vivo* liver uptake of glycoproteins containing terminal GlcNAc is about three to seven times greater than that of the Kupffer cells (Pranning-van Dalen *et al.*, 1987). In support of this, Magnusson and Berg (1989) showed that the endocytic capacity of the Man-receptor system on highly purified rat-liver endothelial cells is very great. This receptor has the highest endocytic rate constant of any yet measured.

As with many of the carbohydrate-receptor systems, the identification of function is a haunting problem. The native ligand(s) for this receptor are unknown. It is certain, however, that if a soluble or particulate glycoconjugate with the typical N-linked oligosaccharide core structure appears in the blood as a result of injury, infection, or normal glycoprotein processing, then its removal and degradation in the liver will be mediated by the nonparenchymal reticuloendothelial Man receptor.

4.3. The Macrophage Man/GlcNAc Receptor

This Man receptor is localized in liver, spleen, and lung, and surprisingly, in leg muscle. Resident and migratory macrophages throughout the body probably contain this receptor (Lee, 1988). However, fresh human-blood monocytes and a number of macrophagelike cell lines do not express this Man receptor, suggesting that it is only expressed in differentiated macrophages. The same result is true for the Gal receptor in hepatocytes. Man-containing ligands such as high-Man oligosaccharides, Man-neoglycoproteins, lysosomal glycosidases, and mannans are degraded after endocytosis mediated by this receptor. An important function seems likely for this receptor considering its extensive activity and its location on an important cell type involved in inflammation, wound healing, antigen processing, and other surveillance functions within the organism.

Expression of Man-receptor activity is modulated in cultured cells by lymphokines and antiinflammatory steroids. Hoppe and Lee (1982) suggested that the receptor facilitates the ability of macrophages to attack yeast or bacteria containing Man in their cell walls. Exposure of macrophages to soluble monosaccharides such as Man, GlcNAc, or Fuc stimulated an increase in the cell-surface binding activity of the receptor. This triggering response may be important in the ability of macrophages to deal with potential microbial pathogens. Subsequently, Ohsumi and Lee (1987) reported that exposure of rabbit alveolar macrophages to Man ligands at 37°C stimulated the cells to secrete lysosomal enzymes in a time- and dose-dependent manner. These workers proposed that occupancy of receptors on the cells may be a trigger for secretion of enzymes that will degrade a yeast bacterial cell wall and enable the macrophage to kill and phagocytose the potential pathogen. Man receptors have also been shown to be comodulated with F_c receptors on macrophages bound to mannan-coated substrata (Sung *et al.*, 1985). This result suggests that the two receptors may be coordinated at least in some aspects of their function and indicates a role for the Man receptor in some immunocompetent process(es).

4.4. The Kupffer Cell Fucosyl (Fuc) Receptor

The receptor is only present in Kupffer cells in the liver (Haltiwanger *et al.*, 1986). Interestingly, this receptor is not found in macrophages from any other tissue source, making it the only carbohydrate receptor other than the hepatic Gal/GalNAc receptor and the hepatic endothelial cell hyaluronan receptor with a unique liver localization. Degradation of FucBSA or other Fuc-glycoconjugates in isolated Kupffer cells has not been characterized but presumably occurs. A likely function of this receptor would be the removal and degradation of Fuc-glycoconjugates from the circulation. As with many of the other carbohydrate receptors, the endogenous ligands are unknown and its function can only be surmised.

4.5. The Macrophage Gal Receptor

This Gal-receptor system is present on resident liver and at least some migratory macrophages, and its involvement has been proposed in regulating the turnover and degradation of aged red blood cells (Aminoff, 1988; Schlepper-Schafer and Kolb-Bauchofen, 1988). It has long been recognized that the sialic acid content on the surface of erythrocytes decreases during the lifetime of these cells. Appearance of the new Gal groups thus exposed would provide a convenient carbohydrate determinant for a system needed to remove old or damaged cells from the body. A large amount of evidence supports this model, although it is not universally accepted. Desialylation of glycoproteins and gangliosides on the red-blood-cell surface would presumably occur spontaneously, reflecting the relative instability of the glycosidic bonds of the sialic acid family, or in response to injury or trauma. Neuraminidases, for example, are often released at a wound site or by some pathogenic organisms or parasites (Prioli *et al.*, 1987). This functional role for the reticuloendothelial Gal receptor is very attractive. As expanded below in Section 5, it could be very important for mammals to remove both soluble and particulate glycoconjugates efficiently when they contain terminal Gal or GalNAc.

This macrophage Gal receptor has a degradative function but, like the macrophage Man receptor, the Gal receptor may also be involved in an immunologically related function. Intraperitoneal coadministration of the simple competing monosaccharide Gal with desialylated sheep red blood cells in mice resulted in an adjuvant activity (Klerx *et al.*, 1986). The adjuvant activity was sugar specific and concentration dependent. Gal, GalNAc, Man-6-P and *D*-Fuc were effective stimulators of the immune response, whereas Glc, Man, fructose, and *L*-Fuc were ineffective. The mechanism of this effect is unknown but involvement of the macrophage Gal/GalNAc receptor is an interesting possibility.

4.6. The Liver Endothelial Cell Hyaluronan/Glycosaminoglycan Receptor

Considering the large endocytic capacity of liver endothelial cells for hyaluronan (McGary *et al.*, 1989; Smedsrod *et al.*, 1984) and the large daily turnover of this glycosaminoglycan in the body (1–4g/day), it is likely that removal of circulating hyaluronan is an important function (Evered and Whelan, 1989). This is the only carbohydrate receptor for which the endogenous ligands are definitely known. The metabolic turnover of chondroitin sulfate or glycosaminoglycan chains other than hyaluronan recognized by this receptor has not been determined but it is clear that this is a generic receptor able to mediate the clearance of this large polysaccharide family. The physiologic consequence of not removing these molecules from the blood is not known but can be imagined. Problems in the microcirculatory system could arise as the viscosity of blood increased with increasing hyaluronan concentration in the blood (Weigel *et al.*, 1986b). Human fibrinogen specifically binds to hyaluronan (LeBoeuf *et al.*, 1986) and the presence of hyaluronan alters the ability of thrombin to induce fibrin clot formation *in vitro* (LeBoeuf *et al.*, 1987). Elevated levels of hyaluronan in the blood could, therefore, affect the normal dynamics of homeostasis and induce thromboses or bleeding.

4.7. The Glucosyl–Advanced Glycation End Products (AGE) Receptor

This receptor is present on macrophages and liver sinusoidal cells (Radoff *et al.*, 1988; Takata *et al.*, 1988). Glycation refers to the nonenzymatic covalent addition of a free reducing sugar such as glucose to a protein by Schiff base formation (Ahmed *et al.*, 1988). Slow rearrangements of these adducts yield stable compounds; the advanced glycation end products. If most or all migratory and resident macrophages in the body have the AGE receptor then it may have an important function; the removal of senescent proteins bearing the age-dependent modification of advanced glycation end products derived from glucose. Patrick *et al.* (1988) documented an excellent correlation in the accumulation of one such end product, carboxymethyl lysine, in the eye lens with age in humans. Because of the phagocytic capacity of macrophages, the AGE receptor would allow removal of “aged” cells as well as soluble proteins.

5. FUNCTION OF THE HEPATIC ASIALOGLYCOPROTEIN RECEPTOR

Mammals express two different Gal/GalNAc receptor systems; one in hepatocytes and one in macrophages (including Kupffer cells). Although the hepatocyte asialoglycoprotein receptor was discovered almost 28 years ago, its physiological role is not clear. The function(s) of both these Gal/GalNAc receptors are unknown. They are generally assumed to remove deleterious agents (Ashwell and Morell, 1974; Steer and Ashwell, 1986; Breitfield *et al.*, 1985) but hypotheses on why or how these agents are deleterious or the consequences of their not being removed have not been presented.

5.1. Previously Proposed Functions of the Hepatic Parenchymal Gal/GalNAc Receptor

5.1.1. Glycoprotein Turnover

Many different desialylated soluble glycoproteins are removed after injection into the blood. Ashwell and Morell (1974) proposed that the function of the receptor was to remove plasma glycoproteins desialylated by endogenous neuraminidases. They suggested that plasma-protein turnover involves loss of sialic acid, either spontaneously or because of circulating neuraminidases, and clearance of the damaged protein by the liver receptor. However, this is not the mechanism by which plasma proteins, such as transferrin, choriogonadotropin, or orosomucoid, turn over (Clarenburg, 1983; Wong *et al.*, 1974; Kuranda and Aronson, 1983; Lefort *et al.*, 1984). Another argument against this primary function for the receptor is its greater affinity for ligands with terminal GalNAc rather than Gal (Ashwell and Harford, 1982; Lee *et al.*, 1982). Since the penultimate sugar on sialylated plasma glycoproteins is Gal, not GalNAc, these are probably not the endogenous ligands.

5.1.2. Removal of Antibody-Antigen Complexes

Mammalian IgGs have Gal/GalNAc-terminal oligosaccharides in the hinge region, yet the molecule is not removed by the Gal/GalNAc receptor. Formation of the IgG-antigen complex could change the antibody conformation so that this oligosaccharide is now recognized by the receptor and the complex can be removed and degraded (Thornburg *et al.*, 1980). While this is an intriguing model, there is at present little evidence to support it.

5.1.3. Cell Adhesion

Cell-cell (Kuhlenschmidt *et al.*, 1982) or cell-matrix (Weigel, 1980; Oka and Weigel, 1986) adhesion can be mediated *in vitro* by the receptor and may be a possible function *in vivo* during metastasis (Springer *et al.*, 1983; Cheingsong-Popov *et al.*, 1983). Since cell surfaces and the extracellular matrix contain Gal/GalNAc-terminal molecules, an adhesive function for the receptor on hepatocytes *in vivo* is possible. It is likely, however, that this adhesion is fortuitously expressed *in vitro* because of the large number of receptors and their mobility.

5.1.4. Cellular Turnover

The presence of cell-surface Gal/GalNAc glycoconjugates can direct a cell to be destroyed. This may be an important function of the macrophage Gal/GalNAc receptor. For example, as erythrocytes age they lose surface sialic acid and increase the number of exposed Gal/GalNAc residues (Aminoff, 1988). These surface saccharides are recognized by the macrophage Gal/GalNAc receptor, which mediates the binding, uptake, and degradation of the cell. Although the hepatocyte Gal/GalNAc receptor may mediate binding to aged erythrocytes, it cannot mediate endocytosis of such large particles. This latter receptor is locked into a coated-pit pathway, whereas the Kupffer- (macrophage-) cell Gal/GalNAc receptor is not. Although their sugar specificities appear almost identical, the Kupffer-cell receptors can bind particles ≥ 11.7 nm in diameter, whereas only particles of < 7.8 nm are bound to hepatocytes (Schlepper-Schafer *et al.*, 1986).

5.1.5. Intracellular Routing of Glycoproteins

Regulation of glycoprotein trafficking is also a possible function. The receptor could bind Gal/GalNAc oligosaccharides of newly made glycoproteins and shuttle them from the Golgi to the cell periphery (Olden *et al.*, 1982). They could then be sialylated, released from the receptor, and exocytosed. Since this receptor is not in other cell types, this function would be unique to hepatocytes, which synthesize large amounts of plasma glycoproteins. In support of this proposed function, Gal/GalNAc receptors are found in Golgi and other organelles between the Golgi and plasma membrane (Breitfield *et al.*, 1985). Also, a small percentage of internalized asialoglycoproteins are resialylated *in vivo* or *in vitro* and returned intact to the medium (Regoeczi *et al.*, 1982). This implies that some receptors shuttle their ligands to a compartment containing active sialyl transferases.

As discussed above, ligand processing and receptor recycling in the Gal/GalNAc receptor system in hepatocytes occur by two different intracellular pathways; a major pathway and a minor pathway (Weigel *et al.*, 1986a; Weigel, 1987). The major pathway is an extremely efficient endocytic- and receptor-recyling pathway, whereas the minor pathway is not. We believe that the recently described minor pathway is involved in the resialylation process and is also responsible for the low level of asialoglycoprotein transcytosis *in vivo*. The function of the major Gal/GalNAc-receptor pathway is, in fact, to remove circulating Gal/GalNAc-containing ligands. These agents are deleterious and their removal is vitally important for the reasons developed below.

5.2. The Galactosyl Homeostasis Hypothesis

We propose that the major function of the two liver Gal/GalNAc receptor systems is to eliminate circulating glycoconjugates. The reasons for this function are embodied in an idea referred to as the *galactosyl homeostasis hypothesis*. The basic tenet of this new hypothesis (Figure 5) is that cell-surface and extracellular Gal/GalNAc glycoconjugates are necessary and important molecules during embryogenesis and organ development. Correct morphogenesis and organization of differentiated tissues relies, in as yet undetermined ways, on these molecules. After development is complete, the maintenance of cell-cell, cell-matrix, and intramatrix interactions and the exertion of normal cell-growth control requires the continuous participation of Gal/GalNAc-specific lectins (either as carbohydrate-binding proteins or as true receptors) with Gal/GalNAc glycoconjugates. These complementary molecules turn over continuously, they are degraded and synthesized, and a constant steady-state concentration of each interacting set of Gal/GalNAc molecules (ligands) and their respective specific lectins (receptors) is maintained in the tissue.

The introduction of soluble or particulate Gal/GalNAc glycoconjugates, for whatever reason, into a tissue will be deleterious because they will compete for binding to the endogenous Gal/GalNAc-specific lectins/receptors in the matrix and on cells. The competing Gal/GalNAc-containing molecules would disrupt the normal organization, growth control, or function maintained by these interactions. This is the reason that soluble or particulate Gal/GalNAc glycoconjugates generated by disease, trauma, cell death, or other causes can injure the organism. Mammals have therefore evolved two efficient systems for removing soluble and particulate Gal/GalNAc ligands. One, the liver parenchymal Gal/GalNAc receptor, is one of the most active endocytic and degradative systems known (Ashwell and Harford, 1982; Weigel, 1987). It removes a wide range of soluble circulating Gal/GalNAc glycoconjugates that would arise in the extracellular spaces in all tissues, and then flow to the lymph and then into the blood. The other system, the macrophage Gal/GalNAc receptor, is present on resident and migratory cells and removes a wide range of Gal/GalNAc-containing particles including whole cells, matrix, and cellular debris (Schlepper-Schaffer *et al.*, 1986). Unlike the liver system, which is fixed, the macrophage system is mobile (except for liver Kupffer cells) and can perform a surveillance, search and destroy, function. Thus, regulation of the synthesis and release of these molecules in tissues, and their removal by these two receptor systems, results in a dynamic Gal/GalNAc

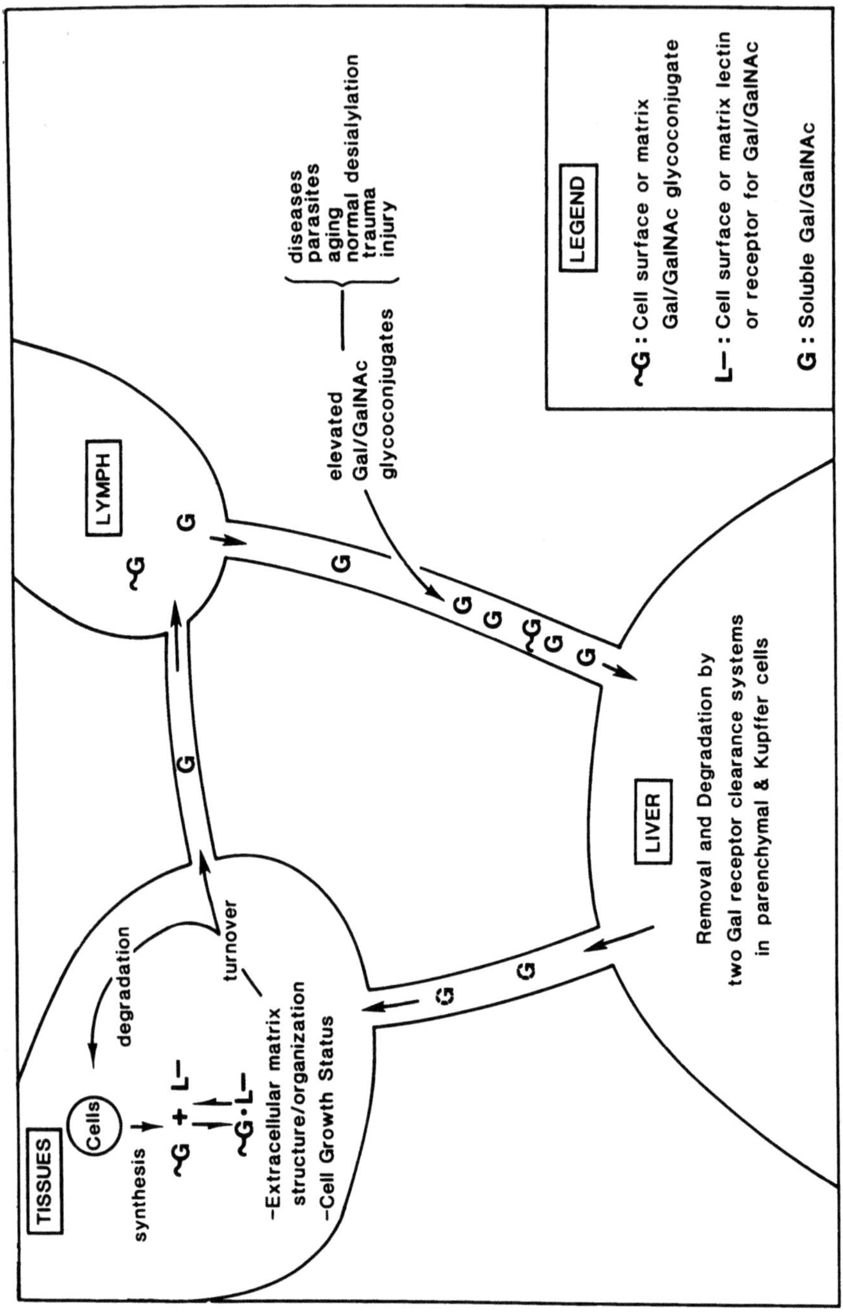
TISSUES
Cells
synthesis
~G + L–
~G·L–
degradation
turnover
–Extracellular matrix structure/organization
–Cell Growth Status
G
LYMPH
~G
G
G
elevated Gal/GalNAc glycoconjugates
diseases
parasites
aging
normal desialylation
trauma
injury
G G ~G G G
LIVER
Removal and Degradation by two Gal receptor clearance systems in parenchymal & Kupffer cells
G G
LEGEND
~G : Cell surface or matrix Gal/GalNAc glycoconjugate
L– : Cell surface or matrix lectin or receptor for Gal/GalNAc
G : Soluble Gal/GalNAc

homeostasis in the body. A brief summary of the evidence in support of this hypothesis is presented below.

5.2.1. Gal-Molecules and Gal/GalNAc Lectins Are Important During Development

The developing mammalian embryo has a high content of Gal/GalNAc glycoconjugates and yet during morphogenesis the two hepatic Gal/GalNAc receptors are not expressed (Dini *et al.*, 1987; Collins *et al.*, 1984; Petell and Doyle, 1985). The Gal/GalNAc receptor is not expressed until birth when it increases dramatically and reaches adult levels within 2 days in rat or 5 days in mouse. The hepatocyte and Kupffer-cell receptors follow parallel patterns of expression. The fetus, therefore, has little or no capacity to remove circulating Gal/GalNAC glycoconjugates, whereas this ability exists shortly after birth. We propose that there is no need to remove these molecules in the embryo; they are part of the normal dynamic developmental process. In fact, their removal during development would be detrimental and result in abnormal or abortive development.

A variety of β-galactoside-binding lectins are regulated during development. Their concentration and distribution pattern change as development of a tissue proceeds (Barondes, 1984). In general, these lectins are soluble proteins and are associated with the extracellular matrix. Cell-surface Gal glycoconjugates are also developmentally regulated. For example, rat-lung macrophages express surface α-linked Gal residues from the 14th day of gestation through the adult (Honda *et al.*, 1989). Cell migration during development also depends on recognition and even modification of Gal-containing molecules in the extracellular matrix (Bayna *et al.*, 1988). Cell-surface Gal/GalNAc transferases are important during cell migration and morula compaction (Bayna *et al.*, 1988).

Galactosemia, high levels of Gal, is associated with premature ovarian failure. Pregnant rats fed diets containing 50% galactose have female offspring with reduced oocyte number (Chen *et al.*, 1981), and development of offspring is

---

FIGURE 5. The Galactosyl Homeostasis Hypothesis. The diagram depicts the dynamic metabolic turnover of Gal/GalNAc glycoconjugates and their removal by the liver. Macrophages in other tissues, especially the reticuloendothelial system (e.g., lymph nodes) will also degrade larger Gal/GalNAc-containing particles or cells. Cells in various tissues synthesize and secrete the Gal/GalNAc lectins and glycoconjugates. As they turn over, some of theses Gal/GalNAc glycoconjugates flow from the extracellular matrix to lymph and then the blood. The two Gal/GalNAc liver receptor systems remove and further degrade these glycoconjugates. The modified Gs leaving the liver represent molecules that might escape degradation if receptor function is impaired or if the influx of Gal/GalNAc molecules (due to diseases, acute phase response, etc.) is so great that the liver's capacity for removal was saturated. These "escaped" glycoconjugates could then exert disruptive and pathologic effects on cell growth or function or the structure/organization of the matrix.

retarded dramatically (30–50% reduced body weight) compared to controls. Gal also induces teratogenesis in mice with defective galactokinase (Shih *et al.*, 1988). Litters of mothers on a Gal diet showed greater fetal loss and 26% of the pups were malformed compared to 0% for control mothers. One interpretation of these results is that a competing ligand, Gal, disrupts the normal lectin-Gal/GalNAc interactions established in the embryo and interferes with normal development. Many studies suggest that Gal/GalNAc lectins and surface Gal/GalNAc glycoconjugates or Gal transferases are developmentally regulated and involved in morphogenesis and early development; periods of active cell migration and reorganization.

5.2.2. A Role for Gal-Molecules and Gal-Lectins in Differentiated Tissues

The extracellular matrix of adult mammalian tissues also contains a variety of soluble Gal/GalNAc-binding proteins. For example, soluble β-Gal lectins can be isolated from brain, and distinct subsets of rat sensory neurons contain both soluble Gal lectins and cell-surface Gal glycoconjugates (Regan *et al.*, 1986). These soluble lectins, which range in size from about 14 to 30 kDa, are thought to be needed for the organization, and perhaps for rigidity or crosslinking, of the matrix (Barondes, 1984). Within the last few years the cDNAs for several of these lectins have been isolated and sequenced (Drickamer, 1988). Other important extracellular molecules, such as the proteoglycan core and lung-surfactant proteins, also have Gal/GalNAc-binding activity and sequence homology to the sugar-binding regions of Gal/GalNAc lectins.

Multiple gene families code for related Gal/GalNAc lectins, and one tissue can express several lectins. Lung has at least three different lectins with similar but distinct carbohydrate specificities (Sparrow *et al.*, 1987). Differentiated tissues have multiple sets of lectin-ligand interactions occurring simultaneously.

5.2.3. Cell Surfaces Contain Gal/GalNAc Glycoconjugates

Although eukaryotic cell surfaces are rich in sialylated glycolipids and glycoproteins, these cells also posssess terminal Gal/GalNAc residues. Some examples are cited above and below. Many mammalian neutral glycosphingolipids contain terminal Gal/GalNAc (Galili *et al.*, 1988). Many pulmonary, urinary, or gastrointestinal pathogenic bacteria establish themselves in the host by binding to these glycolipids. For example, *E. coli* that cause urinary tract infections adhere to Galα1,4Gal- residues on uroepithelial cells (Lund *et al.*, 1987). *Entamoeba histolytica,* a unicellular parasite that infects about 10% of the world's population, adheres to human colonic mucous and epithelial cells by a Gal/GalNAc lectin (Petri *et al.*, 1987).

5.2.4. Gal/GalNAc Glycoconjugates and Receptors Are Involved in Growth Control and Cancer

Soluble Gal/GalNAc lectins are often mitogenic when tested on cells in culture (Lis and Sharon, 1986; Wieser and Oesch, 1988). The phenomenon of contact inhibition of growth, displayed by cultured cells in continuous cell-cell contact, is generally considered to reflect a regulatory process of fundamental importance. Cell-cell contacts via plasma-membrane glycoconjugates with terminal Gal/GalNAc residues mediate this growth-regulatory response in human fibroblasts (Wieser and Oesch, 1988). Cell growth can also be stimulated by β-galactoside treatment. Although the mechanism by which cell-surface Gal/GalNAc regulates growth is unknown, many tumors and tumor cell lines contain endogenous Gal/GalNAc lectins that could be involved in tumor cell growth, adhesion, and metastasis (Raz and Lotan, 1987). Free sialic acid and sialidases active against G_{M3} gangliosides accumulate in a culture medium of fibroblasts and correlate with cell density. These molecules disappear when cells become contact inhibited (Usuki *et al.*, 1987).

Surface Gal/GalNAc groups may regulate cell growth by their interaction with an appropriate surface lectin, which is a growth-factor receptorlike molecule. Binding of the Gal/GalNAc glycoconjugates by some exogenous lectins could release this normal inhibition and stimulate growth. Contact with glycoconjugates on other cells might inactivate the Gal/GalNAc lectin/receptors (Wieser and Oesch, 1988).

Antibodies against cancer cells often react with Gal/GalNAc glycoconjugates, and tumor cells can express large amounts of these molecules (Elices and Goldstein, 1989). The ability of macrophages, natural killer cells, and cytotoxic lymphocytes to kill tumor cells depends on expression of surface Gal and Gal lectins (Silagi *et al.*, 1988). In particular, expression of asialo-G_{M1} on the cytotoxic cells appears necessary for their killing ability. A Gal/GalNAc lectin on these killer cells recognizes tumor cells (Pospisil *et al.*, 1986).

Normal human sera contain a natural anti-Gal IgM that can mediate the complement-dependent lysis of Ehrlich ascites tumor cells (Gross *et al.*, 1988). The sera of cancer patients show reduced titers of these antibodies.

5.2.5. Gal/GalNAc Glycoconjugates and Disease

All normal humans contain a natural serum IgG, regardless of blood type or age, that recognizes α1,3-linked Gal (Galili *et al.*, 1988). A variety of human diseases are accompanied by production of anti-Gal antibodies or by changes in the galactosylation of circulating IgG (Rademacher *et al.*, 1988). Renal injury is accompanied by elevated titers of circulating anti-Gal antibodies. Anti-Gal IgAs against microbial pathogens may be responsible for IgA nephropathy or Henoch-

Schonlein nephritis. Infections of the protozoa *Trypanosoma* lead to host antibodies against terminal Galα1,3Gal-epitopes (Avila *et al.*, 1988). Patients with high titers of these anti-Gal antibodies develop Chagas disease, in which organ degeneration is induced by antibody reaction with blood vessels, endocardium, striated muscle, and peripheral nerves. These anti-Gal antibodies react strongly with the oligosaccharides of the major extracellular basement-membrane glycoprotein, laminin (Avila *et al.*, 1988). Other autoimmune disorders are also associated with the production of anti-Gal antibodies. The presence of anti-Gal antibodies would disrupt the normal glycoconjugates-lectin interactions in a tissue just as a soluble competitor would.

5.2.6. Summary of the Model

This survey of supportive evidence for the central ideas of the Galactosyl Homeostasis Hypothesis is not meant to be exhaustive. From the studies cited, however, we can propose the following scenario. During human fertilization, morphogenesis, and early fetal development Gal/GalNAc glycoconjugates and their respective soluble and cell-surface lectins play important roles. Their concentrations are high and change as development proceeds. Although their functions are not yet clear, they are necessary for successful development. When organogenesis is complete and organ systems are functioning, these sets of glycoconjugate-lectin interactions become stable, and their circulating levels in lymph or blood decreases as expression of the Gal hepatic receptor increases.

The Gal/GalNAc glycoconjugates in differentiated tissues may be so stable that they are essentially cryptic. These molecules may be "seen" only by the very early embryo and then be hidden from the developing immune system before it establishes the program of antigens that will be recognized as self and nonself. The presence of these Gal/GalNAc glycoconjugates on cell surfaces signifies a fetallike condition of cells actively growing and migrating. This situation would appear in the adult only under unusual circumstances, particularly in tumorigenesis and metastasis. Microbial pathogens have evolved strategies to take advantage of these Gal/GalNAc-containing molecules on rapidly dividing epithelial cells in kidney, gut, and lung. During the acute-phase response to inflammation, the circulating levels of Gal/GalNAc glycoconjugates increases dramatically and occupancy of the hepatocyte Gal/GalNAc receptor is more than 50% (Wong and Jamieson, 1979). This result indicates that tissue damage causes release of otherwise cryptic or stable Gal/GalNAc molecules into the blood.

Although these Gal/GalNAc glycoconjugates are cryptic, they are continually degraded and resynthesized in the tissues and these Gal-containing epitopes are transiently but continuously presented to the cellular immune system.

Consequently, anti-Gal antibodies can develop in normal healthy individuals (Galili *et al.*, 1988; Rademacher *et al.*, 1988). These anti-Gal antibodies might be important for control of spontaneous tumors that present these "fetal" Gal markers. However, the level of circulating anti-Gal antibodies can become too high if triggered by infection with microorganisms that contain similar Gal/GalNAc epitopes. In this case the anti-Gal-glycoconjugate antibodies can disrupt the normal stable Gal/GalNAc glycoconjugate-lectin interactions, which will cause lesions and pathological changes in the tissue.

5.2.7. Corollaries of the Galactosyl Homeostasis Hypothesis

Several corollaries arise from the model. First, avian hepatocytes have a similar receptor specific for N-acetylglucosamine (GlcNAc) glycoconjugates (Ashwell and Morell, 1974; Steer and Ashwell, 1986). In birds, GlcNAc lectins and GlcNAc-containing glycoconjugates could function in the way that we propose the Gal/GalNAc system functions in mammals. One could develop a similar N-acetylglucosaminyl homeostasis hypothesis in birds.

Second, if the proposed model is correct, mammals may also have a system to recognize, remove, and degrade circulating Gal/GalNAc lectins. The above discussion has focused on the turnover and clearance of Gal/GalNAc conjugates from tissues but their complementary lectins will also turn over, eventually enter the blood, and need to be removed. Therefore, mammals may have a cellular clearance system in which the clearing receptor contains appropriate Gal/GalNAc residues. No one has yet examined the clearance of Gal/GalNAc lectins from the circulation.

5.2.8. Predictions of the Hypothesis

The premises of the model lead to several testable predictions.

1. There Should Be Elevated Levels of Gal/GalNAc Glycoconjugates in Lymph Relative to Blood. This has not been examined. The reticuloendothelial Gal/GalNAc receptor system may remove the vast majority of Gal/GalNAc glycoconjugates from lymph before they are delivered to blood. The model predicts that the concentration of Gal/GalNAc glycoconjugates in afferent lymph will be higher than in efferent lymph, which will in turn be higher than in blood.

2. If the Liver Parenchymal Gal/GalNAc Receptor is Impaired, the Concentration of Soluble Gal/GalNAc Glycoconjugates in Blood Will Increase. The normal serum concentration of these glycoconjugates is approximately 10 nM (Byrn *et al.*, 1984). Gal/GalNAc-glycoconjugate serum levels increase two to sixfold in cirrhotic patients (Byrn *et al.*, 1984). This could be

because of a defect in receptor function or because damaged liver does not normally sialylate its proteins.

3. Structural Changes. Chronically elevated plasma levels of Gal/GalNAc glycoconjugates should cause structural and/or organizational changes in tissues and have pathological consequences.

These and other related predictions can be tested to understand further the functions of the macrophage and hepatic Gal/GalNAc receptors.

ACKNOWLEDGMENTS. I thank Dr. D. McAbee for helpful discussions and Glenda Romero for help preparing this manuscript. This work was supported by National Institutes of Health grants GM 30218 and GM 35978.

6. REFERENCES

Ahmed, M. U., Dunn, J. A., Walla, M. D., Thorpe, S. R., and Baynes, J. W., 1988, Oxidative degradation of glucose adducts to protein: formation of 3-(N′-lysino)-lactic acid from model compounds and glycated proteins, *J. Biol. Chem.* **263**:8816–8821.

Aminoff, D., 1988, The role of sialoglycoconjugates in the aging and sequestration of red cells from circulation, *Blood Cells* **14**:229–247.

Andersen, T. T., Freytag, J. W., and Hill, R. L., 1982, Physical studies of the rabbit hepatic galactoside-binding protein: effects of calcium and ligands, *J. Biol. Chem.* **257**:8036–8041.

Andersson, G. N., Rissler, P., and Eriksson, L. C., 1988, Asialoglycoprotein receptors in rat liver nodules, *Carcinogenesis* **9**:1623–1628.

Ashwell, G. and Morell, A. G., 1974, The role of surface carbohydrates in the hepatic recognition and transport of circulating glycoproteins, *Adv. Enzymol.* **41**:99–128.

Ashwell, G. and Harford, J., 1982, Carbohydrate-specific receptors of the liver, *Ann. Rev. Biochem.* **51**:531–554.

Avila, J. L., Rojas, M., and Towbin, H., 1988, Serological activity against galactosyl-α(1-3)galactose in sera from patients with several kinetoplastida infections, *J. Clin. Microbiol.* **26**:126–132.

Barondes, S. H., 1984, Soluble lectins: a new class of extracellular proteins, *Science* **223**:1259–1264.

Basu, S. K., Goldstein, J. L., Anderson, R.G.W., and Brown, M. S., 1981, Monensin interrupts the recycling of low density lipoprotein receptors in human fibroblasts. *Cell* **24**:493–502.

Bayna, E. M., Shaper, J. H., and Shur, B. D., 1988, Temporally specific involvement of cell surface β-1,4 galactosyltransferase during mouse embryo morula compaction, *Cell* **53**:145–157.

Berg, T., Blomhoff, R., Naess, L., Tolleshaug, H., and Drevon, C. A., 1983, Monensin inhibits receptor-mediated endocytosis of asialoglycoproteins in rat hepatocytes, *Exp. Cell Res.* **148**:319–330.

Bischoff, J., Libresco, S., Shia, M. A., and Lodish, H. F., 1988, The H1 and H2 polypeptides associate to form the asialoglycoprotein receptor in human hepatoma cells, *J. Cell Biol.* **106**:1067–1074.

Blackshear, P. J., Nairn, A. C., and Kuo, J. F., 1988, Protein kinases 1988: a current perspective, *FASEB J.* **2**:2957–2969.

Breitfeld, P. P., Simmons, C. F., Jr., Strous, G. J. A. M., Geuze, H. J., and Schwartz, A. L., 1985,

Cell biology of the asialoglycoprotein receptor system: a model of receptor-mediated endocytosis, *Internatl. Rev. Cytol.* **97**:47–95.

Byrn, R., Thomas, P., Medrek, P., Spigelman, Z., and Zamcheck, N., 1984, Modified radioassay for measuring asialoglycoprotein in serum, *Clin. Chem.* **30**:1692–1696.

Casey, C. A., Kragskow, S. L., Sorrell, M. F., and Tuma, D. J., 1987, Chronic ethanol administration impairs the binding and endocytosis of asialo-orosomucoid in isolated hepatocytes, *J. Biol. Chem.* **262**:2704–2710.

Casey, C. A., Kragskow, S. L., Sorrell, M. F., and Tuma, D. J., 1990 Effect of chronic ethanol administration on total asialoglycoprotein receptor content and intracellular processing of asialoorosomucoid in isolated rat hepatocytes, *Biochim. Biophys. Acta* **1052**:1–8.

Cheingsong-Popov, R., Robinson, P., Altevogt, P., and Schirrmacher, V., 1983, A mouse hepatocyte carbohydrate-specific receptor and its interaction with liver-metastasizing tumor cells, *Int. J. Cancer* **32**:359–366.

Chen, Y-T., Mattison, D. R., Feigenbaum, L., Fukui, H., and Schulman, J. D., 1981, Reduction in oocyte number following prenatal exposure to a diet high in galactose, *Science* **214**:1145–1147.

Clarenburg, R., 1983, Asialoglycoprotein receptor is uninvolved in clearing intact glycoproteins from rat blood, *Am. J. Physiol.* **244**:G247–G253.

Clarke, B. L., Oka, J. A., and Weigel, P. H., 1987, Degradation of asialoglycoproteins mediated by the galactosyl receptor system in isolated hepatocytes: evidence for two parallel pathways. *J. Biol. Chem.* **262**:17384–17392.

Clarke, B. L. and Weigel, P. H., 1989, Differential effects of leupeptin, monensin, and colchicine on ligand degradation mediated by the two asialoglycoprotein receptor pathways in isolated rat hepatocytes. *Biochem. J.* **262**:277–284.

Collins, J. C., Stockert, R. J., and Morell, A. G., (1984), Asialoglycoprotein receptor expression in murine pregnancy and development, *Hepatology* **4**:80–83.

Dini, L., Conti-Devirgiliis, L., and Russo-Caia, S., 1987, The galactose-specific receptor system in rat liver during development, *Development* **100**:13–22.

Dipaola, M. and Maxfield, F. R., 1984, Conformational changes in the receptors for epidermal growth factor and asialoglycoproteins induced by the mildly acidic pH found in endocytic vesicles, *J. Biol. Chem.* **259**:9163–9171.

Drickamer, K., Mamon, J. F., Binns, G., and Leung, J. O., 1984, Primary structure of the rat liver asialoglycoprotein receptor: structural evidence for multiple polypeptide species, *J. Biol. Chem.* **259**:770–778.

Drickamer, K., 1987, Membrane receptors that mediate glycoprotein endocytosis: structure and biosynthesis, *Kidney Internatl.* **32**:S167–S180.

Drickamer, K., 1988, Two distinct classes of carbohydrate-recognition domains in animal lectins, *J. Biol. Chem.* **263**:9557–9560.

Durand, G., Dodeur, M., Dumont, J-P., Coumoul, S., Agneray, J., and Feger, J., 1984, Surface and total receptors for asialoglycoproteins in hepatocytes from diabetic rats, *Marker Proteins in Inflammation* **2**:655–658.

Elices, M. J. and Goldstein, I. J., 1989, Biosynthesis of bi-, tri-, and tetraantennary oligosaccharides containing α-D-galactosyl residues at their nonreducing termini, *J. Biol. Chem.* **264**:1375–1380.

Evarts, R. P., Marsden, E. R., and Thorgeirsson, S. S., 1985, Modulation of asialoglycoprotein receptor levels in rat liver by phenobarbital treatment, *Carcinogenesis* **6**:1767–1773.

Evered, D. and Whelan, J., eds., 1989, *The Biology of Hyaluronan,* Wiley & Sons, London.

Fiete, D., Brownell, M. D., and Baenziger, J. U., 1983, Evidence for transmembrane modulation of the ligand-binding site of the hepatocyte galactose/N-acetylgalactosamine-specific receptor, *J. Biol. Chem.* **258**:817–823.

Galili, U., Shohet, S. B., Kobrin, E., Stults, C.L.M., and Macher, B. A., 1988, Man, apes, and old world monkeys differ from other mammals in the expression of α-galactosyl epitopes on nucleated cells, *J. Biol. Chem.* **263:**17755–17762.

Gonzalez-Noriega, A., Grubb, J. H., Talkad, V., and Sly, W. S., 1980, Chloroquine inhibits lysosomal enzyme pinocytosis and enhances lysosomal enzyme secretion by impairing receptor recycling, *J. Cell Biol.* **85:**839–852.

Gross, D. J., Simon, E., Szwarc-Bilotynski, L., Margalioth, E., and Gross, J., 1988, An immunochemical assay for natural IgM antibodies with an affinity to galactose and whose titer is reduced in the sera of cancer patients, *Eur. J. Cancer Clin. Oncol.* **24:**363–367.

Halberg, D. F., Wager, R. E., Farrell, D. C., Hildreth J. IV, Quesenberry, M. S., Loeb, J. A., Holland, E. C., and Drickamer, K., 1987, Major and minor forms of the rat liver asialoglycoprotein receptor are independent galactose-binding proteins: primary structure and glycosylation heterogeneity of minor receptor forms. *J. Biol. Chem.* **262:**9828–9838.

Haltiwanger, R. S., Lehrman, M. A., Eckhardt, A. E., and Hill, R. L., 1986, The distribution and localization of the fucose-binding lectin in rat tissues and the identification of a high affinity form of the mannose/N-acetylglucosamine-binding lectin in rat liver. *J. Biol. Chem.* **261:**7433–7439.

Hardy, M. R., Townsend, R. R., Parkhurst, S. M., and Lee, Y. C., 1985, Different modes of ligand binding to the hepatic galactose/N-acetylgalactosamine lectin on the surface of rabbit hepatocytes. *Biochemistry* **24:**22–28.

Harford, J., Bridges, K., Ashwell, G., and Klausner, R. D., 1983a, Intracellular dissociation of receptor-bound asialoglycoproteins in cultured hepatocytes: a pH-mediated nonlysosomal event. *J. Biol. Chem.* **258:**3191–3197.

Harford, J., Wolkoff, A. W., Ashwell, G., and Klausner, R. D., 1983b, Monensin inhibits intracellular dissociation of asialoglycoproteins from their receptor, *J. Cell Biol.* **96:**1824–1828.

Herzig, M. C. S. and Weigel, P. H., 1989, Synthesis and characterization of N-hydroxysuccinimide ester chemical affinity derivatives of asialoorosomucoid that covalently cross-link to galactosyl receptors on isolated rat hepatocytes, *Biochemistry* **28:**600–610.

Herzig, M. C. S. and Weigel, P. H., 1990, Surface and internal galactosyl receptors are hetero-oligomers and retain this structure after ligand internalization or receptor modulation, *Biochemistry* **29:**6437–6447.

Honda, T., Schulte, B. A., and Spicer, S. S., 1989, Glycoconjugate with terminal galactose: a selective property of macrophages in developing rat lung, *Histochem.* **91:**61–67.

Hoppe, C. A. and Lee, Y. C., 1982, Stimulation of mannose-binding activity in the rabbit alveolar macrophage by simple sugars, *J. Biol. Chem.* **257:**12831–12834.

Kaplan, J. and Keogh, E. A., 1981, Analysis of the effect of amines on inhibition of receptor-mediated and fluid-phase pinocytosis in rabbit alveolar macrophages, *Cell* **24:**925–932.

Klerx, J. P. A. M., Molendijk, A. J., Van Dijk, H., Vloet, K. P., and Willers, J. M. N., 1986, Simple sugars with affinity for the macrophage asialoglycoprotein receptor are adjuvants for the humoral immune response to neuraminidase-treated sheep erythrocytes, *J. Immunol.* **136:**73–75.

Kloppel, T. M. (1989). Effects of temperature on the degradation and biliary secretion of asialoorosomucoid by the perfused rat liver: evidence for two intracellular pathways. *J. Cell. Physiol.* **138:**555–560.

Kolset, S. O., Tolleshaug, H., and Berg, T., 1979, The effects of colchicine and cytochalasin B on uptake and degradation of asialo-glycoproteins in isolated rat hepatocytes, *Exp. Cell Res.* **122:**159–167.

Kornfeld, S., 1987, Trafficking of lysosomal enzymes, *FASEB J.* **1:**462–468.

Kuhlenschmidt, M. S., Schmell, E., Slife, C. W., Kuhlenschmidt, T. B., Sieber, F., Lee, Y. C., and Roseman, S., 1982, Studies on the intercellular adhesion of rat and chicken hepatocytes: conditions affecting cell-cell specificity, *J. Biol. Chem.* **257**:3157–3164.

Kuranda, M. J. and Aronson, N. N., Jr., 1983, Tissue locations for the turnover of radioactively labeled rat orosomucoid *in vivo, Arch. Biochem. Biophys.* **224**:526–533.

LeBoeuf, R. D., Raja, R. H., Fuller, G. M., and Weigel, P. H. 1986, Human fibrinogen specifically binds hyaluronic acid, *J. Biol. Chem.* **261**:12586–12592.

LeBoeuf, R. D., Gregg, R. R., Weigel, P. H., and Fuller, G. M., 1987, Effects of hyaluronic acid and other glycosaminoglycans on fibrin polymer formation, *Biochemistry* **26**:6052–6057.

Lee, R. T., Myers, R. W., and Lee, Y. C., 1982, Further studies on the binding characteristics of rabbit liver galactose/N-acetylgalactosamine-specific lectin, *Biochemistry* **21**:6292–6298.

Lee, R. T., Lin, P., and Lee Y. C., 1984, New synthetic cluster ligands for galactose/N-acetylgalactosamine-specific lectin of mammalian liver, *Biochemistry* **23**:4255–4261.

Lee, R. T., and Lee, Y. C., 1988, Rabbit and rat hepatic lectins have two sugar-combining sites per monomeric unit, *Biochem. Biophys. Res. Comm.* **155**:1444–1451.

Lee, Y. C., 1988, Mannose-binding proteins of animal origin, in *The Molecular Immunology of Complex Carbohydrates* (A. M. Wu, ed.) pp. 105–121, Plenum Press, New York.

Lefort, G. P., Stolk, J. M., and Nisula, B. C., 1984, Evidence that desialylation and uptake by hepatic receptors for galactose-terminated glycoproteins are immaterial to the metabolism of human choriogonadotropin in the rat, *Endocrinology* **115**:1551–1557.

Lis, H. and Sharon, N., 1986, Lectins as molecules and as tools, *Ann. Rev. Biochem.* **55**:35–67.

Lund, B., Lindberg, F., Marklund, B-I., and Normark, S., 1987, The PapG protein is the α-D-galactopyranosyl-(1,4)-β-D-galactopyranose-binding adhesin of uropathogenic *Escherichia coli, Proc. Natl. Acad. Sci. USA* **84**:5898–5902.

Magnusson, S. and Berg, T., 1989, Extremely rapid endocytosis mediated by the mannose receptor of sinusoidal endothelial rat liver cell, *Biochem. J.* **257**:651–656.

Maxfield, F. R., 1982, Weak bases and inophores rapidly and reversibly raise the pH of endocytic vesicles in cultured mouse fibroblasts, *J. Cell Biol.* **95**:676–681.

Maynard, Y. and Baenziger, J. U., 1981, Oligosaccharide specific endocytosis by isolated rat hepatic reticuloendothelial cells, *J. Biol. Chem.* **256**:8063–8068.

McAbee, D. D. and Weigel, P. H., 1987, ATP depletion causes a reversible redistribution and inactivation of a subpopulation of galactosyl receptors in isolated rat hepatocytes, *J. Biol. Chem.* **262**:1942–1945.

McAbee, D. D. and Weigel, P. H., 1988, ATP-dependent inactivation and reactivation of constitutively recycling galactosyl receptors in isolated rat hepatocytes, *Biochemistry* **27**:2061–2069.

McAbee, D. D., Oka, J. A., and Weigel, P. H., 1989, Loss of surface galactosyl receptor activity on isolated rat hepatocytes induced by monensin or chloroquine requires receptor internalization via a clathrin-coated pit pathway, *Biochem. Biophys. Res. Commun.* **161**:261–266.

McAbee, D. D., Clarke, B. L., Oka, J. A., and Weigel, P. H., 1990, The surface activity of the same subpopulation of galactosyl receptors on isolated rat hepatocytes is modulated by colchicine, monensin, ATP depletion, and chloroquine, *J. Biol. Chem.* **265**:629–635.

McAbee, D. D., Lear, M. C., and Weigel, P. H., 1991, Total cellular activity and distribution of a subpopulation of galactosyl receptors in isolated rat hepatocytes are differentially affected by microtubule drugs, monensin, low temperature, and chloroquine, *J. Cell. Biochem.* **45**:59–68.

McGary, C. T., Raja, R. H., and Weigel, P. H., 1989, Endocytosis of hyaluronic acid by rat liver endothelial cells: evidence for receptor recycling, *Biochem. J.* **257**:875–884.

Ohsumi, Y. and Lee, Y. C., 1987, Mannose-receptor ligands stimulate secretion of lysosomal enzymes from rabbit alveolar macrophages. *J. Biol. Chem.* **262**:7955–7962.

Oka, J. A. and Weigel, P. H., 1983a, Microtubule-depolymerizing agents inhibit asialo-orosomucoid delivery to lysosomes but not its endocytosis or degradation in isolated rat hepatocytes, *Biochim. Biophys. Acta* **763:**368–376.

Oka, J. A. and Weigel, P. H., 1983b, Recycling of the asialoglycoprotein receptor in isolated rat hepatocytes: dissociation of internalized ligand from receptor occurs in two kinetically and thermally distinguishable compartments, *J. Biol. Chem.* **258:**10253–10262.

Oka, J. A. and Weigel, P. H., 1986, Binding and spreading of hepatocytes on synthetic galactose culture surfaces occur as distinct and separable threshold responses, *J. Cell Biol.* **103:**1055–1060.

Oka, J. A. and Weigel, P. H., 1987, Monensin inhibits ligand dissociation only transiently and partially and distinguishes two galactosyl receptor pathways in isolated rat hepatocytes, *J. Cell. Physiol.* **133:**243–252.

Oka, J. A. and Weigel, P. H., 1988, Effects of hyperosmolarity on ligand processing and receptor recycling in the hepatic galactosyl receptor system. *J. Cell. Biochem.* **36:**169–183.

Oka, J. A. and Weigel, P. H., 1991, Vanadate modulates the activity of a subpopulaton of asialoglycoprotein receptors on isolated rat hepatocytes: Active surface receptors are internalized and replaced by inactive receptors, *Arch. Biochem. Biophys.* **289:**362–370.

Oka, J. A., Christensen, M. D., and Weigel, P. H., 1989, Hyperosmolarity inhibits galactosyl receptor-mediated but not fluid-phase endocytosis in isolated rat hepatocytes, *J. Biol. Chem.* **264:**12016–12024.

Olden, K., Parent, J. B., and White, S. L., 1982, Carbohydrate moieties of glycoproteins: a reevaluation of their function, *Biochim. Biophys. Acta* **650:**209–232.

Patrick, J. S., Thorpe, S. R., and Baynes, J. W., 1988, Oxidative degradation of glycated proteins: age-dependent accumulation of carboxymethyllysine in lens proteins, *FASEB J.* **2:**A1731.

Petell, J. K. and Doyle, D., 1985, Developmental regulation of the hepatocyte receptor for galactose-terminated glycoproteins, *Arch. Biochem. Biophys.* **241:**550–560.

Petri, W. A., Jr., Smith, R. D., Schlesinger, P. H., Murphy, C. F., and Ravdin, J. I., 1987, Isolation of the galactose-binding lectin that mediates the *in vitro* adherence of *Entamoeba histolytica, J. Clin. Invest.* **80:**1238–1244.

Pospisil, M., Kubrycht, J., Bezouska, K., Taborsky, O., Novak, M., and Kocourek, J., 1986, Lactosamine type asialooligosaccharide recognition in NK cytotoxicity, *Immunol. Let.* **12:**83–90.

Pranning-van Dalen, D. P., de Leeuw, A. M., Brouwer, A., and Knook, D. L., 1987, Rat liver endothelial cells have a greater capacity than Kupffer cells to endocytose N-acetylglucosamine and mannose-terminated glycoproteins, *Hepatology* **7:**672–679.

Prioli, R. P., Rosenberg, I., and Pereira, M.E.A., 1987, Specific inhibition of *Trypanosoma cruzi* neuraminidase by the human plasma glycoprotein "cruzin", *Proc. Natl. Acad. Sci. USA* **84:**3097–3101.

Rademacher, T. W., Parekh, R. B., and Dwek, R. A., 1988, Glycobiology, *Ann. Rev. Biochem.* **57:**785–838.

Radoff, S., Vlassara, H., and Cerami, A., 1988, Characterization of a solubilized cell surface binding protein on macrophages specific for proteins modified nonenzymatically by advanced glycosylated end products, *Arch. Biochem. Biophys.* **263:**418–423.

Raz, A. and Lotan, R., 1987, Endogenous galactoside-binding lectins: a new class of functional tumor cell surface molecules related to metastasis, *Cancer and Metastasis Rev.* **6:**433–452.

Regan, L. J., Dodd, J., Barondes, S. H., and Jessell, T. M., 1986, Selective expression of endogenous lactose-binding lectins and lactoseries glycoconjugates in subsets of rat sensory neurons, *Proc. Natl. Acad. Sci. USA* **83:**2248–2252.

Regoeczi, E., Chindemi, P. A., Debanne, M. T., and Charlwood, P. A., 1982, Partial resialylation of human asialotransferrin type 3 in the rat, *Proc. Natl. Acad. Sci. USA* **79:**2226–2230.

Sawyer, J. T., Sanford, J. P., and Doyle, D., 1988, Identification of a complex of the three forms of the rat liver asialoglycoprotein receptor, *J. Biol. Chem.* **263:**10534–10538.

Scarmato, P., Durand, G., Agneray, J., and Feger, J., 1986, Inhibitory effect of sodium arse ite and azide on asialoglycoprotein receptor mediated endocytosis in suspended rat hepatocytes, *Biol. Cell.* **56:**255–258.

Schlepper-Schafer, J., Hulsmann, D., Djovkar, A., Meyer, H. E., Herbertz, L., Kolb, H., and Kolb-Bachofen, V., 1986, Endocytosis via galactose receptors *in vivo:* ligand size directs uptake by hepatocytes and/or liver macrophages, *Exp. Cell Res.* **165:**494–506.

Schlepper-Schafer, J., and Kolb-Bachofen, V., 1988, Red cell aging results in a change of cell surface carbohydrate epitopes allowing for recognition by galactose-specific receptors of rat liver macrophages, *Blood Cells* **14:**259–269.

Schwartz, A. L., 1984, The hepatic asialoglycoprotein receptor, *Critical Rev. Biochem.* **16:**207–233.

Serbource-Goguel, N., Dodeur, M., Borel, B., Feger, J., Agneray, J., and Durand, G., 1985, Study of hepatic binding protein activity in jejuno-ileal bypassed rat hepatocytes, *FEBS lett.* **183:**232–234.

Shih, L. Y., Kuerer, H. M., Chen, T. H., and Deposito, F., 1988, Strain difference in galactokinase level and susceptibility to the teratogenic effect of dietary galactose in mice: I. teratogenic and embryopathic effect, *Teratology* **38:**175–179.

Silagi, S., Dutkowski, R., and Schaefer, A., 1988, *In vivo* requirement for asialo G_{M1} and thyl positive leukocytes for antitumor effect of rIL-2 ± rIFN-γ. *Anticancer Res.* **8:**1265–1270.

Slama, A., Zinbi, H., Feger, J., and Dodeur, M., 1988, Comparative determination of the asialoglycoprotein receptor by ligand and antibody binding in hepatocytes from normal and diabetic rats, *Biol. Cell* **63:**367–369.

Smedsrod, B., Pertoft, H., Eriksson, S., Fraser, J.R.E., and Laurent, T. C., 1984, Studies *in vitro* on the uptake and degradation of sodium hyaluronate in rat liver endothelial cells, *Biochem. J.* **223:**617–626.

Sparrow, C. P., Leffler, H., and Barondes, S. H., 1987, Multiple soluble β-galactoside-binding lectins from human lung, *J. Biol. Chem.* **262:**7383–7390.

Spiess, M. and Lodish, H. F., 1985, Sequence of a second human asialoglycoprotein receptor: conservation of two receptor genes during evolution, *Proc. Natl. Acad. Sci. USA* **82:**6465–6569.

Springer, G. R., Cheingsong-Popov, R., Schirrmacher, V., Desai, P. R., and Tegtmeyer, H., 1983, Proposed molecular basis of murine tumor cell-hepatocyte interaction, *J. Biol. Chem.* **258:**5702–5706.

Steer, C. J., Kempner, E. S., and Ashwell, G., 1981, Molecular size of the hepatic receptor for asialoglycoproteins determined *in situ* by radiation inactivation, *J. Biol. Chem.* **256:**5851–5856.

Steer, C. J. and Ashwell, G., 1986, Hepatic membrane receptors for glycoproteins, in *Progress in Liver Diseases,* vol. 8 (H. Popper and F. Schaffner, eds.), pp. 99–123. Grune and Stratton, New York.

Stein, B. S., Bensch, K. G., and Sussman, H. H., 1984, Complete inhibition of transferrin recycling by monensin in K562 cells, *J. Biol. Chem.* **259:**14762–14772.

Stein, B. S. and Sussman, H. H., 1986, Demonstration of two distinct transferrin receptor recycling pathways and transferrin-independent receptor internalization in K562 cells, *J. Biol. Chem.* **261:**10319–10331.

Stein, M., Braulke, T., Krentler, C., Hasilik, A., and von Figura, K., 1987a, 46-kDa mannose-6-

phosphate-specific receptor: biosynthesis, processing, subcellular location, and topology, *Biol. Chem. Hoppe-Seyler* **368:**937–947.

Stein, M., Zijderhand-Bleekemolen, J. E., Geuze, H., Hasilik, A., and von Figura, K., 1987b, M_r 46 000 mannose-6-phosphate specific receptor: its role in targeting of lysosomal enzymes, *Embo J.* **6:**2677–2681.

Sung, S-S. J., Nelson, R. S., and Silverstein, S. C., 1985, Mouse peritoneal macrophages plated on mannan- and horseradish peroxidase-coated substrates lose the ability to phagocytose by their Fc receptors, *J. Immunol.* **134:**3712–3717.

Takata, K., Horiuchi, S., Araki, N., Shiga, M., Saitoh, M., and Morino, Y., 1988, Endocytic uptake of nonenzymatically glycosylated proteins is mediated by a scavenger receptor for aldehyde-modified proteins, *J. Biol. Chem.* **263:**14819–14825.

Thornburg, R. W., Day, J. F., Baynes, J. W., and Thorpe, S. R., 1980, Carbohydrate-mediated clearance of immune complexes from the circulation: a role for galactose residues in the hepatic uptake of IgG-antigen complexes, *J. Biol. Chem.* **255:**6820–6825.

Tietze, C., Schlesinger, P., and Stahl, P., 1980, Chloroquine and ammonium ion inhibit receptor-mediated endocytosis of mannose-glycoconjugates by macrophages: apparent inhibition of receptor recycling, *Biochem. Biophys. Res. Commun.* **93:**1–8.

Tietze, C., Schlesinger, P., and Stahl, P., 1982, Mannose-specific endocytosis receptor of alveolar macrophages: demonstration of two functionally distinct intracellular pools of receptor and their roles in receptor recycling, *J. Cell Biol.* **92:**417–424.

Tolleshaug, H. and Berg, T. 1979, Chloroquine reduces the number of asialoglycoprotein receptors in the hepatocyte plasma membrane, *Biochem. Phar.* **28:**2919–2922.

Tolleshaug, H., Kolset, S. O., and Berg, T., 1985, The influence of cellular ATP levels on receptor-mediated endocytosis and degradation of asialo-glycoproteins in suspended hepatocytes, *Biochem. Pharm.* **34:**1639–1645.

Usuki, S., Lyu, S-C., and Sweeley, C. C., 1988, Sialidase activities of cultured human fibroblasts and the metabolism of G_{M3} ganglioside, *J. Biol. Chem.* **263:**6847–6853.

von Figura, K. and Hasilik, A., 1986, Lysosomal enzymes and their receptors. *Ann. Rev. Biochem.* **55:**167–193.

Weigel, P. H., 1980, Rat hepatocytes bind to synthetic galactoside surfaces via a patch of asialoglycoprotein receptors, *J. Cell Biol.* **87:**855–861.

Weigel, P. H. and Oka, J. A., 1983, The surface content of asialoglycoprotein receptors on isolated hepatocytes is reversibly modulated by changes in temperature, *J. Biol. Chem.* **258:**5089–5094.

Weigel, P. H. and Oka, J. A., 1984, Recycling of the hepatic asialoglycoprotein receptor in isolated rat hepatocytes: receptor-ligand complexes in an intracellular slowly dissociating pool return to the cell suface prior to dissociation. *J. Biol. Chem.* **259:**1150–1154.

Weigel, P. H., Clarke, B. L., and Oka, J. A., 1986a, The hepatic galactosyl receptor system: two different ligand dissociation pathways are mediated by distinct receptor populations, *Biochem. Biophys. Res. Comm.* **140:**43–50.

Weigel, P. H., Fuller, G. M., and LeBoeuf, R. D., 1986b, A model for the role of hyaluronic acid and fibrin in the early events during the inflammatory response and wound healing, *J. Theor. Biol.* **119:**219–234.

Weigel, P. H., 1987, Receptor recycling and ligand processing mediated by hepatic galactosyl receptors: a two-pathway system, in *Vertebrate Lectins,* K. Olden and J. B. Parent, eds., pp. 65–91, Van Nostrand Reinhold, New York.

Weiser, R. J. and Oesch, F., 1988, Contact-dependent regulation of growth of diploid human fibroblasts is dependent upon the presence of terminal galactose residues on plasma membrane glycoproteins, *Exp. Cell Res.* **176:**80–86.

Wong, K-L., Charlwood, P. A., Hatton, M.W.C., and Regoeczi, E., 1974, Studies of the metabolism of asialotransferrins: evidence that transferrin does not undergo desialylation *in vivo, Clin. Sci. Mol. Med.* **46**:763–774.

Wong, M.W.C. and Jamieson, J. C., 1979, Evidence for reduced uptake of asialo-α_1-acid glycoprotein during the acute phase response to inflammation, *Life Sci.* **25**:827–834.

Yamashiro, D. J. and Maxfield, F. R., 1988, Regulation of endocytic processes by pH, *Trends in Pharm. Sci.* **9**:190–193.

Chapter 6

Characterization of Endocytic Components of Liver Nonparenchymal Cells

Robert Wattiaux, Michel Jadot, Sandra Misquith, and Simone Wattiaux-de Coninck

1. INTRODUCTION

1.1. Liver Heterogeneity

Liver is made up of at least five important cell types: hepatocytes, endothelial cells, Kupffer cells, pit cells, and fat-storing cells. These cells have been broadly classified into two groups: parenchymal cells, or the hepatocytes which comprise almost 65% of the liver cells (Miyai, 1979), and nonparenchymal cells, to which all the other cell types belong.

While the fat-storing cells make up only a very small percentage of the total liver cells, the other three cell types could be roughly divided into 12.8×10^7 hepatocytes, 5.7×10^7 endothelial cells and 1.4×10^7 Kupffer cells per gram of liver. Though the nonparenchymal cells make up 35% of the hepatic cells, in

Robert Wattiaux, Michel Jadot, Sandra Misquith, and Simone Wattiaux-de Coninck Laboratoire de Chimie Physiologique, Facultés Universitaires Notre-Dame de la Paix, B-5000 Namur, Belgium.

Subcellular Biochemistry, Volume 19: Endocytic Components: Identification and Characterization, edited by J. J. M. Bergeron and J. R. Harris. Plenum Press, New York, 1993.

terms of protein, they contribute only 1.3% in comparison with the 98.7% protein that comes from the hepatocytes (Knook, 1974; Blouin, 1977).

1.2. Receptor-Mediated Endocytosis by Liver Cells

Liver cells possess a wide variety of receptors that mediate not only a variety of cellular responses but are also involved in the endocytosis of numerous compounds (reviewed in De Meyts and Hanoune, 1982). In fact, a lot of our information concerning receptor-mediated endocytosis has come from extensive study of the hepatic cells.

Of particular interest are carbohydrate-specific receptors that are present not only on hepatocytes but also on sinusoidal cells. The first of such receptors to be avidly studied was the galactose receptor, which clears desialylated proteins from the plasma. Ever since this observation was made by Morell *et al.* in 1971, this receptor system has received a lot of attention (Morell *et al.*, 1971; Ashwell and Morell, 1974; Wall *et al.*, 1980; Berg *et al.*, 1985). The asialoglycoprotein receptor that according to Stowell and Lee (1978) and Schlessinger *et al.*, (1980) could also be responsible for the removal of proteins covalently bound to glucose can be considered a good example of receptors that direct their ligands to a degradation pathway while they themselves are reutilized thousands of times before degradation (Steer and Ashwell, 1980; Clarke and Weigel, 1985). Initially it was believed that these receptors were exclusively present in hepatocytes, however, extensive work by Kolb-Bachofen and her co-workers have shown the presence of galactose-specific lectins on rat Kupffer cells that differ structurally from those found on the hepatocytes (Schlepper-Schäfer *et al.*, 1982; Kempka and Kolb-Bachofen, 1985; Roos *et al.*, 1985). Electron-microscopic observations suggest that while hepatocytes take up asialoglycoproteins and small galactose particles, Kupffer cells can take up galactose particles of all sizes but seem ineffective in internalizing asialoglycoproteins (Schlepper-Schäfer *et al.*, 1986).

Glycoproteins are also cleared up by the liver through a second type of carbohydrate receptor, involved in the removal of terminal N-acetylglucosaminyl proteins (Stockert *et al.*, 1976). Radiolabeled neoglycoproteins traced these receptors to sinusoidal cells of rat liver and this system came to be known as the mannose or N-acetylglucosamine receptor (Stahl *et al.*, 1976; Achord *et al.*, 1977; Steer and Clarenburg, 1979; Hubbard *et al.*, 1979; Hubbard and Stukenbrok, 1979). According to some authors, the specificity of this receptor could be enlarged to glucose (Summerfield *et al.*, 1982, Schlessinger *et al.*, 1980; Pizzo *et al.*, 1981). A third type of carbohydrate-specific receptor in liver is the one that recognizes fucose-terminated glycoproteins. This receptor has received comparatively little attention and most of our knowledge vis-à-vis this receptor comes from the work of Lehrman and co-workers. (Lehrman *et al.*,

1986a,b; Lehrman and Hill, 1986; Haltiwanger *et al.*, 1986). According to these authors, the fucose receptor is localized in the Kupffer cells. Finally, one should not forget the presence on the hepatic cells of a receptor for mannose-6-phosphate (Man-6-P) residues (Fisher *et al.*, 1980; Geuze *et al.*, 1984b).

Liver is able to remove from the blood proteins that have been modified after reacting with glucose, aldehydes, or low-molecular-weight fatty acids. Receptors involved in the clearance of these proteins have been termed *scavenger receptors* and they have been more precisely located on liver endothelial cells (Nagelkerke *et al.*, 1982; Blomhoff *et al.*, 1984a). In fact, it is now clearly evident that there exists a family of such receptors. For it has been shown that the receptor for formaldehyde-treated bovine serum albumin is distinctly different from the one responsible for removing acetylated LDL (Horiuchi *et al.*, 1985). Subsequently it was shown that malonaldehyde-modified HDL is endocytosed by receptors that are indifferent to both the above-cited modified proteins (Murakami *et al.*, 1986).

2. HOW TO STUDY ENDOCYTIC COMPONENTS IN LIVER NONPARENCHYMAL CELLS

The studies can be performed with the whole liver or with isolated liver cells. Both kinds of material are suitable for morphological studies. In the case of biochemical investigations, especially with centrifugation techniques, the use of the whole organ poses a problem. In homogenates and isolated fractions, subcellular structures coming from the different types of cells are intermingled and it is not easy to establish a distinction between them according to their cellular origin. The experimental approach with isolated cells seems more promising. Cells in suspension or in primary culture are allowed to take up labeled molecules and are fractionated at different times after initiating endocytosis. Another possibility is to inject the substance into an animal and to separate the cells before fractionation.

Although very useful, the use of isolated cells has serious drawbacks. (1) The separation of the different types of hepatic cells takes a long time, particularly when nonparenchymal cells have to be obtained; (2) the procedure requires perfusion with lytic enzymes (collagenase, pronase, etc.) that is probably not without danger for the cells, in particular for their plasma membrane where receptors are located; (3) in primary cultures, liver cells are relatively quickly subjected to changes. For example, asialoglycoprotein endocytosis by hepatocytes is strongly reduced after 1 day of cultivation (Thirion and Wattiaux, 1988); and (4) in the case of nonparenchymal cells, the number of cells recovered may be a limiting factor for centrifugation experiments.

These considerations led us to try, during the last years, to characterize endocytic components of liver nonparenchymal cells by fractionating whole liver homogenates.

3. METHODOLOGICAL CONSIDERATIONS

3.1. Choice of the Endocytosed Molecule

The main difficulty is to locate, in fractions isolated by centrifugation, the subcellular structures belonging to nonparenchymal cells. The preferential association of some receptors with these cells makes it possible to select for injection molecules that will be mainly endocytosed by specific receptors. In our work we have chosen yeast invertase. Because it is a highly mannosylated protein (Trimble and Maley, 1977; Tarentino *et al.*, 1974), it will be chiefly taken up by sinusoidal cells (Madnick *et al.* 1978). According to Hubbard *et al.* (1979) and Praaning-Van Dalen *et al.* (1982) endothelial cells could account for the removal of the bulk of that kind of glycoprotein. We also made use of formaldehyde-treated serum albumin that will be chiefly recovered in the endothelial cells because of the presence of the scavenger receptor on the plasma membrane of these cells (Nagelkerke *et al.*, 1982; Blomhoff *et al.*, 1984).

3.2. Distribution of Nonparenchymal Cell Organelles

In a study performed with the whole liver, a comparison between the distribution of the endocytosed substance and those of organelle markers is not directly possible. There is no information allowing us to characterize the behavior in centrifugation of membrane structures according to their cellular origin. For example, it is not possible to detect in a lysosomal-hydrolase distribution the contribution of sinusoidal cell lysosomes. We have to some extent got around the difficulty thanks to the finding of a procedure permitting us to selectively increase the density of sinusoidal cell lysosomes. Indeed we have found that injecting the rat with a relatively high amount of yeast invertase (Jadot *et al.*, 1985) or of mannan (Wattiaux *et al.*, 1987) causes a change of lysosomal-enzyme distribution after isopycnic centrifugation in a sucrose or Percoll gradient. It consists of the shifting of a proportion of these enzymes to higher-density regions in the gradient because of an accumulation of indigestible mannosylated compounds in lysosomes. It is illustrated in Figure 1, which shows the distribution of several acid hydrolases after injecting a rat with invertase at a concentration of 0.9 mg/100 g of body weight, 15 hr before sacrificing the animal. The distributions are compared with those observed with a noninjected rat. As shown by Madnick *et al.* (1978) and Rodman *et al.* (1978) invertase

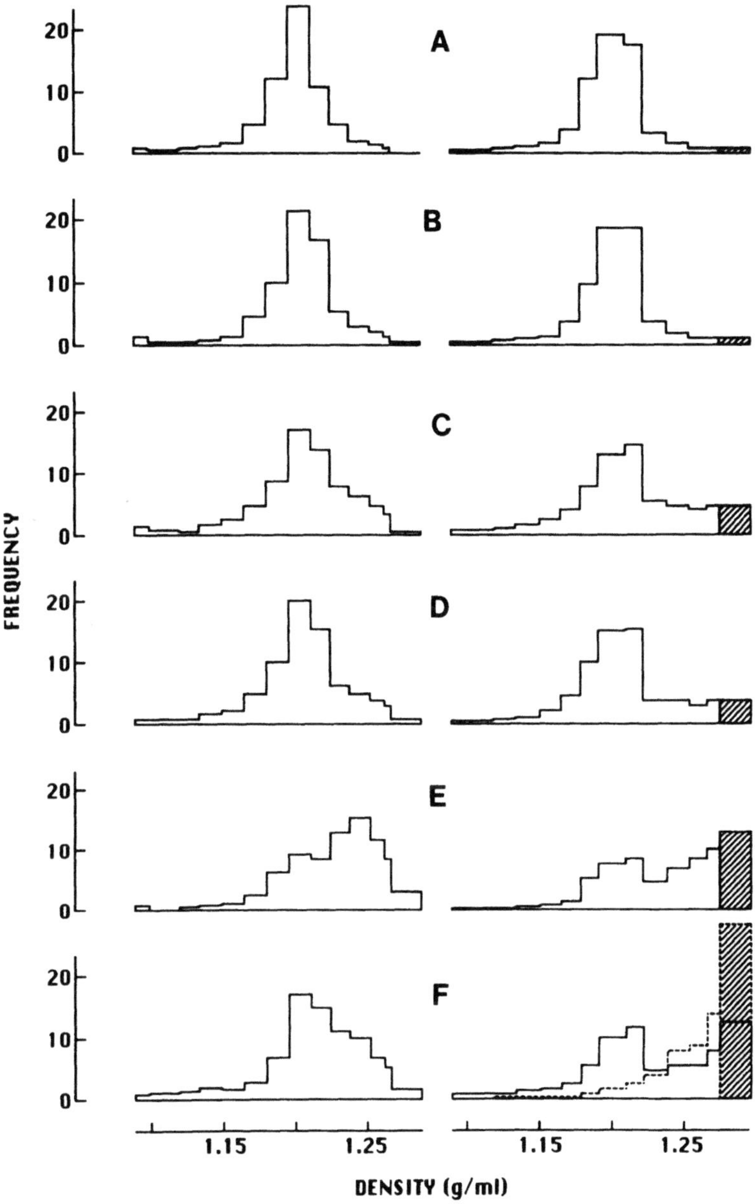

FIGURE 1. Influence of an invertase injection on the distribution of several lysosomal enzymes after isopycnic centrifugation in a sucrose gradient. Left: control rat; right: rat injected with invertase (0.9mg/100 g body weight) and killed 15 hr after injection with (A) acid phosphatase, (B) cathepsin C, (C) β-galactosidase, (D) N-acetylglucosaminidase, (E) arylsulfatase, (F) acid deoxyribonuclease. Dashed lines: invertase (according to Jadot *et al.*, 1985).

Table I
Complementary Data on Experiment in Figure 1: Activity Found in the Bottom Fraction of the Gradients, Expressed as a Percentage of Activity Included in the Gradient. (A) Control Rat (B) Rat Injected with Invertase

	% of activity	
Enzyme	A	B
Acid phosphatase	0.3	1.5
Cathepsin C	0.7	2.8
β-galactosidase	1.0	8.9
N-acetylglucosaminidase	1.5	7.5
Acid deoxyribonuclease	2.8	23.3
Arylsulfatase	6.0	25.6
Invertase	—	54.4

endocytosed by the liver is apparently totally located in nonparenchymal cells. So, the lysosome population that can be individualized by the density change caused by their invertase content must be mainly lysosomes of these cells. The fact that important differences are observed between the percentages of each acid hydrolase shifted to high-density zones is not surprising. Indeed, according to Wattiaux *et al.* (1956), Arborgh *et al.* (1973) and Munthe-Kaas *et al.* (1976), the distribution of lysosomal enzymes between parenchymal and nonparenchymal cells is different for each hydrolase. The proportions of acid hydrolases present in lysosomes whose density is increased by invertase (Table I) are qualitatively similar to those found by Munthe-Kaas *et al.* (1976) in nonparenchymal cell homogenates. That assumption can be substantiated more directly by demonstrating that the distribution profile of molecules known to be recovered in nonparenchymal cell lysosomes after their intravenous injection (e.g., β-glucuronidase, formaldehyde-treated serum albumin, or labeled invertase) is strongly affected by an invertase preinjection. On the contrary, such a treatment is unable to shift the distribution profile of molecules specifically reaching the hepatocyte lysosomes (e.g., asialofetuin, galactosylated albumin, and lactoferrin) (Figure 2).

3.3. Distinction between Prelysosomal and Lysosomal Components

In analytical centrifugation, distinction between endosomal compartment components and lysosomes is relatively easy in hepatocytes because endosome

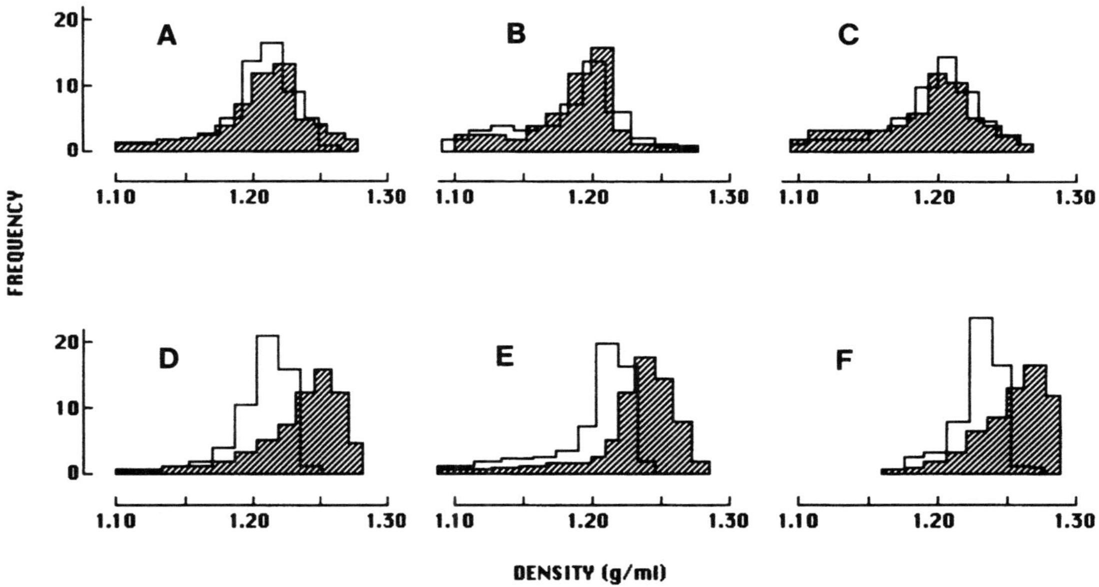

FIGURE 2. Influence of an invertase preinjection on the distribution in a sucrose gradient of molecules endocytosed by parenchymal cells: (A) [^{125}I]asialofetuin, (B) [^{125}I]galactosylated albumin, (C) [^{125}I]lactoferrin; or by sinusoidal cells: (D) [^{125}I]β-glucuronidase, (E) [^{125}I]formaldehyde-treated albumin, or (F) [^{125}I]invertase. Distributions were established 3 hr after injection. Clear blocks: normal rats; shaded blocks: rats injected with invertase (0.9 mg/100 g body weight) 15 hr before sacrificing the animal (according to Wattiaux *et al.*, 1986).

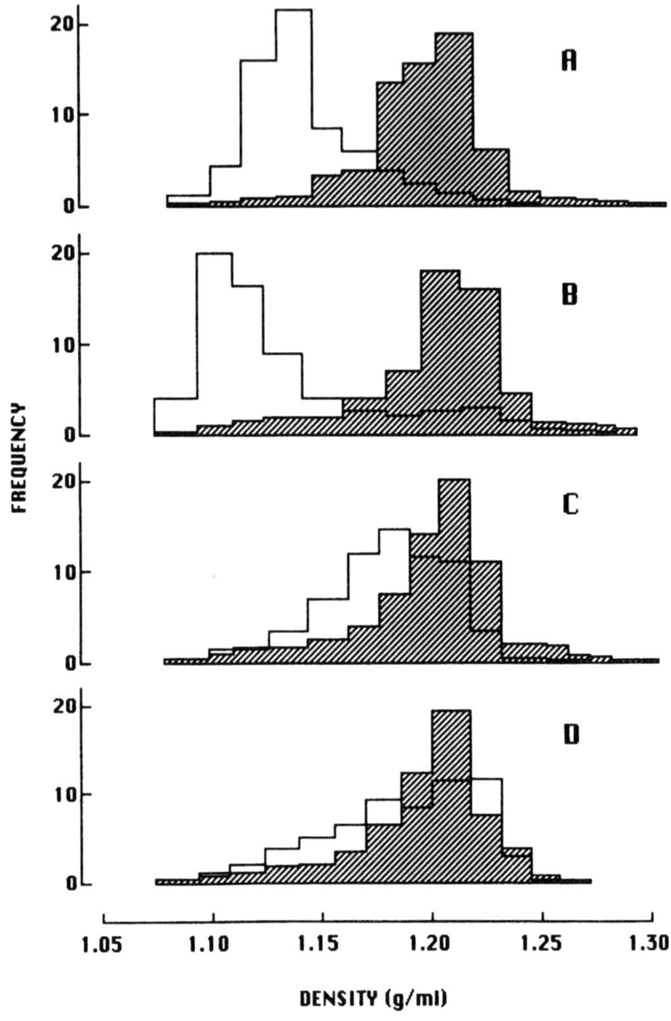

FIGURE 3. Density distribution histograms of asialofetuin, invertase, and cathepsin C after isopycnic centrifugation of a total mitochondrial fraction in a sucrose gradient. The ML fraction was isolated 1 min (A,C) or 10 min (B,D) after injection of 0.050 mg/100 g body weight of [^{125}I]asialofetuin (A,B) or [^{125}I]invertase (C,D). Centrifugations were performed at 50,000 rpm in the VTi50 Spinco rotor. The time integral of the square angular velocity was 270 rad^2/ns. The sucrose gradient extended from 1.09 to 1.32 g/ml. Ordinate represents the frequency (Q/SQ.Δr) where Q represents the activity found in the fraction, SQ the total recovered activity and Δr the increment of density from to to bottom of the fraction. Shaded histograms represent the cathepsin C distributions.

equilibrium density is markedly lower than that of lysosomes, particularly in a sucrose gradient (Figure 3). Unfortunately, such a distinction is much more difficult in nonparenchymal cells where the distribution profile of prelysosomal structures (for example those carrying invertase after a short injection time), considerably overlaps with lysosomes (Figure 3). A specific density perturbant such as Triton WR 1339 could be especially suitable to distinguish lysosomes clearly from prelysosomal structures. It has been shown that 4 days after its intravenous injection, Triton WR 1339 considerably decreases the lysosome density (Wattiaux *et al.*, 1963). As is shown below, such a procedure can be very useful to separate prelysosomal from lysosomal components in sinusoidal cells.

One can also take advantage of methods that do not depend on distribution after centrifugation to distinguish the prelysosomal and the lysosomal structures. Among them, the procedure of Jadot and Wattiaux (1985) is worth mentioning. It allows measurement of the proportion of an endocytosed substance that can be released from the organelles by using a compound that selectively disrupts the lysosomal membrane without affecting the other organelles. It consists of determining the susceptibility to glycyl-L-phenylalanine-2-naphthylamide (Gly-L-Phe-2-NNap) of the structures with which the endocytosed molecule is associated. Gly-L-Phe-2-NNap causes a loss of lysosome integrity as a result of its penetration and its hydrolysis by cathepsin C in lysosomes (Jadot *et al.*, 1984). It is highly specific for lysosomes. Nonlysosomal organelles, which do not contain this hydrolase, will not be affected by Gly-L-Phe-2-NNap. As an example, Figure 4 shows invertase activity unreleasable and releasable by Gly-L-Phe-2-NNap from the structures it is associated with at increasing times after injection of the enzyme to the rat. The amount of invertase not releasable by Gly-L-Phe-2-NNap progressively increases with time until reaching a plateau that is maintained for several minutes; after that, a slow decrease is observed. Invertase releasable by Gly-L-Phe-2-NNap increases with time in a sigmoidal manner; a maximal value is obtained after about 2 hr. Such results allow a precise statement of what proportion of the endocytosed substance is present in a nonlysosomal compartment and in a lysosomal one (Jadot and Wattiaux, 1985).

3.4. Intracellular Degradation of the Endocytosed Compound

Another problem that can complicate interpretation of an endocytosed substance distribution is its hydrolysis in products that leave the subcellular compartment where they are generated. One solution is to make use of molecules that are relatively resistant to intracellular degradation. This is the case for yeast invertase, the amount of which remains unchanged in the liver at least 2 hr after being totally cleared up from the blood (Jadot and Wattiaux, 1985). Another solution is to get the cells to endocytose a labeled compound which, once hydrolyzed intracellularly, yields labeled degradation products that are unable to diffuse

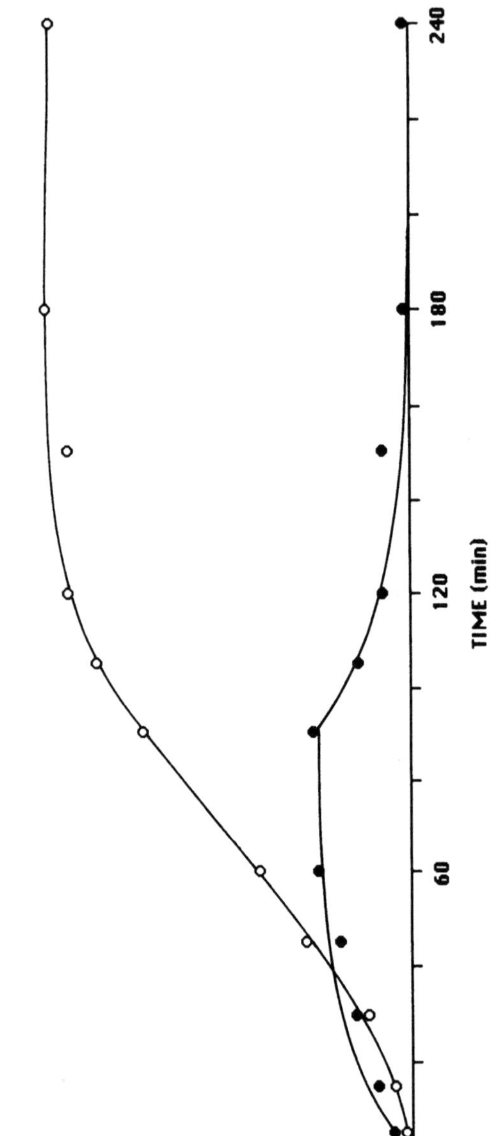

FIGURE 4. Release of endocytosed invertase by Gly-L-Phe-2-NNap. Abscissa: time after injection of yeast invertase (0.9 mg/100 g body weight). Ordinate: invertase activity releasable (○—○) or unreleasable (●—●) by Gly-L-Phe-2-NNap (according to Jadot and Wattiaux, 1985).

through the membrane of organelles where degradation took place. The advantage of the second method is that it allows researchers to spot the structures where hydrolysis of the endocytosed molecule occurs. We made use of that method for the study of formaldehyde-treated serum albumin (fBSA) endocytosis. That molecule is mainly endocytosed by sinusoidal cells. It is readily hydrolysed after endocytosis, giving degradation products that quickly disappear from the liver, probably because they rapidly diffuse through the lysosomal membrane (Figure 5). This drawback can be overcome by linking fBSA to a labeled molecule, [^{125}I] tyramine cellobiose (Pittman *et al.*, 1983), that gives, after intralysosomal hydrolysis of the protein, radioactive degradation products that do not escape from the organelles (Berg *et al.*, 1985). As shown by Figure 5, contrary to what is observed with [^{125}I]-fBSA, the radioactivity of the liver remains constant for several hours after injection, however, with time it becomes more and more acid soluble, indicating that the labeled degradation molecules are produced but do not leave the organ.

4. INTRACELLULAR JOURNEY OF A PROTEIN ENDOCYTOSED BY LIVER SINUSOIDAL CELLS

We will present as an example results obtained after injecting a rat with fBSA labeled with [^{125}I]tyramine cellobiose, abreviated as [^{125}I]-TCfBSA.

4.1. Differential Centrifugation

As shown by Figure 6, at the first minute after injection, almost 50% of [^{125}I]-TCfBSA present in the liver is found in the microsomal fraction P. The amount recovered in P slowly diminishes till, after 60 min, it retains only 5% of the total radioactivity. While P shows a downward decline in its radioactive content, the heavy and light mitochondrial fractions (M and L) exhibit a reverse phenomenon. M is associated with almost 60% at 60 min which remains stable even 180 min after injection. L on the other hand has about 10% after the first minute which increases to 30% and then decreases. The ebb in L's radioactive content corresponds to the rise in that of M. These changes of distribution with time are similar to those described for other substances endocytosed by rat liver; they indicate a progressive transfer of the endocytosed compound from small vesicles (present in the microsomal fraction) to larger vesicles, exhibiting a distribution pattern similar to that of lysosomal enzymes. It is worth remembering that acid-soluble labeled molccules begin to gain significance 5–10 min after injection and then increase up to 30–40 min (see Figure 5). They remain associated with sedimentable components and share common properties of sedimentation with acid-precipitable radioactive molecules.

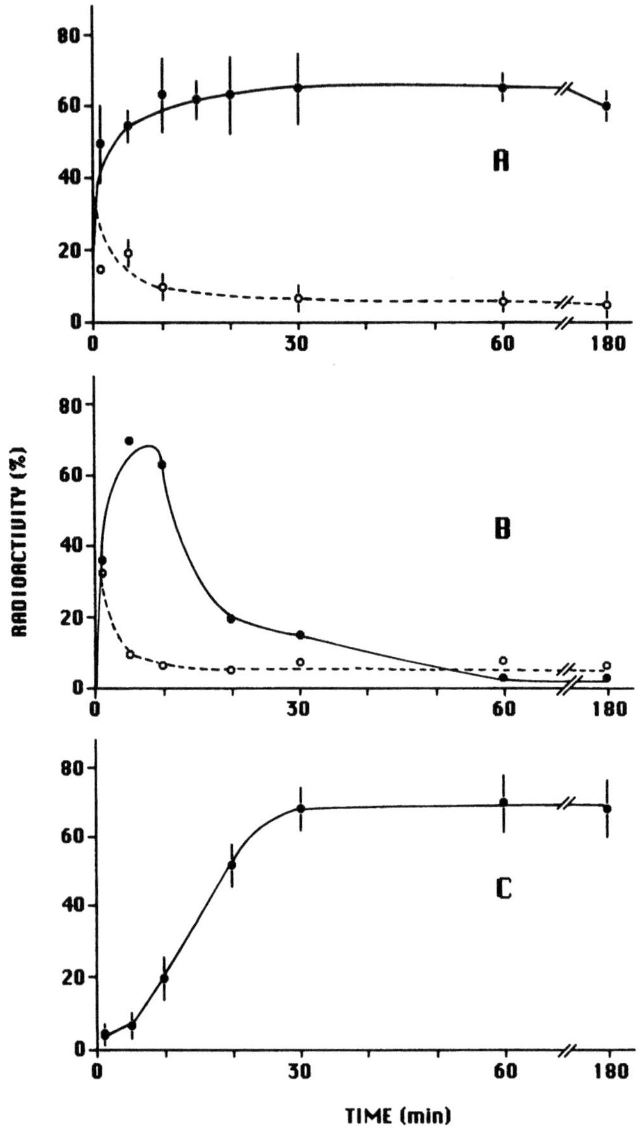

FIGURE 5. Radioactivity recovered in liver and blood after injection of [^{125}I]-TCfBSA (A) or [^{125}I]-fBSA (B). Abscissa: time after injection in minutes. Ordinate: percentages of the injected dose; (●—●) liver, (○—○) blood. In (C) is shown the evolution of acid-soluble radioactivity with time, given as a percentage of the total radioactivity recovered in the homogenate after injection of [^{125}I]-TCfBSA (according to Misquith *et al.*, 1988).

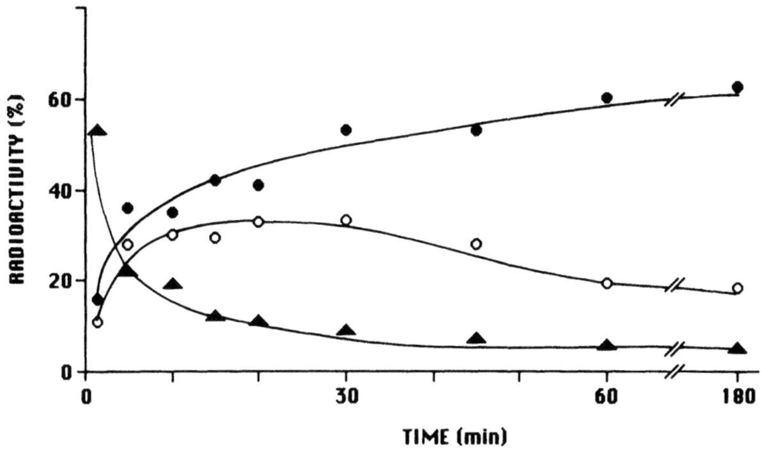

FIGURE 6. Influence of the time after injection on the distribution of TCfBSA after differential centrifugation. Abscissa: time after injection of [^{125}I]-TCfBSA. Ordinate: percentages of the radioactivity present in the homogenate, measured in the microsomal fraction P (▲), in the light mitochondrial fraction L (○), and in the heavy mitochondrial fraction M (●) (according to Misquith *et al.*, 1988).

4.2. Isopycnic Centrifugation

Figure 7 illustrates some distributions observed after isopycnic centrifugation in a sucrose gradient. Within a minute after injection, the organelles containing radioactivity in the microsomal fraction have a distribution that resembles that of alkaline phosphodiesterase, a plasma-membrane marker. Soon after this, there is a shift of the radioactivity to slightly higher density zones of the gradient, in effect, the median equilibrium density increases from a value of 1.16 g/ml at 1 min to 1.19 g/ml at 5 min.

With respect to the total mitochondrial fraction (ML fraction) as its content in acid-soluble radioactivity increases with time, the distribution of acid-precipitable and acid-soluble labeled products are to be presented separately. The density histograms reported in Figure 8 were drawn by taking the percentage of acid-precipitable and acid-soluble radioactivity present in each fraction with respect to the total radioactivity recovered in the sum of the fractions. Such a representation allows researchers to visualize the proportion of each kind of radioactivity associated with the fraction. A considerable overlapping of acid-precipitable and acid-soluble distribution is observed, therefore no clear transfer from one organelle to another could be inferred from such results (Figure 8A).

As in the case of a sucrose gradient, there is no striking difference between the distributions of acid-soluble and acid-precipitable radioactivities when cen-

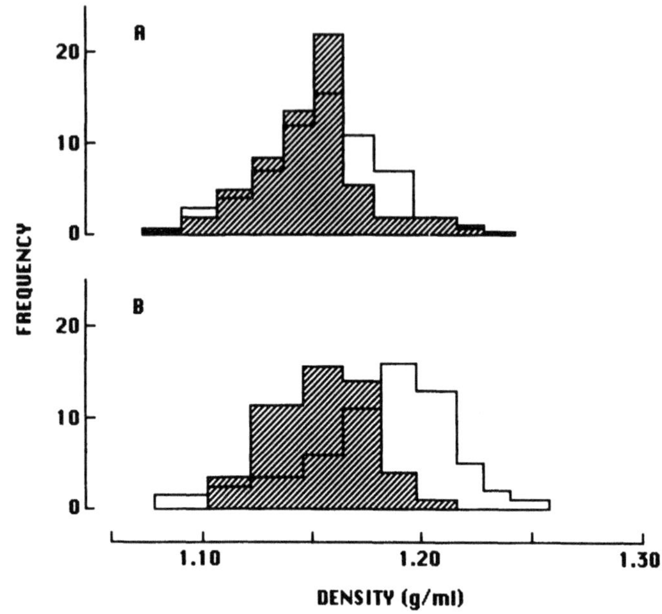

FIGURE 7. Density distribution histograms of radioactivity (clear area) and alkaline phosphodiesterase (shaded area) after isopycnic centrifugation of a microsomal fraction (P) in a sucrose gradient. Distributions were established 1 min (A) or 5 min (B) after injection of [^{125}I]-TCfBSA (according to Misquith *et al.*, 1988).

trifugation is performed in a Percoll gradient with iso-osmotic sucrose as solvent. What is significant is a shift with time of the distributions toward higher density zones of the gradient that occurs as hydrolysis ends, as ascertained by the constant amount of acid-soluble material (Misquith *et al.*, 1988).

Obviously, such results do not indicate if acid-precipitable and acid-soluble radioactivities present in the total mitochondrial fraction ML are located in the same organelles or in different organelles exhibiting a kinship with respect to their behavior in centrifugation. As it can be reasonably supposed that most of the acid-soluble radioactivity is located in lysosomes, this would mean that the acid-precipitable material is quickly transferred from small vesicles derived from the plasma membrane (present in P) to lysosomes, or that the structures with which it becomes associated before reaching lysosomes are similar to these organelles with respect to their size and density.

4.3. Use of Density Perturbants

The use of lysosome-density perturbants, particularly Triton WR 1339 and yeast invertase, may help to answer this question. Triton WR 1339 markedly

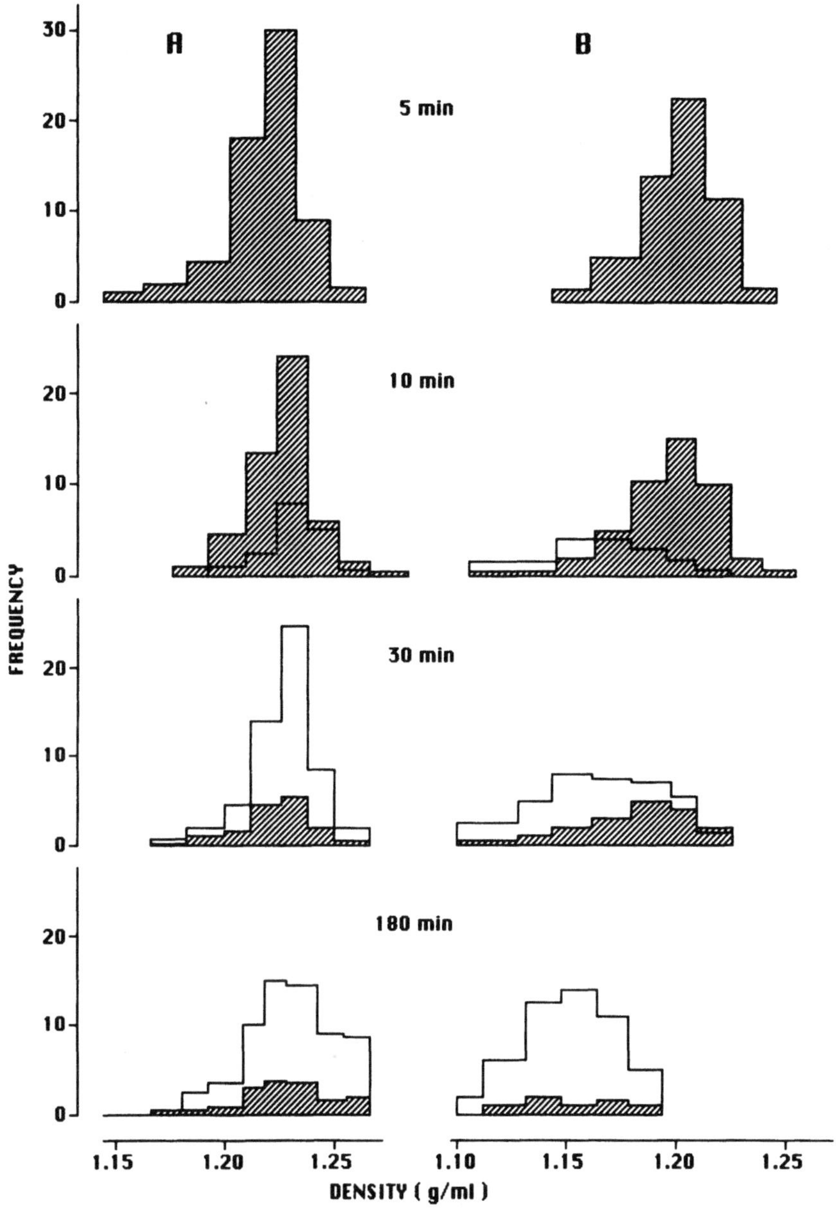

FIGURE 8. Density distribution histograms of acid-soluble (clear area) and acid-precipitable (shaded area) radioactivities after isopycnic centrifugation of an ML fraction in a sucrose gradient. The distributions were established using mitochondrial fractions isolated at increasing time after [^{125}I]-TCfBSA injection, from control animals (A) or from animals preinjected with Triton WR 1339 four days before receiving the [^{125}I]-TCfBSA injection (B) (according to Misquith *et al.*, 1988).

decreases the density of all lysosomes of the liver (Wattiaux *et al.*, 1963) while yeast invertase, as previously mentioned, increases the density of non-parenchymal cell lysosomes only.

Figure 8B exemplifies the effect of Triton WR 1339 treatment on the distribution of acid-soluble and acid-precipitable radioactivities. A clear distinction is now observed between the two distributions because the acid-soluble radioactivity distribution is shifted more markedly toward lower densities than the acid-precipitable one. The decreasing with time of the acid-precipitable radioactivity and the concurrent increase of the acid-soluble one go together with a migration of radioactivity from the high-density region of the gradient to the region where Triton-filled lysosomes equilibrate. The effect of yeast invertase on [^{125}I]-TCfBSA distribution is in agreement with the results obtained after Triton WR 1339 treatment (Misquith *et al.*, 1988).

4.4. Effect of Glycyl-L-Phenylalanine-2-Naphthylamide (Gly-L-Phe-2-NNap)

As shown by Figure 9, whatever the time after injection, most of the acid-soluble radioactivity is released by Gly-L-Phe-2-NNap from the structures it is associated with. The behavior is different for the acid-precipitable material. At 10 min, that form of the molecule that represents most of the radioactivity is almost unreleasable; later when the acid precipitable radioactivity has considerably decreased, it is to a large extent released by Gly-L-Phe-2-NNap.

4.5. Conclusions

An extensive discussion of these observations and of similar ones obtained by following yeast-invertase endocytosis can be found in Jadot *et al.* (1986) and Misquith *et al.* (1988). In summary, with respect to the subcellular structures a molecule taken up by liver sinusoidal cells has to travel through, the next points can be emphasized. (1) The molecule is first trapped in small pinocytic vesicles and recovered in the microsomal fraction but quickly transferred to bigger organelles mainly present in the mitochondrial fraction. (2) The effect of density perturbants and of Gly-L-Phe-2-NNap allows researchers to distinguish at least two kinds of structures among these organelles. A first population becomes involved soon after injection. It carries the acid precipitable radioactivity and is not affected by Gly-L-Phe-2-NNap. Accordingly it does not correspond to lysosomes, but probably represents one or several populations of prelysosomal structures (endosomes). These organelles behave similarly to lysosomes unaffected by density perturbants, with respect to their behavior in centrifugation. (3) Acid-soluble degradation products are always located in lysosomes as ascertained by the effect of density perturbants on their distribution in density gradients and the fact that they are released from organelles by Gly-L-Phe-2-NNap.

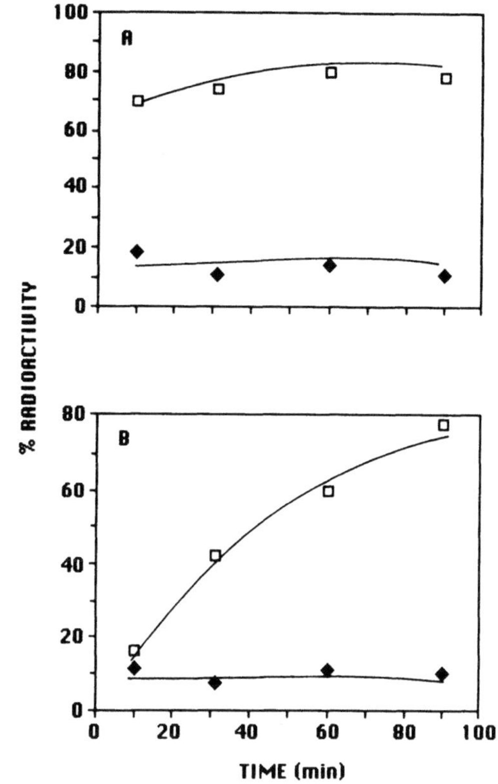

FIGURE 9. Release of endocytosed [^{125}I]-TCfBSA by Gly-L-Phe-2-NNap. Abscissa: time after injection of [^{125}I]-TCfBSA. Ordinate: acid-soluble (A) or acid-precipitable (B) radioactivities (given as percentages of the amount released under the same conditions in the presence of Triton X-100) released from a mitochondrial fraction (ML) after its incubation in the presence (□–□) or absence (♦—♦) of Gly-L-Phe-2-NNap (according to Misquith *et al.*, 1988).

(4) Two populations of lysosomes are probably involved in the intracellular journey of [^{125}I]T-CfBSA. Indeed when hydrolysis of the molecule is terminated, as ascertained by the constant percentage of acid-soluble radioactivity, one observes a change in radioactivity distribution after differential centrifugation (migration from L to M) and after isopycnic centrifugation in Percoll gradient (shift towards higher density).

Accordingly, the subcellular components through which endocytosed [^{125}I]-TCfBSA has to travel would be: pinocytic vesicles to endosomes to lysosomes I to lysosomes II.

5. INTRACELLULAR FATE OF fBSA ENDOCYTOSED BY THE LIVER

To have a more detailed look at the organelles involved in the endocytosis and degradation of fBSA by sinusoidal cells of the liver, the intracellular dis-

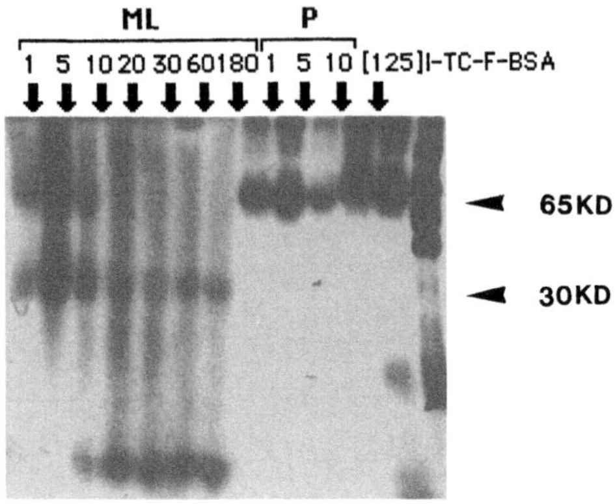

FIGURE 10. SDS-PAGE analysis of mitochondrial (ML) and microsomal (P) fractions. The wells indicate the time after injection of [^{125}I]-TCfBSA (according to Misquith *et al.*, 1988).

tribution of labeled digestion products was investigated by combining centrifugation and SDS-PAGE.

5.1. ML and P Fractions

As already shown, most of the radioactivity is found in the sedimentable fractions ML and P, thus these fractions were the first subjected to SDS-PAGE.

As illustrated by Figure 10, no degradation products are seen in the microsomal fraction P. On the contrary, degradation takes place in the organelles recovered in the mitochondrial fraction ML, and two categories of hydrolysis products have to be considered. A first one of about 30 kD is surprisingly apparent as soon as 1 min after injection, when it is to be remembered that the bulk of radioactivity is still acid precipitable. Later, along with the appearance of acid-soluble radioactivity (see Figure 5), very small degradation products can be identified in SDS-PAGE.

5.2. Fractions Isolated by Gradient Centrifugation

Since the organelles responsible for degradation seemed to be exclusively located in the ML fraction, only that fraction was analyzed in gradient centrifugation. Rats previously treated with Triton WR 1339 were injected with [^{125}I]-TCfBSA and the rats sacrificed at increasing times after injection. ML fractions were isolated and subjected to isopycnic centrifugation in a sucrose gradient, and the distribution of radioactivity was established. The degradation of the endocytosed molecule begins in high-density zones, distinct from those where the Triton-filled lysosomes are located (Misquith *et al.*, 1989). Only when degradation is apparently complete, that is, when small molecules are formed, are they transferred to regions of gradient where Triton-filled lysosomes equilibrate. Thus these results seem to imply that degradation is occurring in organelles that equilibrate in regions of the gradient where endosomes or lysosomes that were not involved in the accumulation of a previously endocytosed compound equilibrate.

These results were confirmed by experiments in which rats preinjected with mannan were used. Once again, as illustrated in Figure 11, the first hydrolysis products are recovered in organelles that are not affected by the density pertur-

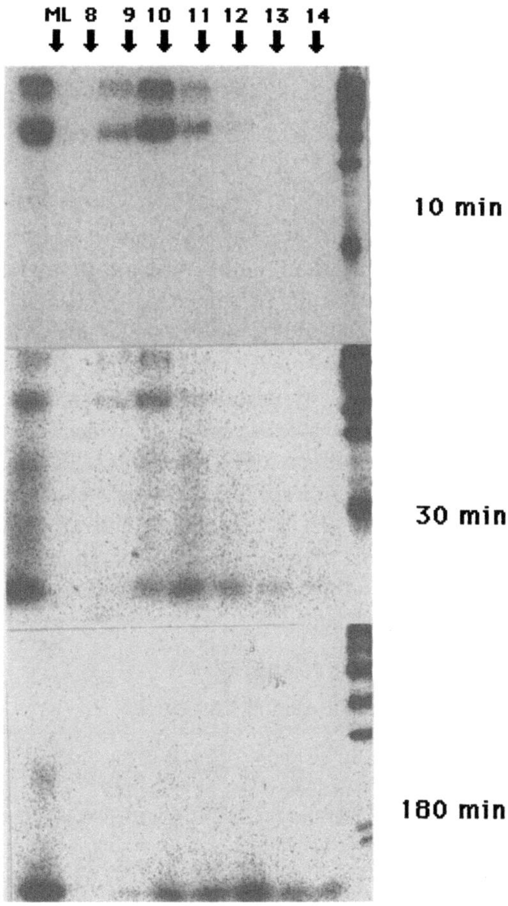

FIGURE 11. SDS-PAGE analysis of [^{125}I]-TCfBSA in sections obtained after isopycnic centrifugation of ML fractions in a sucrose gradient, effect of mannan. The ML fractions were isolated at increasing times after [^{125}I]-TCfBSA injection, from animals injected with mannan 15 hr before receiving the [^{125}I]-TCfBSA injection. The wells bear the numbers corresponding to sections of gradient (according to Misquith *et al.*, 1989).

bant. Only after small degradation products arise are they seen to appear in higher-density zones where mannan-filled lysosomes are present.

Thus, taking the Triton and mannan results together, they seem to indicate that soon after endocytosis, extensive degradation is taking place in organelles that have not been affected by the density perturbants. The question that automatically arises is whether this compartment is made of prelysosomal structures or of lysosomes that have not taken part in accumulation of Triton or mannan. There are now quite a few reports supporting the view that degradation could occur in endosomes, as seen by the work of Diment and Stahl (1985) who have shown the presence of a proteolytic activity in macrophage endosomes, and Hamel *et al.* (1988) who believe that insulin is almost completely degraded in hepatocyte endocytic vesicles. Thus, it seems important to know if the same situation exists in liver sinusoidal cells.

5.3. Purified Organelles

We have tried to purify structures that we suppose to be endosomes with which [^{125}I]-TCfBSA is associated in the ML fraction soon after injection.

Animals were killed 5 or 10 min after injecting [^{125}I]-TCfBSA. These two times were chosen as it was assumed that most of the radioactivity at these times would be associated with prelysosomal compartments. This assumption was based on the observation previously described that at such a short time after injection, the acid-precipitable form of the molecule, which accounts for most of the radioactive material, was unaffected by Triton WR 1339 and more important was not affected by Gly-L-Phe-2-NNap. Four days before, the animals received an injection of Triton WR 1339, to allow the removal of the bulk of the lysosomes by centrifugation in a sucrose gradient. An outline of the procedure is given in Table II. The L fraction was isolated instead of the whole mitochondrial

Table II
Purification of Endosomes

Procedure
Rats injected with Triton WR 1339 to remove bulk of lysosomes
Four days later injected with [^{125}I]-TCfBSA for 5 or 10 min: mainly present in endocytic vesicles.
L fraction isolated to eliminate maximum amount of mitochondria.
Layered on Percoll gradient.
Sections of gradient rapidly analyzed for radioactivity distribution.
Peak of radioactivity pooled and layered on a sucrose gradient.
Choice of gradients
Percoll: good separation of radioactivity from plasma membrane, endoplasmic reticulum, mitochondria.
Sucrose: further separation of lysosomes from endosomes.

Table III
Purification of Endosomes

Substance	Liver	L Fraction	Percoll (7 & 8)	Sucrose (7)	RSA*
Radioactivity	100	21	11.76	5.56	49.85
Glucose-6-phosphatase	100	7	0.11	—	—
Cytochrome oxidase	100	1	0.42	0.10	0.90
Alk. Phosphodiesterase	100	5	0.33	0.05	0.45
Arylsulfatase	100	14	1.67	0.26	2.33
Cathepsin C	100	16	1.69	0.18	1.61
Galactosyl transferase	100	12	6.60	3.26	29.66
Proteins	100	3	0.78	0.11	

*RSA is given as the ratio of the percentage of the compound present in a fraction to the amount of protein found in the same fraction.
Note: Values are expressed as percentages found in the total liver homogenate. The degree of purification is expressed as the relative specific activity (RSA) of the compound in the purified fraction.

fraction (ML) to eliminate as much of the mitochondria as possible. The fraction was then subjected to isopycnic centrifugation in a Percoll gradient; this medium was chosen as it has been shown to be a good one for separating lysosomes from plasma membrane, endoplasmic reticulum, and mitochondria. Table III assesses the yield and the purification reached after each step. The maximum purification for radioactivity and thus for the structure it is associated with is 50 times, with a yield of 5.5%. As ascertained by galactosyltransferase activity, the main contaminant organelle is the Golgi apparatus. These results are similar to those obtained by Marsh *et al.* (1987) who purified hepatocyte endosomes by free flow electrophoresis. Fractions of the two gradients containing most of the endocytosed compound were analyzed by SDS-PAGE. No proteolytic products are revealed in the radiogram, particularly in the preparation that exhibits the highest purification, which we suppose to be made up chiefly of endosomes (Misquith *et al.*, 1989).

5.4. *In Vitro* Degradation of fBSA

That these organelles are not able to split fBSA is confirmed because, when incubated in the presence of [^{125}I]-fBSA, they do not generate degradation products as ascertained by SDS-PAGE analysis (Figure 12). These results lead us to speculate that the first proteolysis event that we detected after fBSA endocytosis takes place in another kind of particle (see Figure 10). The characteristics of these lytic organelles are that they are unaffected by lysosome-density perturbants (Triton WR 1339, invertase), are not disrupted by Gly-L-Phe-2-NNap, and are able to cause only a partial proteolysis of fBSA. They could be considered as

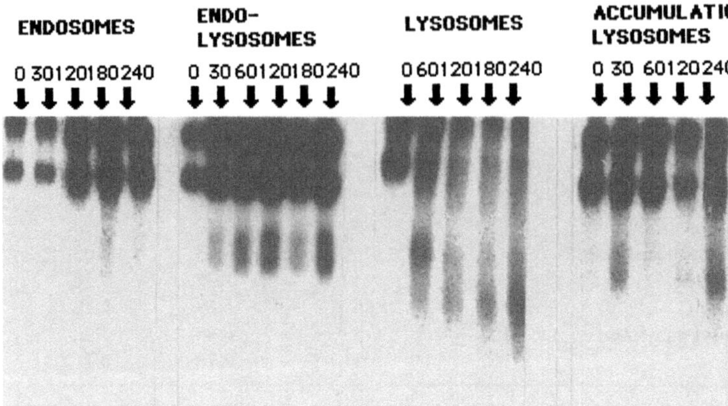

FIGURE 12. SDS-PAGE analysis of *in vitro* degradation products found when [^{125}I]-fBSA is incubated with different organelles. The wells indicate the different incubation times (according to Misquith *et al.*, 1989).

intermediary components between endosomes and lysosomes (late endosomes, endo-lysosomes).

On the other hand, as we have seen, centrifugation results suggest that among lysosomes, two categories of organelles are involved: lysosome I (transfer and digestion) and lysosome II (accumulation of digestion products). An additional distinction between these two types of lysosomes can be found by assessing their hydrolytic power toward [^{125}I]-fBSA. Lysosome II can be purified by injecting invertase or mannan to rats and recovering the shifted granules in a sucrose gradient. On the other hand, a representative sample of the total lysosome population can be obtained by the method of Wattiaux *et al.* (1978). The organelles were incubated with [^{125}I]-fBSA and the amount of acid-soluble products measured. Moreover, aliquots of the digestion mixture were analyzed by SDS-PAGE. For comparison, [^{125}I]-fBSA hydrolysis by endosomes and endo-lysosomes was also assayed. Endosomes were purified as described in Section 5.3. To obtain what we suppose to be endo-lysosomes, an ML fraction originating from a Triton WR 1339–injected rat was centrifuged in a sucrose gradient. Fractions located between 1.19 and 1.22 g/ml density, devoid of lysosomes shifted by Triton WR 1339, were collected. It is in these fractions that the radioactivity present in ML during the first 10 min after injection is mostly recovered.

Figure 12 represents the radiogram obtained after electrophoresis. There are no detectable acid-soluble products in the medium, even after a long incubation, when endosomes are used (Misquith *et al.*, 1989) and no proteolytic products are observable after SDS-PAGE. With endo-lysosomes, even after 4 hr, more than 90% of [^{125}I]-fBSA remains precipitable by TCA (Misquith *et al.*, 1989). However, electrophoresis shows that a first degradation product ($\pm$ 30 kD) appears during incubation. It is to be remembered that this product is seen *in vivo* in the ML fraction even at 1 min. If we recall results obtained with Gly-L-Phe-2-NNap, we become aware that this product is located in organelles that are unaffected by the lysosome labilizer, for at 10 min Gly-L-Phe-2-NNap is unable to release any acid-precipitable radioactivity. These results seem to signify that, as we have proposed, these organelles may be intermediates between endosomes and fully mature secondary lysosomes. They could be formed by the fusion of endosomes with new primary lysosomes that do not contain a large number of hydrolases, but do have a sufficient number to cause the first split that leads to an ignition of the degradation process as proposed by Opresko and Karpf (1987). Lysosomes purified by the metrizamide method cause an extensive degradation of [^{125}I]-fBSA (Figure 12). On the other hand, lysosomes isolated after mannan injection, where labeled degradation products finally accumulate after [^{125}I]-TCfBSA endocytosis, are apparently less efficacious as ascertained by the measurement of acid-soluble labeled products (Misquith *et al.*, 1989) and the SDS-PAGE analysis (Figure 12). This implies that these lysosomes have limited proteolytic activity.

5.5. Identification of Proteolytic Activities

Preceding results have shown that fBSA is degraded in a two-step process, in intact organelles after its intravenous injection as well as in an *in vitro* assay. The appearance of a first 30 kD fragment in endo-lysosomes is followed by extensive degradation in lysosomes. To probe the enzymes involved in degradation, we made use of the purified enzymes cathepsin B and cathepsin D and of their respective specific inhibitors leupeptin and pepstatin. We also assessed the requirement for dithiothreitol (DTT) in the incubation medium. Our choice of enzymes and inhibitors was based on the report of Stahl and co-workers that a prelysosomal compartment contained cathepsin D (Diment and Stahl, 1985; Diment *et al.*, 1988) and the report of Mego (1984) that the cathepsin B family was responsible for the degradation of fBSA.

[^{125}I]-fBSA degradation was assessed by SDS-PAGE after its incubation either with lysosomal extracts in the presence of an inhibitor or purified lysosomal enzymes (Figure 13). Results demonstrate that DTT is not required for the first split in the molecule but is absolutely necessary for its complete degradation. In the presence of leupeptin the first split is prominant, however, there is little subsequent degradation. In the case of pepstatin, there is comparatively little of the 30 kD fragment, but a large proportion of what is split is degraded by 4 hr. The use of purified enzymes shows that cathepsin D leads mainly to the first split while cathepsin B completely degrades the molecule. These observations,

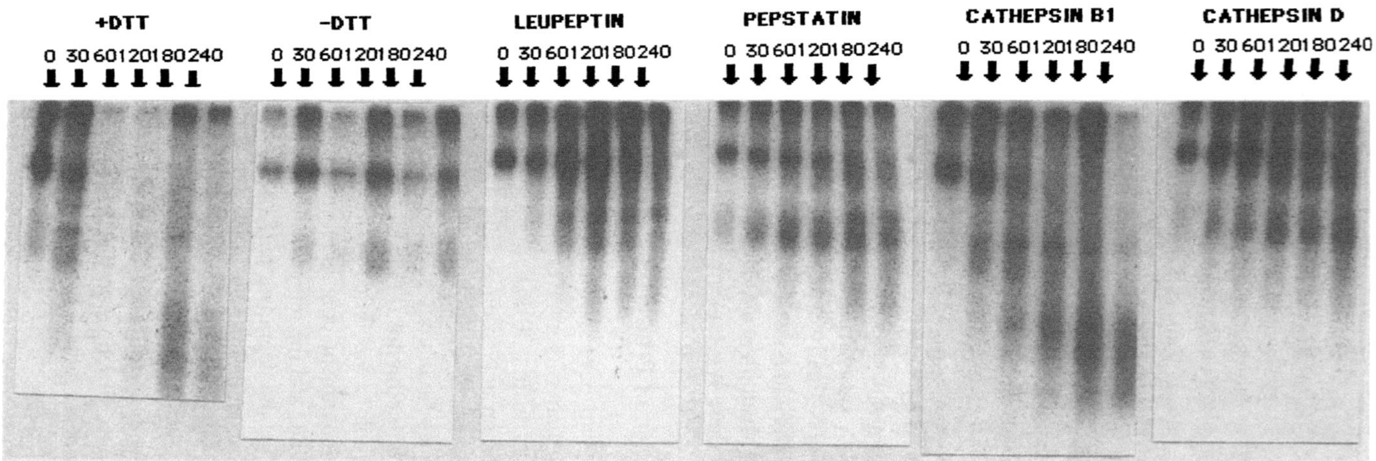

FIGURE 13. SDS-PAGE analysis of *in vitro* degradation products formed when [^{125}I]-fBSA is incubated with either lysosomal extracts in the presence of different inhibitors or purified lysosomal enzymes. [^{125}I]-fBSA was incubated in the presence of acetate buffer pH 4.5 prepared with 0.25 M sucrose containing 2 mM DTT unless specified to the contrary and 2% Triton X-100. Concentrations used for these assays were leupeptin: 3.3×10^{-5} M; pepstatin: 4.8×10^{-6} M; purified cathepsin B1: 0.01 mg lysosomal protein/test; purified cathepsin D: 0.01 mg lysosomal protein/test. The wells indicate the different incubation times.

showing that the protease responsible for the first split is not a thiol-dependent one and that cathepsin D is able to catalyze the ignition of proteolysis—a process blocked by pepstatin—provide an indication that the protease present in endolysosomes is cathepsin D or a cathepsin D–like enzyme.

6. CONCLUSIONS

Table IV features some characteristics of the organelles that we believe to be involved in the intracellular transport and degradation of an endocytosed molecule by the endothelial liver cells. The proposed pathway involving five different organelles (pinosomes, endosomes, endolysosomes, transfer lysosomes, and accumulation lysosomes) to which the endocytosed substance is successively associated bears close resemblance to the endocytic process recently described in other cell types. For example, morphological as well as biochemical investigations lead to the discovery of different endosomal populations (Wall and Hubbard, 1985; Mueller and Hubbard, 1986; Geuze *et al.*, 1983) and to the separation of some of these structures (Baenziger and Fiete, 1986; Evans and Flint, 1985). On the other hand, the involvment of two populations of lysosomes in the nonparenchymal cell endocytic pathway reminds researches of the report of Pertoft *et al.* (1978) showing two lysosomal populations taking part in the endocytosis of asialoglycoproteins in hepatocytes, and the demonstration by England *et al.* (1986) of two classes of lysosomes involved in the transport of asialofetuin in hepatocytes; classes that can be distinguished on the basis of their density in a Nycodenz gradient. It is also worth noticing that by combining the use of centrifugation, SDS-PAGE, and preinjections of Triton WR 1339 we were able to demonstrate that asialofetuin labeled with [^{125}I]-TC follows, once endocytosed by hepatocytes, an intracellular pathway involving, in succession endosomes, a digestive compartment not affected by Triton WR 1339, and lysosomes filled with the detergent (Wattiaux *et al.*, 1989). Nevertheless, the endosomal compartment identified in sinusoidal cells exhibits at least one outstanding difference from its hepatocytic counterpart: an equilibrium density that is markedly higher in a sucrose gradient, making distinguishing between that compartment and lysosomes difficult.

Endosomes have frequently been defined on the basis of their very low equilibrium density in isopycnic centrifugation (see, for example, Kornfeld and Mellman, 1989). Our experiments show that such an oversimplified assumption (Hubbard, 1989) is not valid at least as far as sinusoidal cells are concerned, and therefore should be used very carefully. It is to be noted, however, that we cannot exclude the existence of a very-low-density endosomal compartment in that cell type. That putative structure would have to be crossed extremely rapidly by the endocytosed molecule because the percentage of radioactivity present in the low-density region of the gradient may almost be ignored at any time after injection.

TABLE IV
Some Characteristics of the Organelles Involved in the Endocytic Process of [^{125}I]TCfBSA

Organelles	Fraction (highest proportion)	Density in sucrose gradient	Effect of Triton WR 1339 (density shift)	Effect of invertase (density shift)	Effect of Gly-L-Phe-2-NNap	Presence of degraded products
Pinosomes	P	1.16g/ml	Unaffected	Unaffected	Unaffected	Not observed
Endosomes	P & ML	1.19g/ml	Unaffected	Unaffected	Unaffected	Not present
Endolysosomes	ML	1.21g/ml	Unaffected	Unaffected	Unaffected	First proteolytic product only
Transfer lysosomes	ML	1.22g/ml	Unaffected	Unaffected	Affected	Whole range of degradation products
Accumulation lysosomes	ML	1.22g/ml	Affected 1.15g/ml	Affected 1.245g/ml	Affected	Presence of low mol. wt. radiolabeled compounds

That observation illustrates also the risk of relying exclusively on a comparison between a hydrolase distribution and an endocytic-marker distribution to assess the lysosomal location of that marker. In the nonparenchymal cells, such a comparison alone would lead to the misguided impression that the endocytic marker, still located in prelysosomal structures after a short injection time, is already present in lysosomes, or in other words, that the transfer to lysosomes is much faster in sinusoidal cells than in hepatocytes as proposed by Quintart *et al.* (1983).

The specific disruption of lysosomes by Gly-L-Phe-2-NNap previously described can be particularly useful in assessing the station of a compound in the lysosomal compartment. In our opinion, the main advantage of the method [which can be applied to study an endocytic process (Oude Elferink *et al.*, 1985; Jadot *et al.*, 1985; Misquith *et al.*, 1988) as well as to follow the processing of lysosomal hydrolases (Oude Elferink *et al.*, 1986)] is that it allows a functional definition of lysosomes, a definition that could be stated as organelles containing enough active cathepsin C to account for their disruption by Gly-L-Phe-2-NNap.

The discovery of such a highly complicated intracellular pathway, involving intermediate organelles between endosomes and lysosomes, raises a further question: Are these multiple compartments preexisting in the cell when we get it to endocytose the marker or do they represent multiple steps on a temporal scale? The second hypothesis, generally referred as the *maturation model*, could very well fit our experimental results. Such a model supposes that an endosome is progressively transformed into an intermediate compartment (endolysosome) by its fusion with primary lysosomes that contain enough hydrolases to account for the ignition of proteolysis (leading to the appearance of the first split products). Cytochemical observations have demonstrated such a fusion process between primary lysosomes and endosomes (De Bruyn *et al.*, 1983a,b; Jose-Vu *et al.*, 1986). On the other hand, studies of the lysosomal enzyme biogenesis seem to be in agreement with the proposed model. Indeed, these works show that newly synthesized lysosomal enzymes bound to the Man-6-P receptor are discharged in an endosomal compartment (Brown *et al.*, 1986; Griffiths *et al.*, 1988; Duncan and Kornfeld, 1988) where they meet the different compounds internalized by endocytosis. On the basis of these observations, it is tempting to speculate that the late endosomal compartment, where sorting between the hydrolase and its receptor occurs (Schmid *et al.*, 1989), could be assimilated to the endolysosomes identified in the sinusoidal cells. That structure would be the site where the discharge of the bulk of lysosomal enzymes (including cathepsin C) would occur, thereby transforming that prelysosomal compartment into a fully mature lysosome. It is to be noted that such a model that presupposes the existence of prelysosomal structures containing a limited amount of hydrolases substantiates the claim of Stahl and co-workers that macrophage endosomes contain protease activity (Diment and Stahl, 1985; Diment *et al.*, 1988) as well as the report of Hamel *et al.* (1988) showing that insulin degradation would occur in endosomes.

Results obtained in the course of the study of the sinusoidal-cell endocytic process are fully consistent with the proposed maturation model of Stoorvogel *et al.* (1989). According to that hypothesis, newly formed pinosomes budding from the plasma membrane fuse with all endosomes, but the probability of such a fusion event decreases from the early to the late endosomal compartment. On the other hand, a preferential fusion of vesicles arising from the trans-Golgi network (TGN) and carrying newly synthesized lysosomal enzymes, along with late endosomes, would account for the progressive maturation of the endosomal apparatus.

Our observations suggest that the last step in the intracellular pathway is the transfer of the endocytosed molecule from a first population of lysosomes (lys I) exhibiting a powerful degradative activity, to a second population of these organelles (lys II) mainly concerned with the storage of the hydrolysis products. The way the transfer takes place again favors the hypothesis of a maturation model. In the hypothesis of a massive discharge of lys I in lys II, one should expect the establishment of a bimodal distribution for acid-soluble radioactivity (see Figure 8). On the contrary, the distribution of the acid-soluble compounds moves progressively to the equilibrium density of Triton-filled lysosomes. That observation could be best explained by assuming that small vesicles arising, for example, from the fission of tritosomes are carrying small density material to lys I, thereby inducing the maturation of that population of organelles by decreasing its density. In that respect, we should keep in mind that extensive fusions between secondary lysosomes have been described and that fission of secondary lysosomes has been observed in the course of the phagocytic process (Piao and Ogawa, 1985).

Nevertheless, keeping this maturation model in mind, we cannot exclude the existence of continuous tubular structures, instead of discrete vesicular organelles. Such a cytoplasmic "digestive tract" that would be transformed in vesicles by the homogenization procedure is strongly reminiscent of the tubular lysosomes described by Swanson *et al.* (1987) in macrophages.

ACKNOWLEDGMENTS. Our laboratory has been supported by grants from the Fonds National de la Recherche Scientifique, the Fonds de la Recherche Scientifique Médicale (contracts No. 3.45 32.87 and No. 3.452397), the ASBL Air Escargot and the Mr. and Mrs. Detilleux donation.

7. REFERENCES

Achord, D. T., Brot, F. E., and Sly, W. S., 1977, Inhibition of the rat clearance system for agalacto-orosomucoid by yeast mannans and by mannose, *Biochem. Biophys. Res. Commun.* **77**:409–415.

Arborgh, B., Berg, T., and Ericsson, J. L. E., 1973, Quantitation of acid phosphatase and arylsulfatase in rat hepatic parenchymal and Kupffer cells. *FEBS Lett.*, **35**:51–53.

Ashwell, G. and Morell, A. G., 1974, The role of surface carbohydrates in the hepatic recognition and transport of circulating glycoproteins, *Adv. in Enzymol.* **41**:99–128.

Baenziger, J. U. and Fiete, D., 1986, Separation of two populations of endocytic vesicles involved in receptor-ligand sorting in rat hepatocytes, *J. Biol. Chem.* **261**:7445–7454.

Berg, T., Kindberg, G. M., Ford, T., and Blomhoff, R., 1985, Intracellular transport of asialoglycoproteins in rat hepatocytes, *Exp. Cell Res.* **161**:285–296.

Bloomhoff, R., Eskild, W., and Berg, T., 1984a, Endocytosis of formaldehyde-treated bovine serum albumin via scavenger pathway in liver endothelial cells, *Biochem. J.* **218**:81–86.

Blouin, A., 1977, Morphometry of liver sinusoidal cells, in *Kupffer Cells and Other Liver Sinusoidal Cells.* (E. Wisse and D. L. Knook, eds.) pp. 61–71, Elsevier, New York.

Brown, W. J., Goodhouse, J., and Farquhar, M. G., 1986, Mannose-6-phosphate receptors for lysosomal enzymes cycle between the Golgi complex and endosomes, *J. Cell Biol.* **103**:1235–1247.

Clarke, B. L., and Weigel, P. H., 1985, Recycling of the asialoglycoprotein receptor in isolated rat hepatocytes, *J. Biol. Chem.* **260**:128–133.

De Bruyn, P. P. H., Cho, Y., and Michelson, S., 1983a, *In vivo* endocytosis by bristle coated pits of protein tracers and their intracellular transport in the endothelial cells lining the sinuses of the liver. I. The endosomal disposition, *J. Ultrastruct. Res.* **85**:272–289.

De Bruyn, P. P. H., Cho, Y., and Michelson, S., 1983b, *In vivo* endocytosis by bristle coated pits of protein tracers and their intracellular transport in the endothelial cells lining the sinuses of the liver. II. The endosomal-lysosomal transformation. *J. Ultrastruct. Res.* **85**:290–299.

De Meyts, P. and Hanoune, J., 1982, Plasma membrane receptors and function, in *The Liver: Biology and Pathobiology,* (I. Arias, H. Popper, D. Schachter, & D. A. Schafritz, eds.) pp. 551–580, Raven Press, New York.

Diment, S., and Stahl, P., 1985, Macrophage endosomes contain proteases which degrade endocytosed protein ligands, *J. Biol. Chem.* **260**:15311–15317.

Diment, S., Leech, M. S., and Stahl, P., 1988, Cathepsin D is membrane-associated in macrophage endosomes, *J. Biol. Chem.* **263**:6901–6907.

Duncan, J. R., and Kornfeld, S., 1988, Intracellular movement of two mannose-6-phosphate receptors: return to the Golgi apparatus, *J. Cell Biol.* **106**:617–628.

England, I. G., Naess, L., Blomhoff, R., and Berg, T., 1986, Uptake, intracellular transport, and release of [^{125}I]-Poly(vinylpyrrolidone) and [^{14}C]-sucrose-asialofetuin in rat liver parenchymal cells, *Biochem. Pharma.* **35**:201–208.

Evans, W. H. and Flint, N., 1985, Subfractionation of hepatic endosomes in Nycodenz gradients and by free flow electrophoresis, *Biochem. J.* **232**:25–32.

Fisher, H. D., Gonzalez-Noriega, A., Sly, W. S., and Morre, D. J., 1980, Phosphomannosyl-enzyme receptors in rat liver, *J. Biol. Chem.* **255**:9608–9615.

Geuze, H. J., Slot, W. J., and Strous, G. J. A. M., 1983, Intracellular site of asialoglycoprotein receptor-ligand uncoupling: double label immunoelectron microscopy during receptor mediated endocytosis, *Cell* **32**:277–287.

Geuze, H. J., Slot, J. W., Strous, G. J. A. M., Hasilik, A., and Von Figura, K., 1984, Ultrastructural localization of the mannose-6-phosphate receptor in rat liver, *J. Cell Biol.* **98**:2047–2054.

Griffiths, G., Hoflack, B., Simons, K., Mellman, I., and Kornfeld, S., 1988, The mannose-6-phosphate receptor and the biogenesis of lysosomes, *Cell* **52**:329–341.

Haltiwanger, R. S., Lehrman, M. A., Eckhardt, A. E., and Hill, R. L., 1986, The distribution and localization of the fucose-binding lectin in rat tissues and the identification of a high affinity form of the mannose/ N-acetylglucosamine-binding lectin in rat liver, *J. Biol. Chem.* **261**:7433–7439.

Hamel, F. G., Posner, B. I., Bergeron, J., Frank, B., and Duckworth, W., 1988, Isolation of insulin degradation products from endosomes derived from intact rat liver, *J. Biol. Chem.* **263:**6703–6708.

Horiuchi, S., Takata, K., Maeda, H., and Morino, Y. 1985, Scavenger function of sinusoidal liver cells, *J. Biol. Chem.* **259:**53–56.

Hubbard, A. L., Wilson, G., Ashwell, G., and Stukenbrok, H., 1979, An electron microscope autoradiographic study of the carbohydrate recognition systems in rat liver I. Distribution of ^{125}I-ligands among the liver cell types, *J. Cell Biol.* **83:**47–64.

Hubbard, A. L., and Stukenbrok, H., 1979, An electron microscope autoradiographic study of the carbohydrate recognition systems in rat liver. II. intracellular fate of ^{125}I-ligands, *J. Cell Biol.* **83:**65–81.

Hubbard, A. L., 1989, Endocytosis, *Current Opinion in Cell Biology,* **1:**615–616.

Jadot, M., Colmant, C., Wattiaux-De Coninck, S., and Wattiaux, R., 1984, Intralysosomal hydrolysis of glycyl-L-phenylalanine 2-Naphthylamide, *Biochem. J.* **219:**965–970.

Jadot, M. and Wattiaux, R., 1985, Effect of glycyl-L-phenylalanine-2-naphthylamide on invertase endocytosed by rat liver, *Biochem. J.* **225:**645–648.

Jadot, M., Wattiaux-De Coninck, S., and Wattiaux, R., 1985, Effect on lysosomes of invertase endocytosed by rat liver, *Eur. J. Biochem.* **151:**485–488.

Jadot, M., Misquith, S., Dubois, F., Wattiaux-De Coninck, S., and Wattiaux, R., 1986, Intracellular pathway followed by invertase endocytosed by rat liver, *Eur. J. Biochem.* **161:**695–700.

Jost-Vu, E., Hamilton, R. L., Hornick, C. A., Belcher, J. D., and Havel, R. J., 1986, Multivesicular bodies isolated from rat hepatocytes, *Histochemistry* **85:**457–466.

Kempka, G. and Kolb-Bachofen, V., 1985, Galactose-specific receptors on liver cells. I. Hepatocyte and liver macrophage receptors differ in their membrane anchorage, *Biochim. Biophys. Acta.* **847:**108–114.

Knook, D. L., 1974, Distribution of lysosomal enzyme activities between parenchymal and non-parenchymal cells from rat liver, *Hoppe Seyler's Physiol. Chem.* **255:**1217–1219.

Kornfeld, S. and Mellman, I., 1989, The biogenesis of lysosomes, *Annu. Rev. Cell Biol.* **5:**483–525.

Lehrman, M. A. and Hill, R. L., 1986, The binding of fucose-containing glycoproteins by hepatic lectins—Purification of a fucose-binding lectin from rat liver, *J. Biol. Chem.* **261:**7419–7425.

Lehrman, M. A., Pizzo, S. V., Imber, M. J., and Hill, R. L., 1986a, The binding of fucose-containing glycoproteins by hepatic lectins—Re-examination of the clearance from blood and the binding to membrane receptors and pure lectins, *J. Biol. Chem.* **261:**7412–7418.

Lehrman, M. A., Haltiwanger, R. S., and Hill, R. L., 1986b, The binding of fucose-containing glycoproteins by hepatic lectins—The binding specificity of the rat liver fucose lectin. *J. Biol. Chem.* **261:**7426–7432.

Madnick, H. M., Winckler, J. R., and Segal, H. L., 1978, Uptake of yeast invertase by rat liver cells *in vivo* and *in vitro*. *Arch. Biochem. Biophys.* **191:**385–392.

Marsh, M., Schmid, S., Kern, H., Harms, E., Male, P., Mellman, I., and Helenius, A., 1987, Rapid analytical and preparative isolation of functional endosomes by free flow electrophoresis, *J. Cell Biol.* **104:**875–886.

Mego, J. L., 1984, Role of thiols, pH and cathepsin D in the lysosomal catabolism of serum albumin, *Biochem. J.* **218:**775–783.

Misquith, S., Wattiaux-De Conick, S., and Wattiaux, R., 1988, Uptake and intracellular transport in rat liver of formaldehyde-treated bovine serum albumin labeled with ^{125}I-l-tyramine-cellobiose, *Eur. J. Biochem.* **174:**691–697.

Misquith, S., Wattiaux-De Coninck, S., and Wattiaux, R., 1989, Intracellular degradation by liver endothelial cells, *Mol. Cell. Biochem.* **91:**63–74.

Miyai, K., 1979, Ultrastructural basis for toxic liver injury, in *Toxic Injury of the Liver,* (E. Farber and M. M. Fisher, eds.) pp. 59–154, Marcel Dekker New York.

Morell, A. G., Gregoriadis, G., Sceinberg, I. H., Hickman, J., and Ashwell, G., 1971, The role of

sialic acid in determining the survival of glycoproteins in the circulation, *J. Biol. Chem.* **246:**1461–1467.
Mueller, S. C. and Hubbard, A. L., 1986, Receptor-mediated endocytosis of asialoglycoproteins by rat hepatocytes: receptor-positive and receptor-negative endosomes, *J. Cell Biol.* **102:**932–942.
Munthe-Kaas, A. C., Berg, T., and Seljelid, R., 1976, Distribution of lysosomal enzymes in different types of rat liver cells, *Expt. Cell Res.* **99:**146–154.
Murakami, M., Horiuchi, S., Takata, K., and Morino, Y., 1986, Scavenger receptor for malondialdehyde-modified high-density lipoprotein on rat sinusoidal liver cells, *Biochem. Biophys. Res. Comm.* **137:**29–35.
Nagelkerke, J. F., Barto, K. P., and van Berkel, T.J.C., 1982, Receptor-mediated endocytosis in rat liver non-parenchymal cells. An evaluation of two recently developed cell isolation procedures, in *Sinusoidal Liver Cells.* (D. L. Knook and E. Wisse, eds.), pp. 319–326, Elsevier Biomedical Press, Amsterdam.
Opresko, L. K., and Karpf, R. A., 1987, Specific proteolysis regulates fusion between endocytic compartments in *Xenopus* oocytes, *Cell* **51:**557–568.
Oude Elferink, R. P. J., Van Doorn-Van Wakeren, J., Hendriks, T., Strijland, A., Reuser, A. J. J., and Tager, J., 1985, Biosynthesis and intracellular transport of a-glucosidase and cathepasin D in normal and mutant human fibroblasts, *Eur. J. Biochem.* **153:**55–63.
Oude Elferink, R. P. J., Van Doorn-Van Wakeren, J., Hendriks, T., Strijland, A., and Tager, J., 1986, Transport and processing of endocytosed lysosomal a-glucosidase in cultured human skin fibroblasts, *Eur. J. Biochem.* **158:**339–344.
Pertoft, H., Wärmegard, B., and Höök, M., 1978, Heterogeneity of lysosomes originating from rat liver parenchymal cells, *Biochem. J.* **174:**309–317.
Piao, Y. J. and Ogawa, K., 1985, Ultrastructural and cytochemical observations on heterophagy and autophagy of macrophage in mouse thymus, *Acta Histochem. Cytochem.* **18:**615–632.
Pittman, R. C., Careco, J. E., Glass, C. K., Green, S. R., Taylor, C. A., and Attie, A. D., 1983, A radioiodinated, intracellularly trapped ligand for determining the sites of plasma protein degradation *in vivo, Biochem. J.* **212:**791–800.
Pizzo, S. V., Lehrman, M. A., Imber, M. J., and Guthrow, C. E., 1981, The clearance of glycoproteins in diabetic mice. *Biochem. Biophys. Res. Commun.* **101:**704–708.
Praaning-Van Dalen, D. P., De Leeuw, M., Brouwer, A., De Ruiter, C. F., and Knook, D. L., 1982, Ultrastructural and biochemical characterization of endocytic mechanisms in rat liver Kupffer and endothelial cells, in *Sinusoidal Liver Cells* (D. L. Knook and E. Wisse, eds.), pp. 271–278, Elsevier Biomedical Press, Amsterdam.
Quintart, J., Courtoy, P. J., Limet, J. N., and Baudhuin, P., 1983, Galactose specific endocytosis in rat liver, *Eur. J. Biochem.* **131:**105–112.
Rodman, J. S., Schlessinger, P., and Stahl, P., 1978, Rat plasma clearance of horseradish peroxidase and yeast invertase is mediated by specific recognition, *FEBS Lett.* **85:**345–348.
Roos, P. H., Hartman, H-J., Schlepper-Schafer, J., Kolb, H., and Kolb-Bachofen, V., 1985, Galactose-specific receptors on liver cells. II. characterization of the purified receptor from macrophages reveals no structural relationship to the hepatocyte receptor, *Biochim. Biophys. Acta.* **847:**115–121.
Schlepper-Schäfer, J., Kolb-Bachofen, V., Holl, N., Friedrich, E., and Kolb, H., 1982, Galactose-specific lectin on rat Kupffer cells: localization and function, in *Sinusoidal Liver Cells* (D. L. Knook & E. Wisse, eds.), pp. 279–286, Elsevier Biomedical Press, Amsterdam.
Schlepper-Schäfer, J., Hülsmann, D., Djovkar, A., Meyer, H. E., Herbertz, L., Kolb, H., and Kolb-Bachofen, V., 1986, Endocytosis via galactose receptors *in vivo, Expl. cell res.* **165:**494–506.
Schlessinger, P. H., Rodman, J. S., Doebber, T. W., Stahl, P. D., Lee, Y. C., Stowell, C. P., and Kuhlenschmidt, T. B., 1980, The role of extra-hepatic tissues in the receptor-mediated plasma clearance of glycoproteins terminated by mannose or N-acetylglucosamine, *Biochem. J.* **192:**597–606.

Schmid, S., Fuchs, R., Kielian, M., Helenius, A., and Mellman, I., 1989, Acidification of endosome subpopulations in wild-type Chinese hamster ovary cells and temperature-sensitive acidifcation-defective mutants, *J. Cell Biol.* **108:**1291–1300.

Stahl, P., Schlessinger, P. H., Rodman, J. S., and Doebber, T., 1976, Recognition of lysosomal glycosidases *in vivo* inhibited by modified glycoproteins, *Nature* **264:**86–88.

Steer, C. J. and Ashwell, G., 1980, Studies on a mammalian hepatic binding protein specific for asialoglycoproteins, *J. Biol. Chem.* **255:**3008–3013.

Steer, C. J. and Clarenburg, R., 1979, Unique distribution of glycoprotein receptors on parenchymal and sinusoidal cells of rat liver, *J. Biol. Chem.* **254:**4457–4461.

Stockert, R. J., Morell, A. G., and Scheinberg, H., 1976, The existence of a second route for the transfer of certain glycoproteins from the circulation into the liver, *Biochem. Biophys. Res. Comm.* **68:**988–993.

Stoorvogel, W., Geuze, H. J., Griffith, J. M., Schwartz, A. L., and Strous, G. J., 1989, Relations between the intracellular pathway of the receptors for transferrin, asialoglycoprotein, and mannose-6-phosphate in human hepatoma cells, *J. Cell Biol.* **108:**2137–2148.

Stowell, C. P. and Lee, Y. C., 1978, The binding of D-glucosyl-neoglycoproteins to the hepatic asialoglycoprotein receptor, *J. Biol. Chem.* **253:**6107–6110.

Summerfield, J. A., Vergalla, J., and Jones, E. A., 1982, Modulation of a glycoprotein recognition system on rat hepatic endothelial cells by glucose and diabete mellitus, *J. Clin. Invest.* **69:**1337–1347.

Swanson, J., Bushnell, A., and Silverstein, S. C., 1987, Tubular lysosome morphology and distribution within macrophages depend on the integrity of cytoplasmic microtubules, *Proc. Natl. Acad. Sci. USA.* **84:**1921–1925.

Tarentino, A. L., Plummer, T. H., Jr, and Maley, F., 1974, The release of intact oligosaccharides from specific glycoproteins by endo-B-N-acetylglucosaminidase H., *J. Biol. Chem.* **249:**818–824.

Thirion, J. and Wattiaux, R., 1988, Endocytic modifications occuring in primary culture of rat hepatocytes, *4th Int. Congr. Cell Biol.*, Montreal. Abstracts of papers, p. 240.

Trimble, R. B. and Maley, F., 1977, Subunit structure of external invertase from saccharomyces cerevisiae, *J. Biol. Chem.* **252:**4409–4412.

Wall, D. A., Wilson, G., and Hubbard, A. L., 1980, The galactose-specific recognition system of the mammalian liver: the route of ligand internalization in rat hepatocytes, *Cell* **21:**79–93.

Wall, D. A. and Hubbard, A. L., 1985, Receptor mediated endocytosis of asialoglycoproteins by rat liver hepatocytes: biochemical characterization of the endosomal compartments, *J. Cell Biol.* **101:**2104–2112.

Wattiaux, R., Baudhuin, P., Berleur, A. M., and De Duve, C., 1956, Tissue fractionation studies. 8. cellular localization of bound enzymes, *Biochem. J.* **63:**608–610.

Wattiaux, R., Wibo, M., and Baudhuin, P., 1963, Influence of the injection of Triton WR 1339 on the properties of rat liver lysosomes, in *Ciba Foundation Symposium on Lysosomes.* (A. V. de Reuck and M. P. Cameron, eds.), pp. 176–200, J. & A. Churchill Ltd. London.

Wattiaux, R., Wattiaux-De Coninck, S., Ronveaux-Dupal, M. F., and Dubois, F., 1978, Isolation of rat liver lysosomes by isopycnic centrifugation in a metrizamide gradient, *J. Cell Biol.* **78:**349–368.

Wattiaux, R., Jadot, M., Misquith, S., Dubois, F., and Wattiaux-De Coninck, S., 1986, Differences in the cellular location of substances endocytosed by rat liver as observed from the distribution patterns obtained after isopycnic centrifugation in a sucrose gradient, *Biochem. Biophys. Res. Commun.* **136:**504–509.

Wattiaux, R., Wattiaux-De Coninck, S., Dubois, F., and Braun, A., 1987, Influence of the injection of mannan on rat liver lysosomes, *Biochem. Soc. Transac.* **15:**435.

Wattiaux, R., Misquith, S., Wattiaux-De Coninck, S., and Dubois, F., 1989, Fate of asialofetuin endocytosed by rat liver, *Biochem. Biophys. Res. Commun.* **158:**313–318.

Chapter 7

The Endocytic Compartments of Normal and Regenerating Liver

Carlos Enrich and W. Howard Evans

1. INTRODUCTION

The liver is pivotally positioned to remove a wide range of molecules circulating in the blood, thereby acting as a key organ in the regulation of blood composition. The basolateral plasma membrane, especially the blood-facing sinusoidal domain of the hepatocyte, contains a large number of various receptors that account for the liver's ability to selectively and efficiently endocytose and metabolize a variety of ligands (see Table I). Endocytotic uptake of ligands from the blood circulating in the space of Disse is a multistep process. Receptor-ligand complexes assembled at coated regions and probably also at morphologically undifferentiated regions of the plasma membrane are internalized within minutes to a membrane-bound compartment described variously as compartment for uncoupling of receptors and ligands (CURL), receptosomes, diacytosomes, endosomes, prelysosomal compartment, and so forth. In this chapter we will use the composite term *endocytic compartment* to describe the complex membrane networks present in all animal cells where the intracellular sorting of internalized

---

Carlos Enrich Departmento de Biologia Celular, Facultad Medicina, Universitat de Barcelona, Spain. **W. Howard Evans** National Institute for Medical Research, Mill Hill, London NW7 1AA, United Kingdom.
Subcellular Biochemistry, Volume 19: Endocytic Components: Identification and Characterization, edited by J. J. M. Bergeron and J. R. Harris. Plenum Press, New York, 1993.

ligands and receptors occurs. Subcellular fractions prepared from liver homogenates and composed largely of membrane vesicles originating from the endocytic compartment will be termed *endosomes*.

In hepatocytes, several endocytic routes are delineated. The major route delivers the receptor-ligand complexes directly to the endocytic compartment and after a pH-induced dissociation of the complex the receptors are recycled back to the plasma membrane and the liberated ligands either gain entry into the cytoplasm or are transferred onward to lysosomes where they are degraded. In transcytosis the receptor-ligand complexes are transported through the endocytic compartment to the apical pole of the hepatocyte (the bile-canalicular plasma membrane) where the ligand and the receptor (a truncated form in the case of the secretory-component receptor) may be released in the biliary spaces. In specific endocytic circumstances (e.g., during ligand overload) ligands and associated receptors are transferred in tandem to lysosomes for degradation (e.g., the epidermal growth factor–ligand receptor complex). Other routes operationally facilitate the return from the endocytic compartment of undissociated ligand-receptor complexes to the originating plasma membrane domain, for instance, transferrin and its receptor during iron delivery into cells. In this special case, the route is restricted mainly to a more superficial (or early) region of the endocytic compartment located underneath the sinusoidal plasma membrane where the pH, although inappropriate for dissociation of the receptor-transferrin complex, allows iron release. Trafficking between the plasma membrane and the endocytic compartment may also occur in the absence of bound ligands. For example, the endocytosis and release into bile of secretory component can proceed in the absence of circulating polymeric IgA (Mullock *et al.*, 1980) and receptor recycling in the absence of ligand also appears to occur with the low-density lipoprotein, asialoglycoprotein, α_2-macroglobulin, and insulinlike growth factor (IGF) II receptors (Oka and Czech, 1986).

These routes comprise the main arteries of the endocytic compartment, but this description is probably an oversimplification of the overall trafficking networks in operation. For example, resialylation (repair) of ligands (e.g., asialotransferrin, as seen in Regoeczi *et al.*, 1982) as well as receptors (transferrin receptor, see Snider and Rogers, 1985; mannose-6-phosphate receptor, see Duncan and Kornfeld, 1988) occurs, although at a longer transit timescale to that of conventional trafficking, suggesting that the routes followed probably involve the Golgi apparatus. Indeed, increasing complexity of the Golgi apparatus is seen at its immature "cis" side, where interaction with salvage pathways that involve vesicles linking the endoplasmic reticulum to the Golgi occurs, as well as in the trans-Golgi networks. On the other hand, it should not be assumed that all plasma-membrane constituents are endocytosed to lysosomes via the endocytic compartment. An exception appears to be the connexin proteins that form the gap junctional plaques. Connexin 32, the member of the connexin superfamily predominant in rat liver, is detected by specific antibodies in lateral and sinusoidal plasma membranes and, although relatively abundant in lysosomes, is not present in endosomes (Rahman *et al.*, unpublished work). Thus, a direct endocytic route from the lateral plasma membrane, especially from gap junctional plaques to lysosomes, may be operational. Indeed, all these additional features contribute to an emerging picture of a highly regulated network of intracellular trafficking routes.

Liver provides a useful experimental epithelial-tissue model to study endocytosis. In normal adult liver, the hepatocytes are highly active metabolically, and endocytic process are correspondingly accentuated. In regenerating liver, the alterations in endocytosis occurring as the hepatocytes divide in a synchronous fashion are open to exploration. Livers perfused with specific radiolabeled ligands can be subjected to temperature manipulation to block, for example, transport between endosomes and lysosomes, leading to the accumulation of ligands in endosomes (Courtoy *et al.*, 1984, 1985; Jones *et al.*, 1985; Luzio and Stanley, 1983; Evans and Flint, 1985). In livers of adult animals, hepatocytes account for most of the endocytic activity, but other cells, especially endothelial and Kupffer cells (Wisse, 1977), undoubtedly contribute specifically to the overall activity. In general, particles larger than 0.1 μm are phagocytosed by Kupffer cells, and the endothelial cells contain their own repertoire of receptors (Steer and Clarenburg, 1979). In the subcellular fractionation approaches used to study endocytosis, the use of liver homogenates first filtered through fine nylon gauze to remove unbroken smaller cells has helped to reduce substantially the contribution of nonhepatocyte membranes in the fractions subjected to biochemical analysis. In hepatocytes, endosome structures account for about 1% of the total cell membrane; in contrast, the plasma membrane accounts for 2%, the endoplasmic reticulum for 35%, and the Golgi apparatus for 7% (Blouin *et al.*, 1977).

At the biochemical level, discrete membrane components feature in mediating the functions carried out in the endocytic compartment. In this chapter, the biochemical properties of hepatic endosomes are described. Immunological approaches toward identifying endosome-specific components are summarized, as well as biochemical studies in which changes in the properties of the endocytic compartment in regenerating liver have been examined.

2. MORPHOLOGICAL FEATURES OF THE HEPATIC ENDOCYTIC COMPARTMENT

Detailed models have been advanced showing the anatomical intricacies of the endocytic compartment, especially in studies carried out using cultured cells (Marsh *et al.*, 1986). Undoubtedly, the endocytic compartments of epithelial cells or tissues are more morphologically and functionally complex.

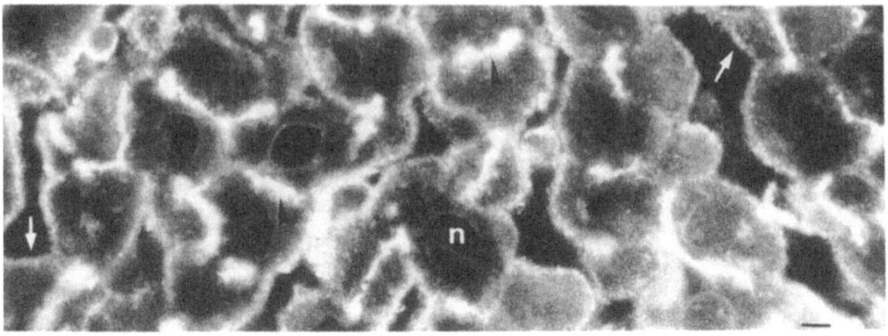

FIGURE 1. Frozen rat-liver section stained by immunofluorescent endosomal antibodies. Sections were treated with antibodies generated in rabbits to the integral membrane proteins of rat hepatic endosomes followed by a fluorescent second antibody. Arrows point to an endocytic region located below the sinusoidal plasma membrane. Black arrowhead points to intensive staining around the bile canaliculus. n = nucleus. Bar = 10μm.

The endocytic compartment of hepatocytes is evident in specific regions of the cytoplasm. The more peripheral endocytic structures in hepatocytes comprise a complex network of anastomosing tubular elements located below the sinusoidal plasma membrane and multivesicular bodies that are present mainly near the canalicular plasma membrane (Enrich and Evans, 1989b; Evans and Enrich, 1989; Geuze *et al.*, 1987). These morphological details have emerged through the application to thin liver sections of antibodies raised to the integral membrane proteins and receptors of hepatic endosome fractions. These antibodies (the immunodominant antigen being the asialoglycoprotein receptor) have mapped out in liver tissue the two major regions of the endocytic compartment (Figure 1) (Enrich and Evans, 1989a). Frozen sections of rat liver stained with an anti-endosome antiserum showed higher-intensity labeling at the pericanalicular region with less intense fluorescence underneath the sinusoidal surface area (see Figure 1). When peroxidase-labeled antibodies were used to stain liver sections, further details were discerned, especially a region running closely parallel to the lateral plasma membrane, a region that could represent an endocytic pathway connecting the sinusoidal and canalicular endocytic zones and thus provide the transcytotic route operating in liver tissue (Enrich and Evans, unpublished work). It is noteworthy that the staining patterns observed with antibodies generated to proteins resident in endosomal membranes were different from the patterns obtained, for example, with antibodies to a Golgi-specific 135 kDa protein (Burke *et al.*, 1982), to a lysosomal-specific 100 kDa protein (Lippincott-Schwartz and Fambrough, 1987), to liver plasma membranes (Tauber *et al.*, 1983), and to clathrin (Enrich *et al.*, 1989). Immunoelectron microscopy has revealed further details of the nature of the endocytic structures (Figure 2). In Lowicryl sections of rat liver, gold particles were associated with vacuolar structures, especially multivesicular bodies present at the biliary pole of the hepatocyte. These membrane vesicles with figure-eight-like shapes were similar in morphology to those described in liver cryosections by Geuze *et al.* (1983, 1984) and to structures observed when cell-free systems were used to analyze endocytosis (Mayorga *et al.*, 1989). When isolated hepatocytes and HepG2 cells were stained with the same anti–liver endosome antibodies, the fluorescence was, however, more diffusely distributed in the cells (Enrich and Evans, 1989a), and the overall pattern of these cells with a less pronounced polarity was generally similar to that observed in fibroblasts (Willingham and Pastan, 1985).

In contrast to conventional models in which geographically and functionally defined endocytic compartments are described, tubular reticulum models have also been proposed in which there is a membrane continuity between, for example, the early and late compartments in liver (Evans and Flint, 1985). Indeed, recent studies using fluorescent microscopy and video-enhancement techniques show the endocytic compartment of epidermoid carcinoma cells as an extensive network of tubular cisternae in which are contained swellings that may correspond to multivesicular bodies; these fluorescent bodies were observed to move toward the periphery of the endocytic tubules (Hopkins *et al.*, 1990). Clearly, morphological models of the endocytic compartment are currently in a most fluid state.

3. BIOCHEMICAL PROPERTIES OF HEPATIC ENDOSOMES

In view of the morphological complexity of the endocytic compartment, it is not surprising that isolation and biochemical characterization of this pleimorphic organelle from tissue homogenates is a major task. Since clear-cut biochemical markers have proved difficult to identify, endocytic vesicles have been isolated on the basis of the presence of high concentrations, relative to cell/tissue homogenates, of radioiodinated ligands internalized by cells. These ligand-enclosing vesicles are predominantly of light buoyant density when fractionated, for example, in sucrose, Ficoll, dextran, Percoll, or Nycodenz gradients. The application of density-perturbation techniques has also contributed to the isolation of endocytic vesicles relatively free of other membrane vesicles of low buoyant densities (Quintart *et al.*, 1984, 1989). Free flow electrophoresis has also been used to isolate a single endosome vesicle fraction from liver (Debanne *et al.*, 1982) but this methodology appears unable to subfractionate hepatic endosomes (Evans and Flint, 1985; Fuchs *et al.*, 1989). In most studies aimed at the isolation and biochemical characterization of endocytic vesicles, it has been convenient to categorize endosomes kinetically according to the timepoint of preparation when maximal accumulation of radioligands occurs in various membrane fractions

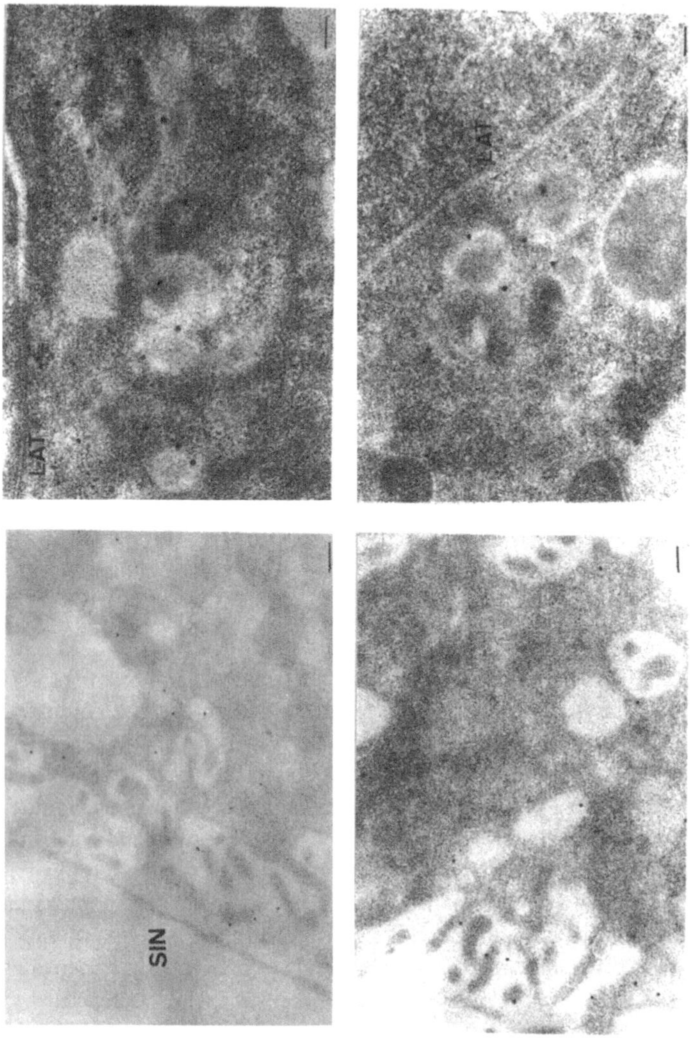

FIGURE 2. Lowicryl-fixed rat-liver sections stained with antibodies to integral membrane proteins of hepatic endosomes followed by Protein A 5 nm gold. In A, a region below sinusoidal plasma membrane and the lateral plasma membrane is shown. These endocytic structures are mainly vesicotubular in appearance. In B, the region surrounding the bile canaliculus contains many immunolabeled multivesicular bodies with characteristic figure-eight-like structures evident.

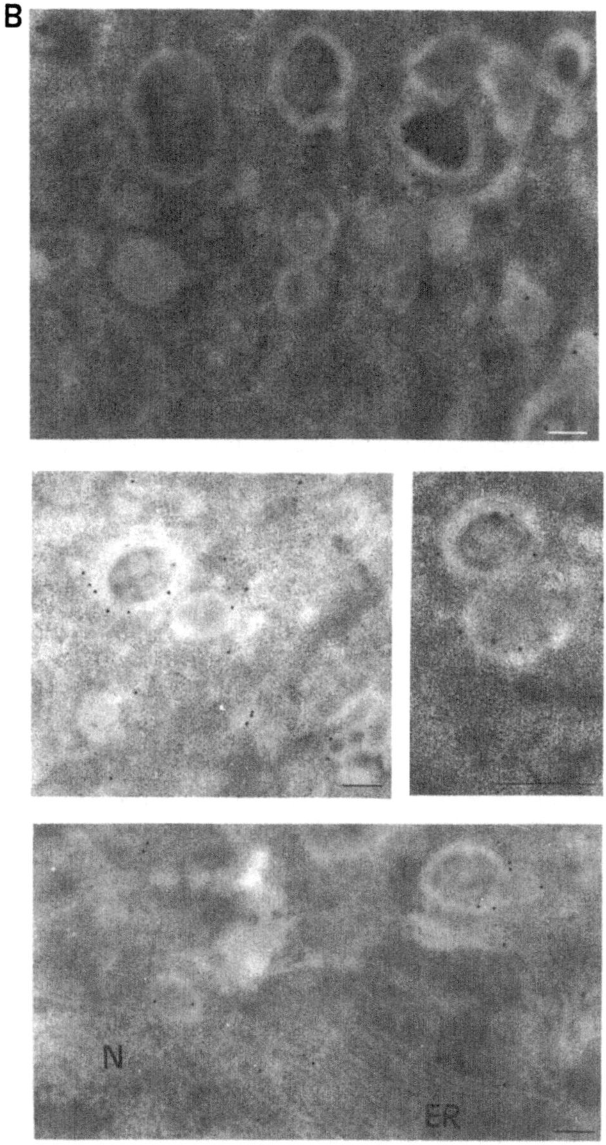

FIGURE 2. *(Continued)*

assigned to the endocytic compartment after organ perfusion with ligands or their addition to the medium surrounding cells. In this chapter we describe the biochemical properties of two ligand-encapsulating vesicular fractions designated *early-* and *late endosomes* prepared at 2- and at 10- to 20-min intervals, respectively, after the uptake of a wide range of ligands by the perfused liver, (Evans and Flint, 1985). A third endosomal fraction, heavier than the others, is also obtained by this methodology when endosome fractions are resolved in sucrose followed by iso-osmotic Nycodenz gradients, (Evans and Flint, 1985; Traub *et al.*, 1990). Fractions with similar properties have been isolated from liver homogenates by other workers (Runquist and Havel, 1991; Jackle *et al.*, 1991a,b).

3.1. Proteins of the Liver Plasma Membrane and Endosomes: Establishing a Database

To analyze the protein profiles in the subcellular fractions enriched in liver plasma membranes domains and in early and late endosomes, a comprehensive electrophoretic resolution of proteins across the molecular mass range 20–200 kDa and the pH range 3.5 to 9.0 was carried out. The resolved proteins were detected by Coomassie blue staining as well as by autoradiography of fractions derived from rat livers metabolically prelabeled with ^{35}S-methionine (Enrich *et al.*, 1990). The glycoproteins in the fractions were detected after transfer of gels to nitrocellulose sheets by a lectin overlay technique using radioiodinated Concanavalin A, wheat germ agglutin, or a sialic-acid-recognizing lectin from *Limax flavus*. These combined approaches have revealed clear differences in the overall protein and glycoprotein compositions of the sinusoidal, lateral, and canalicular plasma membrane fractions (Figure 3). Extending the comparison to include early and late endosomes indicated a high degree of similarity in protein composition between the sinusoidal plasma membranes and early and late endosomes. This applied to the Coomassie blue stained proteins resolved in the pH range 3.5 to 9.0 and especially to the glycoproteins identified using the three lectins (Figure 4). These results point to the conclusion that many of the major proteins and glycoproteins identified in the subcellular fractions by these approaches may correspond to the receptors (see Table I) that are internalized from the plasma membranes to the endocytic compartment; many of these receptors are then recycled back to the plasma membrane. Indeed, the asialoglycoprotein and divalent cation-sensitive mannose-6-phosphate receptors, for example, were specifically identified by immunoblotting techniques in the sinusoidal plasma membranes and the two endosome fractions. This overall similarity between the early and late endosomes prepared from liver tissue was also evident after staining the fractions using two radiolabeled lectins (Figure 5) and reinforces other work on liver endosomal fractions (Baenziger and Fiete, 1986; Belcher *et al.*, 1987). Overall, the data are best interpreted in terms of maturation models of the

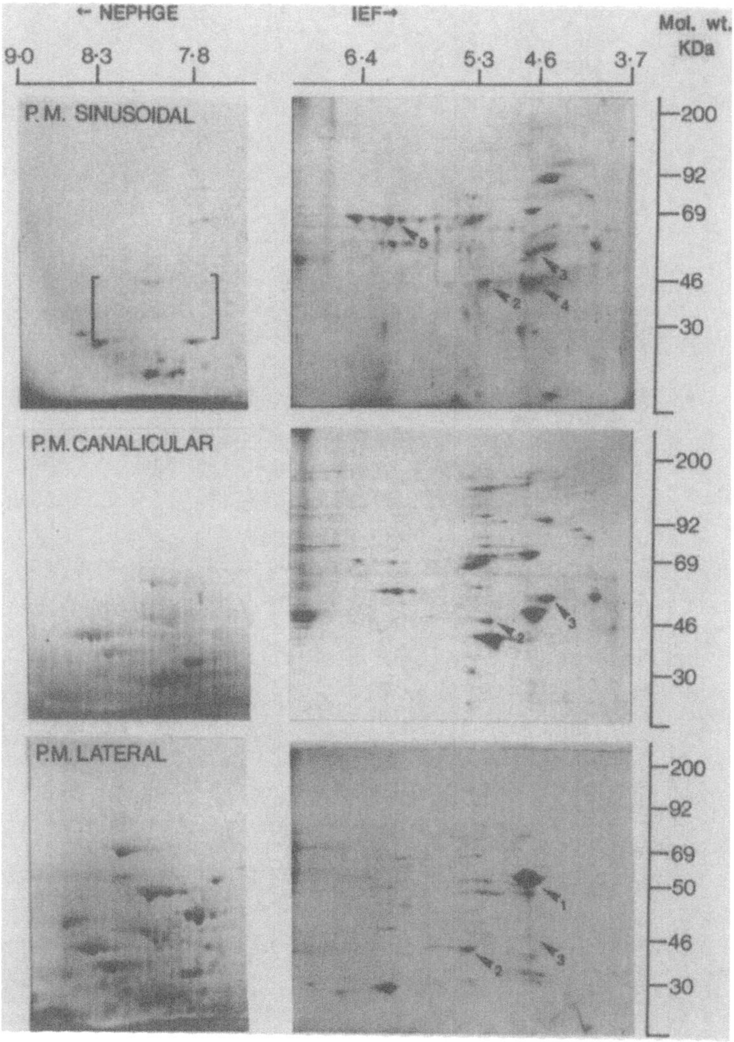

FIGURE 3. Two-dimensional gel electrophoresis of Coomassie blue stained polypeptides of three rat-liver plasma membrane domains resolved by the O'Farrell (1975) procedure. Labeled polypeptides are: (1) a 55 kDa nonglycosylated protein, (2) actin, (3) tubulin, (4) asialoglycoprotein receptor, (5) 5′-nucleotidase, (7) calmodulin binding protein. NEPHGE, nonequilibrium pH gel electrophoresis, IEF isoelectrofocusing. Bracketed area indicates area for comparison with Figure 4.

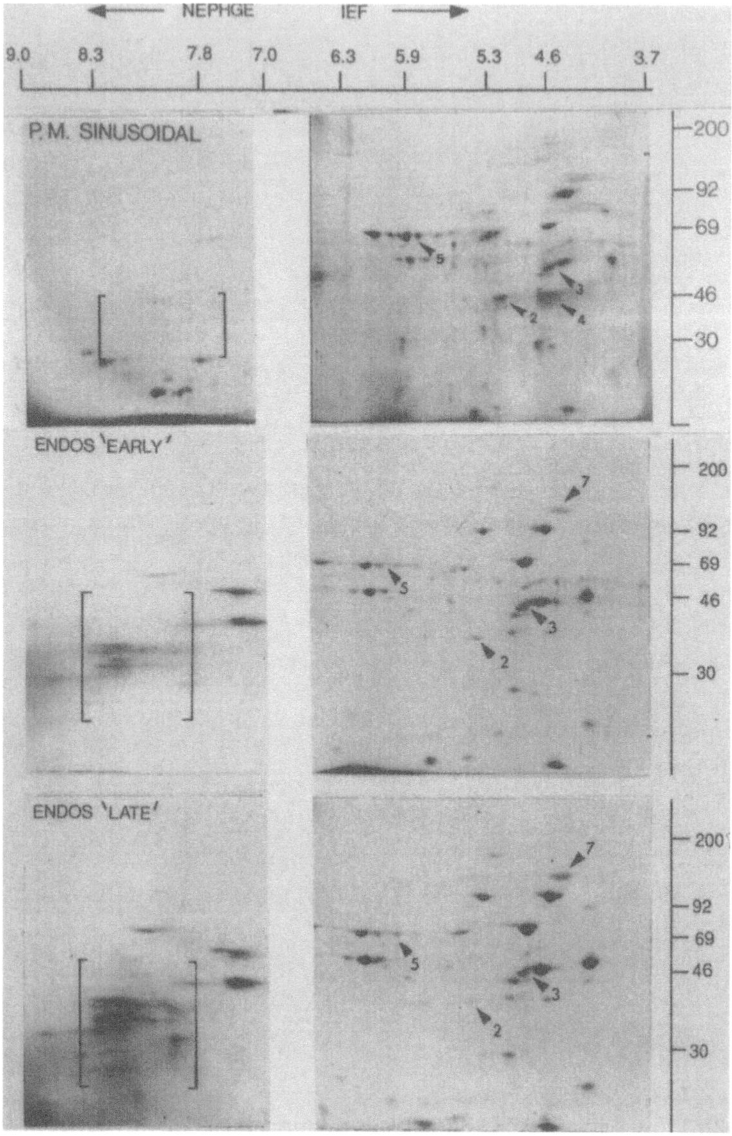

FIGURE 4. Two-dimensional gel electrophoretic analysis of the polypeptides of sinusoidal plasma membranes, early and late endosomes. Numbered bands are labeled as in Figure 3. Bracketed area shows increasing amounts of low-molecular-weight basic polypeptides.

Table I
Receptors Located Predominantly at the Sinusoidal Aspect of the Hepatocyte Plasma Membrane

Receptors	References
For metabolites	
Chylomicrons	Carrella and Cooper, 1979
Low density lipoproteins	Chao *et al.*, 1981
Fatty acids	Renaud *et al.*, 1978
Bile acids	Levy and von Dieppe, 1989
Organic anions	Wolkoff and Chung, 1980
Asialoglycoproteins	Ashwell and Morrell, 1974
IgG (Fc portion)	Frommel and Rachman, 1979
Carcinoembryonic antigen	Thomas, 1980
Immune complexes	Hopf *et al.*, 1981
IgA (polymeric)	Fisher *et al.*, 1979
Transcobalamin	Nexo and Hollenberg 1980
α_2-Macroglobulin	Ashcom *et al.*, 1990
Transferrin	Goldenberg *et al.*, 1990
Hemoglobin-haptoglobin	Kino *et al.*, 1980
Kallikreins	Kouyoumdjian *et al.*, 1989
Plasminogen activator (tissue type)	Owensby *et al.*, 1988
For polypeptide hormones, growth factors, and transmitters	
Insulin	Bergeron *et al.*, 1985
Glucagon	Sperling *et al.*, 1980
Growth hormone	Fouchereau-Peron *et al.*, 1980
Prolactin	Bergeron *et al.*, 1983
Parathyroid hormone	Neuman and Schneider, 1980
Vasopressin	Cantau *et al.*, 1980
Thyroxine	Gharbi and Torresani, 1979
Estradiol	Pietras and Szego, 1980
Prostaglandins	Okamura and Terayama, 1977
α-Adrenergic	El-Refau *et al.*, 1979
β-Adrenergic	Hadjiivanova *et al.*, 1984
Epidermal growth factor	Dunn *et al.*, 1986; Jackle *et al.*, 1991b
Hepatic growth factor	Nakamura *et al.*, 1989
Transforming growth factors α,β	Mead and Fausto, 1989
Integrins β1, α5, and α1	Johannson *et al.*, 1987; Forsberg *et al.*, 1990

liver endocytic compartment in which there is some degree of membrane compositional similarity between the regions through which ligands and receptors traverse.

Noteworthy differences between sinusoidal plasma membranes and early and late endosomes were, however, discerned by extending the pH range in which the proteins were examined (see Figure 4). The two-dimensional elec-

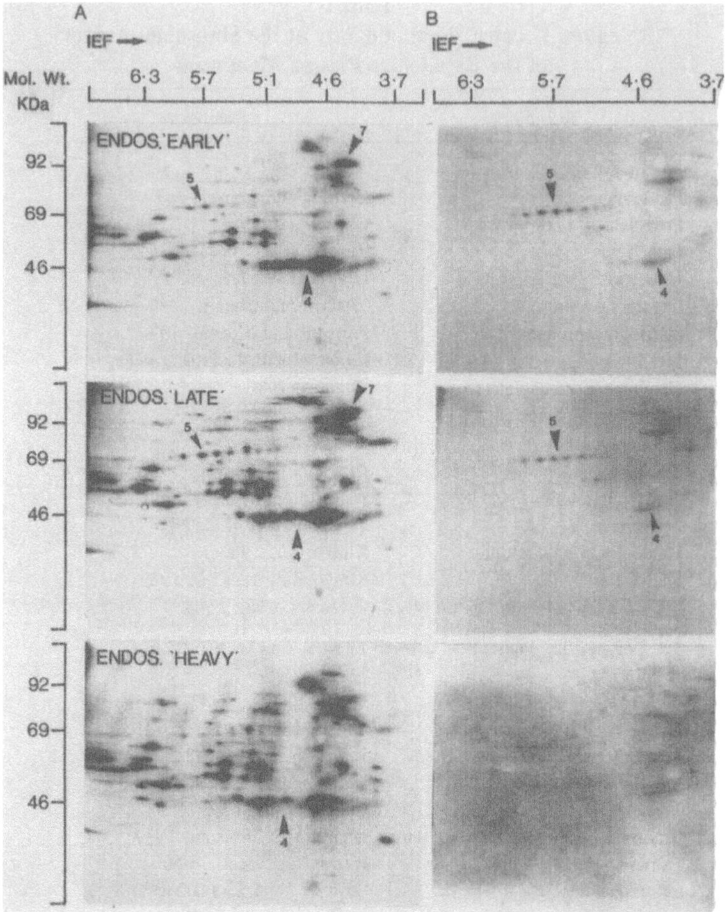

FIGURE 5. Two-dimensional analysis of the glycoproteins of early, late, and heavy endosomes stained after transfer to nitrocellulose paper with radioiodinated Concanavalin (A) and wheat germ agglutinin (B). Numbered bands correspond to those indicated in Figure 3.

trophoretic analysis of the fractions indicated major differences between sinusoidal plasma membranes and endosomes in polypeptides of molecular mass 20–25 kDa and pH 7.5 to 8.5. These basic low-molecular-mass polypeptides were shown, by Coomassie blue staining, to be more evident in late endosomes. Table II summarizes some of the differences found between hepatic early and late endosomes.

Table II
Differences Noted between Early and Late Hepatic Endosome Fractions

	Early endosomes	Late endosomes	Reference
Morphology and location	Vesicotubular, subsinusoidal p.m. region	Mainly multivesicular bodies, pericanalicular p.m. region	Enrich and Evans, 1989
Time taken for ligands to reach fractions after perfusion into portal vein	2–5 min	15 min	Evans and Flint, 1985
Density (in Nycodenz gradients)	1.090 g/ml	1.050 g/ml	Evans and Flint, 1985
pH (in situ)	6.0–6.2	5.5–6.0	Geisow and Evans, 1984; Fuchs *et al.*, 1989
Enzymes			
Acid hydrolases	+	+	Runquist and Havel, 1991; Evans (unpublished); Quintart *et al.*, 1989; Evans and Flint, 1985; Saermark *et al.*, 1985
Na^+ K^+-ATPase	+	n.d.	
5′-nucleotidase	+	++	
Alkaline phosphodiesterase	+	++	
Ionophore-activated Mg^{2+}-ATPase	+	++	
Other properties			
Mannose-6-phosphate receptors	++	++	Enrich *et al.*, 1990; Runquist and Havel, 1991
Asialoglycoprotein receptor (RHL-1)	+++	+	Enrich *et al.*, 1992
downregulated in regenerating liver			Evans (unpublished)
Protein kinase activity	++	+	Strong and Evans, 1987
K^+-activated Ca^{2+}-channels	++	+	Ali and Evans, 1990b
Phospho-inositide anchored glycoproteins	n.d.	110 kDa, 35 kDa	Ali *et al.*, 1989b, 1990
Trimeric and low mol. wt G proteins	+	++	

n.d., not detected; +, relative amounts of the various constituents measured; p.m., plasma membrane.

Note: In addition, a third "receptor-enriched" endosome fraction is also recovered in Nycodenz gradients. This fraction, with high insulin and asialoglycoprotein binding activity, is depleted in the enzymes described in section 5, but is characterised by a high mol. wt G protein (~100kDa) (Traub *et al.*, 1990).

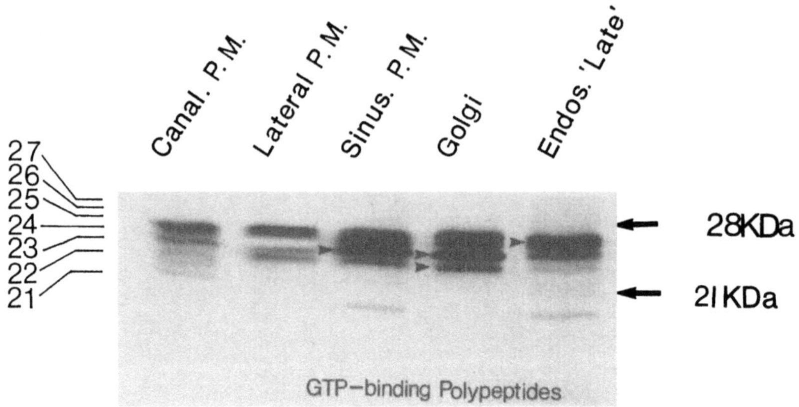

FIGURE 6. Distribution of GTP-binding polypeptides in rat-liver subcellular fractions. Membrane fractions were resolved in 9–18% polyacrylamide gels, transferred to nitrocellulose sheets and incubated with ^{35}S-labeled GTP-γ-S. The molecular masses of the GTP-binding proteins in the various fractions is indicated. Equal amounts of protein were applied to each channel.

3.2. Distribution in Liver Membranes of Trimeric and Low-Molecular-Weight G Proteins

As part of a continuing search for endosome-specific proteins, the possibility was examined that basic low-molecular mass polypeptides present in higher amounts in endosomes corresponded to specific categories of GTP-binding (G) proteins. Support emerged from a one-dimensional comparative analysis of the GTP-binding and *ras* oncoproteins present in these fractions. Figure 6 shows that, using radioactive GTP-gamma S to label nitrocellulose blots of the proteins present in various liver subcellular fractions resolved by SDS-polyacrylamide gel electrophoresis, the relative distribution of the 20–30 kD low-molecular-weight GTP-binding proteins was different. A more specific compartmentation of proteins that bind GTP was evident when the fractions resolved by electrophoresis and transferred to nitrocellulose sheets were probed with a panel of antibodies to the *ras* oncoproteins, a class of GTP-binding proteins mainly associated with plasma membranes. Figure 7 shows that two monoclonal antibodies to p21$^{N\text{-}ras}$, as well as antipeptide polyclonal antibody to an amino terminal sequence of p21$^{N\text{-}ras}$ recognized an approximate 28 kDa polypeptide in endosomal fractions, but not in plasma membranes and Golgi fractions, where the major polypeptide recognized generally by the antibodies was a 21–23 kDa product (Ali and Evans, 1990a). These low-molecular-weight G proteins in hepatic endosomes may correspond to *rab* 5 and 7 proteins identified in polarized

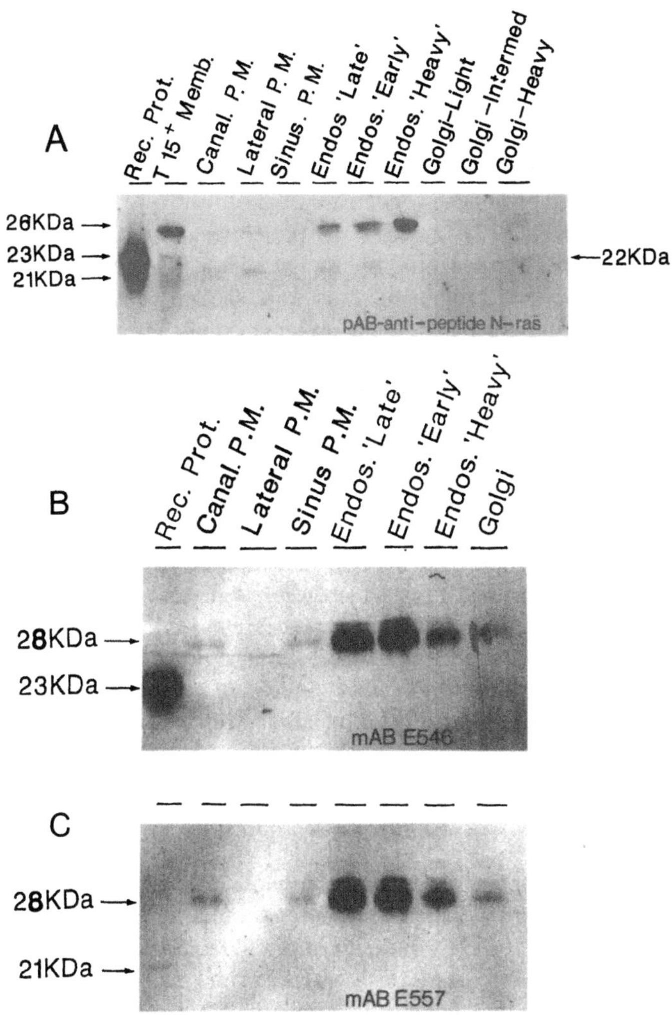

FIGURE 7. Western blots showing the distribution of *ras*-related polypeptides in various liver fractions. Panel A was exposed to an antipeptide (N-terminal sequence) antibody. Panels B and C were exposed to two monoclonal antibodies raised to the ras oncoprotein. Rec. Prot. = recombinant protein purified from E Coli; T15+ Memb = a membrane fraction from cultured cells; other abbreviations are self-explanatory. Equal amounts of protein were applied to each channel.

Madine-Darby Canine Kidney (MDCK) cells (Chavrier *et al.*, 1990). The high concentration of specific categories of low-molecular-weight G proteins in endosomes is in agreement with current views that they are involved in the regulation of trafficking through the endocytic pathways. A role for low-molecular-weight G proteins in membrane trafficking is supported by independent evidence utilizing cell-free systems in which G proteins allow specific membrane-fusion events underlying endocytosis to occur (Mayorga *et al.*, 1989).

The results showing the identification of approximately 26–28 kDa *ras*-type polypeptides in liver endosomes contrast with those showing the distribution of the trimeric G proteins, 21–23 kDa *ras* proteins, and enzymes hydrolyzing inositol triphosphate (IP_3) and tetrakisphosphate (IP_4) (Evans and Ali, 1991). For example, the α and β G-protein subunits were shown, by antibody and toxin labeling, to be widely distributed, being present in sinusoidal plasma membranes and in endosomes, but the highest specific activity (on a semiquantitative immunostaining basis) was recorded in the bile-canalicular plasma membrane fraction (Ali *et al.*, 1989a,b). This distribution of the trimeric G proteins between the hepatic plasma membrane and endosome fractions closely mirrored that of 21–23 kDa *ras*-type polypeptides and the enzymes involved in phosphatidyl inositol metabolism, especially the inositol tris- and tetrakisphosphatases located at the inner (cytoplasmic) surface of the bile-canalicular plasma membrane (Shears *et al.*, 1988). Taken together, the results suggest that these membrane constituents migrate from the sinusoidal plasma membrane via the endocytic compartment to the bile-canalicular plasma membrane domain. The functions of G proteins and enzymes of phosphoinositol metabolism that feature in transmembrane signaling and are located at the bile-canalicular membrane remain obscure at the present time.

3.3. Targeting to Plasma Membrane Domains

The plasma membrane of each of the hepatocyte's functional domains is characterized by a specific protein, glycoprotein, and lipid composition (Evans 1981; Whetton *et al.*, 1983; Evans and Enrich, 1989; Enrich *et al.*, 1990). This raises the question of the nature of the mechanisms ensuring that each surface domain of the hepatocyte maintains its distinctive composition. These questions were first addressed in studies of the kinetics of incorporation of glycoproteins metabolically labeled *in vivo* into the plasma membrane domains, and indicated a delayed arrival of glycoproteins into the bile-canalicular plasma membrane, despite its close proximity to the Golgi apparatus (Evans *et al.*, 1980; Buchsel *et al.*, 1980). These results were interpreted to indicate that an indirect route to the canalicular plasma membrane was followed, implicating the endocytic apparatus (Evans 1981); this interpretation was undoubtedly influenced by the first isolation of intracellular ligand-enclosing endocytic vesicles from rat liver (Debanne *et al.*, 1982). Further biochemical characterization of endosome fractions in liver (Evans and Flint, 1985; Evans and Hardison, 1985; Saermark *et al.*, 1985) together with extensive studies on other tissues and cells have laid the foundation of current knowledge of the membrane trafficking routes underlying the structural and functional polarity of hepatocytes and epithelial cells in general, and indicated that a route for indirect trafficking of glycoproteins from the sinusoidal to the bile-canalicular pole of the hepatocyte existed, especially in relation to the mechanisms and regulation of bile secretion. The application of antibodies generated to specific bile-canalicular glycoproteins allowed the indirect route to be further characterized (Hubbard *et al.*, 1989) and has led to the view that all plasma membrane proteins are initially transported to the sinusoidal (basolateral) domain of the hepatocyte before being removed and then translocated to the canalicular domain. However, further work has identified important exceptions to this generalization, and points to a functional difference between early and late endosomes.

By taking advantage of a major improvement in the methodology for preparing domain-identified liver plasma membranes (Ali *et al.*, 1990), the application of a technique for separating and identifying phosphoinositide-anchored glycoproteins (Lisanti *et al.*, 1990) showed that six such components were confined to the bile-canalicular membrane (Ali and Evans, 1990b). Because no phosphoinositide-anchored glycoproteins were identified in the sinusoidal plasma membrane, it was concluded that the indirect route was unlikely to be operational, a view reinforced when an examination of endosomes identified two of these components exclusively in late endosomes. Thus, in common with many epithelial cell lines (Lisanti *et al.*, 1990; Lisanti and Rodriguez-Boulon, 1991), these results suggest that hepatocytes target phosphoinositide-anchored glycoproteins directly to the apical bile-canalicular plasma membrane domain. Also, the late endosomal compartment emerges as the locus in the hepatocyte where exocytic (i.e., phosphoinositide-anchored glycoproteins arriving directly from the Golgi apparatus) and endocytic (endocytosed glycoproteins arriving indirectly via the blood-sinusoidal plasma-membrane domain) converge as shown in Figure 8. These observations also provide an important trafficking distinction between early and late, endosomes of liver. Two of the phosphoinositide-anchored glycoproteins were identified as 5′-nucleutidase and alkaline phosphatase; both of these ecto-enzymes are highly enriched at the bile-canalicular plasma membrane domain of the hepatocyte (Ali and Evans, 1990b).

3.4. Polydispersity of Liver Endosomal Fractions

The complexity of the endocytic compartment is evident in biochemical approaches toward isolation and characterization. Many investigators have prepared several endosome subfractions from liver homogenates (e.g., Evans and

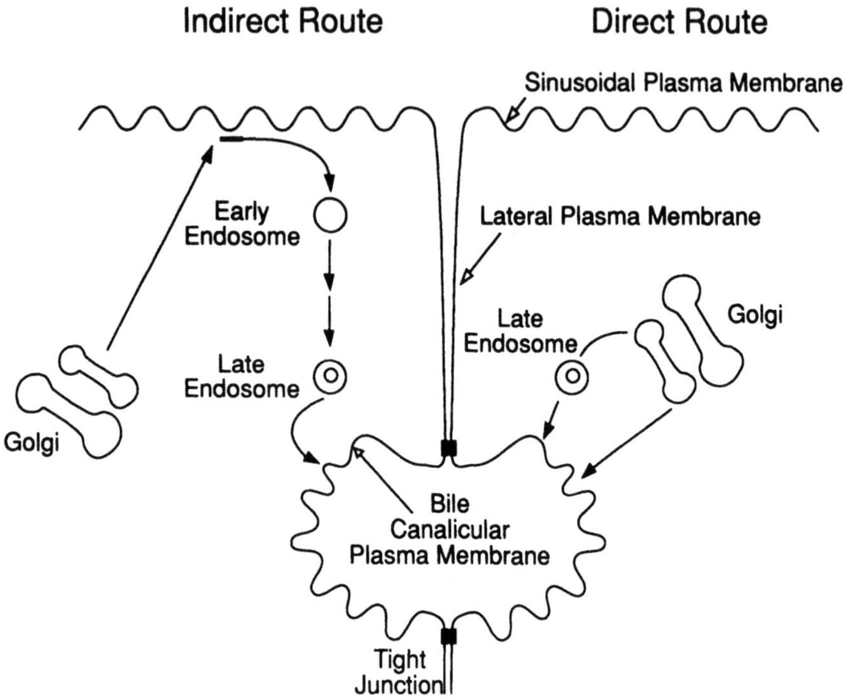

FIGURE 8. Schematic representation of indirect and direct routes of glycoproteins to the hepatocytes' bile-canalicular plasma membrane. The indirect route involves passage from the Golgi to the sinusoidal plasma membrane followed by transcytosis probably involving the early and late endocytic compartment. The direct route followed by phosphoinositide-anchored glycoproteins appears to involve the late endocytic compartment (Ali and Evans, 1990b).

Flint, 1985; Belcher *et al.*, 1987). A third endosomal fraction, obtained by the procedure utilizing Nycodenz gradients (Evans and Flint, 1985), was shown on the basis of the specific binding of the ligands to the vesicles to contain high levels of asialoglycoprotein and insulin receptors. Further work, however, using specific antibodies shows that the distribution of the asialoglycoprotein receptors is more heterogeneous in the fractions designated early, late, and receptor-enriched or heavy endosome fractions (Enrich *et al.*, 1990). In immunocytochemical studies, antibodies generated to the integral membrane proteins of the receptor-enriched heavy endosome fraction located primarily to a region beneath the sinusoidal plasma membrane, suggesting that these endocytic vesicles may be en route back to the plasma membrane, that is, they may be categorized as retroendocytic receptor carriers. (Enrich and Evans, unpublished work). This receptor-enriched endosome fraction contained a 100 kDa protein identified by antibodies to the α subunit of G proteins (Ali *et al.*, 1989a,b) that was absent from all the other liver subcellular fractions examined. Intriguingly, this 100 kDa protein was also specifically recognized by antibodies to transducin and may correspond to a new category of G protein involved in regulating membrane traffic to and from the plasma membrane (Traub *et al.*, 1990, 1991).

4. ASSOCIATION OF ENDOSOMES WITH CYTOSKELETAL COMPONENTS, CALMODULIN, AND EXTRACELLULAR MATRIX

Microtubules have been implicated in the intracellular movement of endocytic structures (Klausner, 1989) thus prompting an examination of the possible association of actin, myosin, tubulin, clathrin, and microtubule-associated proteins (MAPs) with liver endosomes and plasma membranes. All these proteins can be conveniently identified by Western-blotting techniques. For example, they were all detected in lateral plasma membranes, but sinusoidal plasma membranes contained only actin, clathrin, and low amounts of MAPs (mainly the high-molecular-weight MAP). In contrast, these proteins were not detected in endosomes. However, endosomes contained high levels of *tau* proteins (60–65 kDa MAPs) (Enrich *et al.*, unpublished work) with higher levels being detected in late endosomes. These observations, made by analysis of the protein composition of subcellular fractions, are consonant with the view that early steps in endocytosis do not require microtubules, whereas onward trafficking to late endosomes and to lysosomes is microtubule-dependent (Gruenberg and Howell, 1989).

A comparison of the Ca^{2+} and calmodulin-binding proteins of liver plasma membranes and endosomes has also highlighted further differences between these liver-membrane networks. Four Ca^{2+}-binding proteins (180, 92, 90, and 55 kDa) were shown to be mainly located in liver endosomes (Enrich and Evans, 1989b). The identification and purification of a 130-kDa calmodulin-binding protein in plasma membranes that shows amino-terminal sequence homology with a major 115-kDa calmodulin-binding protein in endosomes (Enrich *et al.*, 1988) suggests that trafficking of specific calmodulin-binding proteins from the plasma membrane to the endocytic compartment may be a protease-regulated step.

Finally, the close relationship between the plasma membrane and the endocytic compartment is emphasized in studies of the distribution of extracellular matrix components. These indicate a role for the endocytic compartment in the modulation of fibronectin (Enrich and Evans, 1987), laminin and nidogen (Smedsrod *et al.*, 1989), heparin sulfate (Ishihara *et al.*, 1986), and the integrin receptor (Raub and Kuentzel, 1989). Indeed, it is possible that the liver is the major organ responsible for the clearance of laminin from the circulation (Smedsrod *et al.*, 1989).

5. IMMUNOLOGICAL APPROACHES TOWARD IDENTIFYING ENDOSOME-SPECIFIC PROTEINS

Polyclonal and monoclonal antibodies generated to endosome fractions are being used to identify proteins specific to the endocytic compartment (Ahluwalia *et al.*, 1988; Enrich and Evans, 1987). A major problem to be confronted is that many of the major immunogenic proteins present in endosomal fractions correspond to receptors that recycle extensively between the plasma membrane and the endocytic compartment, a fact well illustrated in the two-dimensional analysis of the proteins in the fractions described above. Indeed, when the integral membrane proteins of liver endosomes (corresponding to those hydrophobic proteins partitioning into Triton X-114 in two-phase systems) were used to generate antibodies in rabbits, the immunodominant antigen detected in endosomes was the 42-kDa asialoglycoprotein receptor subunit (Enrich and Evans, 1987). Antibodies to calmodulin-binding proteins, shown to be mainly associated with liver endosomes (Enrich *et al.*, 1988) identified proteins of 130 and 68 kDa in endosome fractions; immunofluorescence studies of liver sections using antibodies to these calmodulin-binding proteins have confirmed an intracellular location (Enrich *et al.*, unpublished work). These results highlight the candidacy of these calmodulin-binding proteins as endosome-specific constituents.

6. ENDOCYTOSIS IN REGENERATING LIVER

Liver growth, induced by partial hepatectomy of the organ, is a precisely regulated process during which a radical reorganization of metabolism occurs as the hepatocytes become committed to cycles of cell division (Michalopoulos, 1990). In the context of receptor-mediated endocytosis, vesicular trafficking in the cell in general is arrested (Warren, 1985) and recycling of receptors between the plasma membrane and the endocytic compartment of transferrin receptors is greatly diminished (Warren *et al.*, 1984).

Using antibodies generated to the integral membrane proteins of liver endosomes (Enrich and Evans, 1989a), immunocytochemical studies of liver sections have indicated that 24 h after hepatectomy, a general diminution in the intensity of staining of the overall endocytic compartment occurred, but the zone of the endocytic compartment located at the biliary pole was less affected (Enrich *et al.*, 1992). By immunoblotting of endocytic proteins resolved by gel electrophoresis, a diminution in the staining intensity of the asialoglycoprotein receptor (42 kDa) and calmodulin-binding protein (130 kDa), but an increase in the staining of *tau* proteins (60–65 kDa) was observed, especially in early endosomes. These results indicate that, in the regenerative period prior to the triggering of DNA synthesis and the onset of mitotic activity, the early endocytic compartment is the more distressed of the two endocytic compartments. These changes in the expression of specific proteins are, nevertheless, subtle, for no overall differences are seen when the protein profiles of endosomes prepared from normal and regenerating livers are compared.

The complexity of the changes that occur in the endocytic compartment in receptor distribution, recycling, and oligomeric composition during liver regeneration was illustrated when studies were focused on the asialoglycoprotein receptor. This receptor features in the endocytic uptake of galactose and N-acetylgalactosamine-terminating glycoproteins circulating in the blood. The receptor in rat liver is composed of three subunits, RHL-1, -2, and -3, with RHL-1 (42 kDa) being the major receptor subunit, located mainly in the endocytic compartment. The expression of the asialoglycoprotein receptor is down-regulated in livers regenerating after a partial hepatectomy (Bartles *et al.*, 1990) and this reduced expression (assessed by Western blotting using subunit-specific antibodies) was most evident in the RHL-1 subunit present in endosomes (Figure 9). Agents that influence adenosine 3′,5′-monophosphate (cAMP) production, for example β-adrenergic antagonists such as dl-propanolol, significantly diminished the reduction in the level of expression of the RHL-1 receptor, whereas the RHL-2 and -3 subunits were largely unaffected (Enrich *et al.*, 1992). The specific modulation of the asialoglycoprotein RHL-1 receptor subunit was greater in early than in the late endosome fractions prepared from 24 h regenerating livers. Although the reasons for the presence of such substantial levels of asialoglycoprotein receptors in hepatocytes is still unclear, because receptor numbers and capacity are in excess of that necessary to remove damaged glycoproteins in normal liver functioning, the above observations illustrate, in the context of a multimeric membrane receptor, the differential regulation of subunit expression that occurs in the endocytic compartment during liver regeneration. Since the primary amino-acid sequences of the receptor subunits are different (Spiess, 1990) the regulation probably occurs at the transcriptional and translational levels (Stockert and Morrell, 1990; Huber *et al.*, 1986) and possibly involves signaling between endocytic compartment components and the nucleus.

A further model worthy of consideration in which the endocytic traffic is perturbed is the alcoholic liver. Recent studies (Casey *et al.*, 1987; Tuma and Sorrell, 1988) have emphasized that ethanol-induced damage to liver profoundly affects the trafficking of proteins to the various plasma-membrane domains. Experimental models such as the regenerating and neoplastic liver (Eriksson and Anderson, 1992) and the liver in alcohol disease should prove useful in unravel-

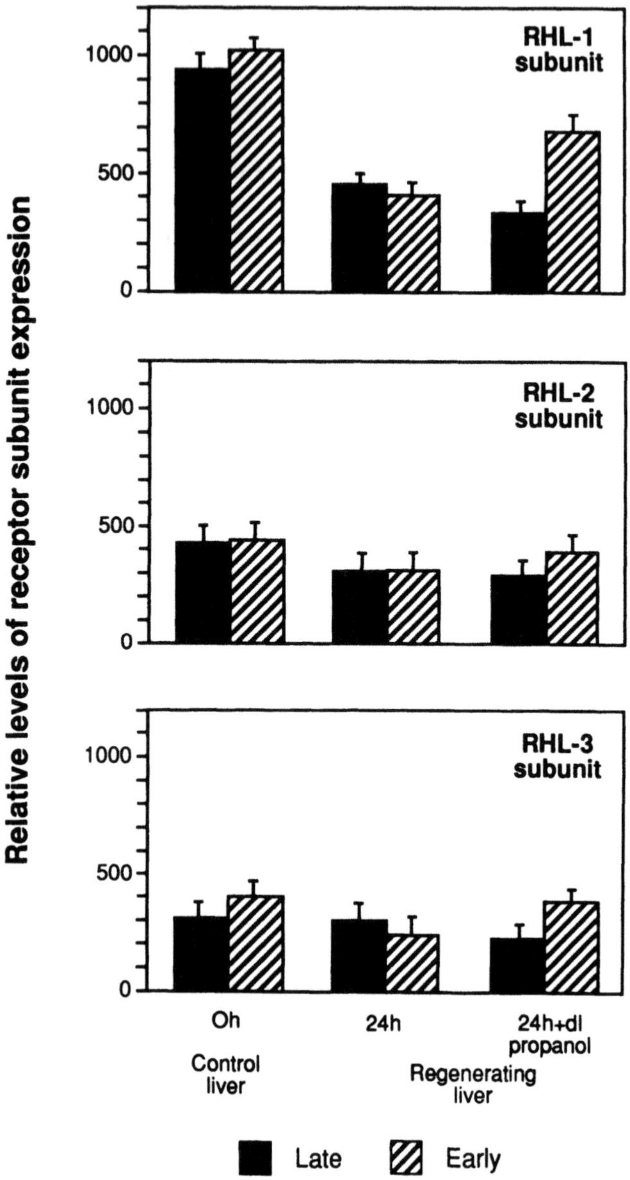

FIGURE 9. Quantitative comparison of the expression of the RHL-1, -2, and -3 asialoglycoprotein receptor subunits determined by densitometry of Western blots of early and late hepatic endosome fractions using antibodies that stain specifically the various receptor subunits. The expression levels in control livers, regenerating livers 24 h after hepatectomy and in regenerating livers treated with dl-propanolol are shown. The figure shows the diminution in the expression of the RHL-1 subunit but recovery of receptor expression in early endosomes, in rats treated with dl-propanolol (Enrich *et al.*, 1992).

ing the intricate trafficking networks that occur in the hepatocyte and that implicate the endocytic compartments described in this review.

ACKNOWLEDGMENTS. We thank the Acciones Integradas scheme for the collaboration that made the writing of this review possible in Barcelona and London.

7. REFERENCES

Ahluwalia, J. P., Doherty, J. J., Troulis, M., Posner, B. I., and Bergeron, J. J. M., 1988, Identification of antigen(s) specific to a "late" endosome fraction, in *Cell-Free Analysis of Membrane Traffic*, (D. J. Morré, K. E. Howell, G. M. W. Cook, and W. H. Evans, eds.) pp. 411–415. Alan R. Liss, Inc., New York.

Ali, N., and Evans, W. H., 1990a, Distribution of polypeptides binding guanosine 5'[γ ^{35}S-thiol]-triphosphate and anti-(ras proteins) antibodies in liver subcellular fractions. Evident for endosome-specific components, *Biochem. J.* **271**:179–183.

Ali, N. and Evans, W. H., 1990b, Priority targeting of glycosyl-phosphatidylinositol anchored proteins to the bile canalicular (apical) plasma membrane of hepatocyte. Involvement of "late" endosomes, *Biochem. J.* **271**:193–199.

Ali, N., Milligan, G., and Evans, W. H., 1989a, Distribution of G-proteins in rat liver plasma membrane domains and endocytic pathways, *Biochem. J.* **261**:905–912.

Ali, N., Milligan, G., and Evans, W. H., 1989b, G-proteins of rat liver membranes. Subcellular compartmentation and disposition in the plasma membrane, *Molec. Cellular Biochem.* **91**:75–84.

Ali, N., Aligué, R., and Evans, W. H., 1990, Highly purified vesicles and plasma membranes isolated from rat liver on Nycodenz gradients, *Biochem J.* **271**:185–192.

Ashcom, J. D., Tiller, S. E., Dickerson, K., Cravens, J. L., Argraves, W. S., and Strickland, D. K., 1990, A human α2-macroglobulin receptor: identification of a 420-kD cell surface glycoprotein specific for the activated conformation of α2-macroglobulin, *J. Cell. Biol.* **110**:1041–1048.

Ashwell, G. and Morell, A. G., 1974, The role of surface carbohydrates in the hepatic recognition and transport of circulating glycoproteins, *Adv. Enzymol.* **41**:99–128.

Baenziger, J. U. and Fiete, D., 1986, Separation of two populations of endocytic vesicles involved in receptor-ligand sorting in rat hepatocytes, *J. Biol. Chem.* **261**:7445–7454.

Bartles, J. R., Zhang, L. Q., Verheyen, E. M., Hospodar, K. S., Nehme, C. L., and Fayos, B. E., 1990, Decreases in the relative concentrations of specific hepatocyte plasma membrane proteins during liver regeneration: down regulation or dilution? *Develop. Biol.* **143**:258–270.

Belcher, J. D., Hamilton, R. L., Brady, S. E., Hornick, C. A., Jaeckle, S., Schneider, W. J., and Havel, R. J., 1987, Isolation and characterization of three endosomal fractions from the liver of estradiol-treated rats, *Proc. Natl. Acad. Sci. USA* **84**:6785–6789.

Bergeron, J. J. M., Resch, L., Rachubinski, R., Patel, B. A., and Posner, B. I., 1983, Effect of colchicine in internalization of prolactin in female rat liver: an *in vivo* radioautographic study, *J. Cell Biol.* **96**:875–886.

Bergeron, J. J. M., Cruz, J., Khan, M. N., and Posner, B. I., 1985, Uptake of insulin and other ligands into receptor-rich endocytic components of target cells: the endosomal apparatus, *Ann. Rev. Physiol.* **47**:383–403.

Blouin, A., Bolender, R. P., and Weigel, E. R., 1977, Distribution of organelles and membranes between hepatocytes and non-hepatocytes in the rat liver parenchyma. A stereological study, *J. Cell Biol.* **72**:441–455.

Buchsel, R., Berger, D., and Reutter, W., 1980, Routes of fucoproteins in plasma membrane domains, *FEBS Lett* **113**:95–98.

Burke, B., Griffiths, G., Reggio, H., Louvard, D., and Warren, G., 1982, A monoclonal antibody against a 135-kD Golgi membrane protein, *EMBO Journal* **1**:1621–1628.

Cantau, B., Keppens, S., de Wulf., and Jand, S., 1980, ^{3}H-vasopressin binding to isolated rat hepatocytes and liver membranes. Regulation by GTP and relation to glycogen phosphorylase activation, *J. Receptor Res.* **1**:137–168.

Carrella, M. and Cooper, A. D., 1979, High affinity binding of chylomicron remnants to rat liver plasma membranes, *Proc. Natl. Acad. Sci. USA* **76**:338–342.

Casey, C. A., Kragskow, S. L., Sorrell, M. F., and Tuma, D. J., 1987, Chronic ethanol administration impairs the binding and endocytosis of asialo-orosomucoid in isolated hepatocytes, *J. Biol. Chem.* **262**:2704–2710.

Chao, Y-S., Jones, A. L., Hradek, G. T., et al., 1981, Autoradiographic localization of the sites of uptake, cellular transport and catabolism of low density lipoproteins in the liver of normal and estrogen-treated rats, *Proc. Natl. Acad. Sci. USA* **78**:597–601.

Chavrier, P., Parton, R. G., Hauri, H. P., Simons, K., and Zerial, M., 1990, Localization of low molecular weight GTP binding proteins to exocytic and endocytic compartments, *Cell* **62**:317–329.

Courtoy, P. J., Quintart, J., and Baudhiun, P., 1984, Shift of equilibrium density induced by 3,3′-diaminobenzidine cytochemistry: a new method for the analysis and purification of peroxidase-containing organelles, *J. Cell Biol.* **98**:870–876.

Courtoy, P. J., Quintart, J., Limet, J. N., de Roe, C., and Baudhiun, P., 1985, Polymeric IgA and galactose-specific pathways in rat hepatocytes: evidences for intracellular ligand sorting, in *Endocytosis* (I. Pastan and M. C. Willingham, eds.) pp. 163–194, Plenum Press, New York and London.

Debanne, M. T., Evans, W. H., Flint, N., and Regoeczi, E., 1982, Receptor-rich intracellular membrane vesicles transporting asialotransferrin and insulin in liver, *Nature* **298**:398–400.

Duncan, J. R. and Kornfeld, S., 1988, Intracellular movement of the two mannose-6-phosphate receptors: return to the Golgi apparatus, *J. Cell Biol.* **106**:617–628.

Dunn, W. A., Connolly, T. P., and Hubbard, A. L., 1986, Receptor-mediated endocytosis of epidermal growth factor by rat hepatocytes: receptor pathway, *J. Cell. Biol.* **102**:24–36.

El-Refau, M. F., Blackmore, P. F., and Exton, J. H., 1979, Evidence for 2 α-adrenergic binding sites in liver plasma membranes. Studies with ^{3}H-epinephrine and ^{3}H-dihydroergocryptone, *J. Biol. Chem.* **254**:4375–4386.

Enrich, C. and Evans, W. H, 1987, Evidence for a role of the hepatic endocytic compartment in the modulation of the extracellular matrix, *Exp. Cell Res.* **173**:99–108.

Enrich, C. and Evans, W. H., 1989a, Antibodies to hepatic endosomes. Identification of two endosome antigens. *Europ. J. Cell Biol.* **48**:344–352.

Enrich, C. and Evans, W. H., 1989b, Calcium and calmodulin-binding proteins of liver endosomal and plasma membrane fractions, in *Biochemical Approaches to Cellular Calcium. Methodological Surveys in Biochemistry and Analysis, Vol. 19,* (E. Reid, G. M. W. Cook, and J. P. Luzio, eds.) pp. 53–58. The Royal Society of Chemistry.

Enrich, C., Bachs, O., and Evans, W. H., 1988, A 115 kDa calmodulin-binding protein is located in rat liver endosome fractions, *Biochem. J.* **255**:999–1005.

Enrich, C., Jones, G. R., Bordas, J., and Evans, W. H., 1989, A 220 kDa polypeptide immunolocalised to epithelial tight junctions is associated with brain clathrin preparations, *Eur. J. Cell Biol.* **50**:304–312.

Enrich, C., Tabona, P., and Evans, W. H., 1990, A two-dimensional electrophoretic analysis of the proteins and glycoproteins of liver plasma membrane domains and endosomes. Implications for endocytosis and transcytosis, *Biochem. J.* **271**:171–178.

Enrich, C., Vergés, M., and Evans, W. H., 1992, Differential expression of asialoglycoprotein receptor subunits in the endocytic compartment during liver regeneration, *J. Cell. Physiol.* **150**:344–352.

Eriksson, L. C. and Andersson, G. N., 1992, Membrane biochemistry and chemical hepatocarcinogenesis, *Critical Revs. in Biochem. and Molec. Biol.* **27**:1–55.

Evans, W. H., 1981, Membrane traffic at the hepatocyte's sinusoidal and canalicular surface domains, *Hepatology*. **5**:452–457.

Evans, W. H. and Ali, N., 1991, Subcellular distribution of trimeric and low molecular weight G-proteins in liver, in *Cell Signalling: Experimental Strategies* (E. Reid, G. M. W. Cook, and J. P. Luzio, eds.) pp. 423–428. Royal Society of Chemistry, London.

Evans, W. H. and Enrich, C., 1989, Liver plasma membrane domains and endocytic trafficking, *Biochem. Soc. Trans.* **17**:619–622.

Evans, W. H. and Flint, N., 1985, Subfractionation of hepatic endosomes in Nycodenz gradients and by free flow electrophoresis. Separation of ligand-transporting and receptor-enriched membranes, *Biochem. J.* **232**:25–32.

Evans, W. H. and Hardison, W. G. M., 1985, Phospholipid, cholesterol, polypeptide and glycoprotein composition of hepatic endosome subfractions. *Biochem. J.* **232**:33–36.

Evans, W. H., Flint, N., and Vischer, P., 1980, Biogenesis of membrane domains. Incorporation of $[^3H]$ fucose into plasma membrane and Golgi apparatus glycoproteins, *Biochem. J.* **192**:903–910.

Fisher, M. M., Nagy, B., Bazin, H., and Underdown, B. J., 1979, Biliary transport of IgA: role of secretory component, *Proc. Natl. Acad. Sci. USA*. **76**:2008–2012.

Forsberg, E., Paulsson, M., Timpl, R., and Johansson, S., 1990, Characterization of a laminin receptor on rat hepatocytes, *J. Biol. Chem.* **265**:6376–6381.

Fouchereau-Peron, M., Broer, M., and Rosselin, G., 1980, Growth hormone and insulin binding to isolated hepatocytes in the genetically dwarf mouse, *Biochem. Biophys. Acta* **631**:451–462.

Frommel, D. and Rachman, F., 1979, Receptor for the Fc portion of IgG on the plasma membrane of hepatocyte. *Ann. Immunol. (Paris)* **130c**:553–560.

Fuchs, R., Male, P., and Mellman, I., 1989, Acidification and ion permeabilities of highly purified rat liver endosomes, *J. Biol. Chem.* **264**:2212–2220.

Geisow, M. J. and Evans, W. H., 1984, pH in the endosome, *Exp. Cell Res.* **150**:36–46.

Geuze, H. J., Slot, J. W., Strous, G. J. A. M., Lodish, H. F., and Schwartz, A. L., 1983, Intracellular site of asialoglycoprotein receptor-ligand uncoupling: double-label immunoelectron microscopy during receptor-mediated endocytosis, *Cell* **32**:277–287.

Geuze, H. J., Slot, J. W., Strous, G. J. A. M., Peppard, J., von Figura, K., Hasilik, A., and Schwartz, A. L., 1984, Intracellular receptor sorting during endocytosis: comparative immunoelectron microscopy of multiple receptors in rat liver. *Cell* **37**:195–204.

Geuze, H. J., Slot, J. W., and Schwartz, A. L., 1987, Membranes of sorting organelles display lateral heterogenicity in receptor distribution, *J. Cell Biol.* **104**:1715–1723.

Gharbi, J. and Torresani, J., 1979, High affinity thyroxine binding to purified rat liver plasma membranes, *Biochem. Biophys. Res. Commun.* **88**:170–177.

Goldenberg, H., Seelos, C., Chatwani, S., and Pumm, R., 1990, Uptake and endocytic pathways of transferrin and iron in perfused rat liver. *Biochim. Biophys. Acta.* **1067**:145–152.

Gruenberg, J. and Howell, K. E., 1989, Membrane traffic in endocytosis: insights from cell-free systems. *Ann. Rev. Cell Biol.* **5**:453–481.

Hadjiivanova, N., Flint, N., Evans, W. H., Dix, C., and Cooke, B. A., 1984, Endocytosis of β-adrenergic ligands by rat liver, *Biochem. J.* **222**:749–754.

Hopf, U., Schaefer, H. E., Hess, G. and Meyer Zum Büschenfelde, K. H., 1981, *In vivo* uptake of immune complexes by parenchymal and nonparenchymal liver cells in mice, *Gastroenterology* **80**:250–259.

Hopkins, C. R., Gibson, A., Shipman, M., and Miller, K., 1990, Movement of internalized ligand-receptor complexes along a continuous endosomal reticulum. *Nature* **346:**335–339.

Hubbard, A. L., Stieger, B., and Bartles, J. R., 1989, Biogenesis of endogenous plasma membrane proteins in epithelial cells, *Ann. Rev. Physiol.* **51:**755–770.

Huber, B. E., Glowinski, I. B., and Thorgeisson, S. S., 1986, Transcriptional and post-transcriptional regulation of the asialoglycoprotein receptor in normal and neoplastic rat liver, *J. Biol. Chem.* **261:**12400–12407.

Ishihara, M., Fedarko, N. S., and Conrad, H. E., 1986, Transport of heparan sulphate into nuclei of hepatocytes, *J. Biol. Chem.* **261:**13575–13580.

Jäckle, S., Runquist, E., Brady, S., Hamilton, R. L., and Havel, R. J., 1991a, Isolation and characterisation of three endosomal fractions from the liver of normal rats after lipoprotein loading, *J. Lipid Res.* **32:**485–498.

Jäckle, S., Runquist, E. A., Miranda-Brody, S., and Havel, R. J., 1991b, Trafficking of the epidermal growth factor receptor and transferrin in three hepatocyte endosomal fractions, *J. Biol. Chem.* **266:**1396–1402.

Johansson, S., Forsberg, E., and Lundgren, B., 1987, Comparison of fibronectin receptors from rat hepatocytes and fibroblasts, *J. Biol. Chem.* **262:**7819–7824.

Jones, A. L. and Burwen, S. J., 1985, Hepatic receptors and their ligands: problems of intracellular sorting and vectorial movement, *Sem. Liver Dis.* **5:**136–146.

Kino, K., Tusnoo, H., Higa, Y., Takami, M., Hamagushi, H., and Nakajima, H., 1980, Hemoglobin-haptoglobin receptor in rat liver plasma membranes. *J. Biol. Chem.* **255:**9616–9620.

Klausner, R. D., 1989, Sorting and traffic in the central vacuolar system, *Cell* **57:**703–706.

Kouyoumdjian, M., Borges, D. R., Prado, E. S., and Prado, J. L., 1989, Identification of receptors in the liver that mediate endocytosis of circulating tissue kallikreins. *Biochim. Biophys. Acta.* **980:**299–304.

Levy, D. and von Dieppe, P., 1989, Identification of bile acid transport protein in hepatocyte sinusoidal plasma membranes, *Methods Enzymol.* **174:**25–31.

Lippincott-Schwartz, J. and Fambrough, D. M., 1987, Cycling of the integral glycoprotein LEP100, between plasma membrane and lysosomes: kinetic and morphological analysis, *Cell* **49:**669–677.

Lisanti, M. P., Rodriguez-Boulan, E., and Saltiel, A. R., 1990, Emerging functional roles for GP-1 membrane protein anchor, *J. Membrane Biol.* **113:**155–167.

Lisanti, M. P. and Rodriguez-Boulan, E., 1991, Polarized sorting of GP-1 linked proteins in epithelial and membrane microdomains, *Cell Biol. Int. Rep.* **15:**1023–1049.

Luzio, J. P. and Stanley, K. K., 1983, The isolation of endosome-derived vesicles from rat hepatocytes, *Biochem. J.* **216:**27–36.

Marsh, M., Griffiths, G., Dean, G. E., Mellman, I., and Helenius, A., 1986, Three-dimensional structure of endosomes in BHK-21 cells. *Proc. Natl. Acad. Sci. USA* **83:**2899–2903.

Mayorga, L. S., Diaz, R., Colombo, M. I., and Stahl, P. D., 1989, GTPγS stimulation of endosome fusion suggests a role for a GTP-binding protein in the priming of vesicles before fusion, *Cell Regulation* **1:**113–124.

Mead, J. E. and Fausto, N., 1989, Transforming growth factor α may be a physiological regulator of liver regeneration by means of an autocrine mechanism, *Proc. Natl. Acad. Sci. USA* **86:**1558–1562.

Michalopoulos, G. K., 1990, Liver regeneration: molecular mechanisms of growth control, *FASEB Journal* **4:**176–187.

Mullock, B. M., Jones, R. S., and Hinton, R. H., 1980, Movement of endocytic shuttle vesicles from the sinusoidal to the bile canalicular face of hepatocytes does not depend on occupation of receptor sites, *FEBS. Lett.* **113:**201–205.

Nakamura, T., Nishizawa, T., Hagiya, M., Seki, T., Shimonishi, M., Sugimura, A., Tashiro, K., and Shimizu, S., 1989, Molecular cloning and expression of human hepatocyte growth factor, *Nature* **342:**440–443.

Neuman, W. F. and Schneider, N., 1980, The parathyroid hormone-sensitive adenylate cyclase system in plasma membranes of rat liver, *Endocrinology* **107:**2082–2087.

Nexo, E. and Hollenberg, M. D., 1980, Characterization of the particulate and soluble receptor for transcobalamin II from human placenta and rabbit liver. *Biochim. Biophys. Acta.* **628:**190–200.

O'Farrell, P. H., 1975, High resolution two-dimensional electrophoresis of proteins, *J. Biol. Chem.* **250:**4407–4021.

Oka, Y and Czech, M. P., 1986, The type II insulin-like growth factor receptor is internalized and recycles in the absence of ligand, *J. Biol. Chem.* **261:**9090–9093.

Okamura, N. and Terayama, H., 1977, Prostaglandin receptor-adenylate cyclase system in plasma membranes of rat liver and ascites hepatomas, and the effect of GTP upon it, *Biochim. Biophys. Acta.* **465:**54–67.

Owensby, D. A., Sobel, B. E., and Schwartz, A. L., 1988, Receptor-mediated endocytosis of tissue-type plasminogen activator by the human hepatoma cell line Hep G2, *J. Biol. Chem.* **263:**10587–10594.

Pietras, R. J. and Szego, C. M., 1980, Partial purification and characterization of estrogen receptors in subfractions and hepatocyte plasma membranes. *Biochem. J.* **191:**743–760.

Quintart, J., Courtoy, P. J., and Baudhiun, P., 1984, Receptor-mediated endocytosis in rat liver, and enzymic characterization of low density organelles involved in the uptake of galactose-exposing proteins, *J. Cell Biol.* **98:**877–884.

Quintart, J., Baudhuin, P., and Courtoy, P. J., 1989, Marker enzymes in rat liver vesicles involved in transcellular transport, *Eur. J. Biochem.* **184:**567–574.

Raub, T. J. and Kuentzel, S. L., 1989, Kinetic of morphological evidence for endocytosis of mammalian cell integrin receptors by using an anti-fibronectin receptor β-subunit monoclonal antibody, *Exp. Cell Res.* **184:**407–426.

Regoeczi, E., Chindemi, P. A., Debanne, M. T., and Charlwood, P. A., 1982, Partial resialylaton of human asialoglycoprotein type 3 in the rat. *Proc. Natl. Acad. Sci.* **79:**2226–2230.

Renaud, G., Foliot, A., and Infante, R., 1978, Increased uptake of fatty acids by the isolated rat liver after raising the fatty acid binding protein concentration with clofibrate, *Biochem. Biophys. Res. Commun.* **80:**327–334.

Runquist, E. A. and Havel, R. J., 1991, Acid hydrolase in early and late endosome fractions from rat liver, *J. Biol. Chem.* **266:**22557–22563.

Saermark, T., Flint, N., and Evans, W. H., 1985, Hepatic endosome fractions contain an ATP-driven proton pump, *Biochem. J.* **225:**51–58.

Shears, S. B., Evans, W. H., Kirk, C. J., and Michell, R. H., 1988, Preferential localization of rat liver D-*myo*-inositol 1,4,5-triphosphate/1,3,4,5-tetrakisphosphate 5-phosphatase in bile-canalicular plasma membrane and "late" endosomal vesicles, *Biochem. J.* **256:**363–369.

Smedsrod, B., Paulsson, M., and Johannsson, S., 1989, Uptake and degradation *in vivo* and *in vitro* of laminin and nidogen by rat liver cells, *Biochem. J.* **261:**37–42.

Snider, M. D. and Rogers, O. C., 1985, Intracellular movement of cell surface receptors after endocytosis. resialylation of asialotransferrin receptor in human erythroleukemia cells, *J. Cell. Biol.* **100:**826–834.

Sperling, M. A., Ganguli, S., Voina, S., Kaptein, V. E., and Nicoloff, J. T., 1980, Modulation by thyroid status of the glucagon receptor adenyl cyclase system in rat liver plasma membrane, *Endocrinology* **107:**684–690.

Spiess, M., 1990, The asialoglycoprotein receptor: a model for endocytic transport receptors, *Biochemistry* **29:**10009–10018.

Steer, C. J. and Clarenburg, R., 1979, Unique distribution of glycoprotein receptors on parenchymal and sinusoidal cells of rat liver, *J. Biol. Chem.* **254:**4457–4461.

Stockert, R. J. and Morell, A. G., 1990, Second messenger modulation of asialoglycoprotein receptor, *J. Biol. Chem.* **265:**1841–1846.

Strong, P. N. and Evans, W. H., 1987, Receptor-mediated endocytosis of apamin by liver cells, *Eur. J. Biochem.* **163:**267–273.

Tauber, R., Park, C-S., and Reuter, W., 1983, Structural carbohydrates from liver and hepatoma, in *Structural Carbohydrates in the Liver: Falk Symposium 34,* (H. Popper, W. Reutter, F. Gudat, and E. Kotgen, eds.) pp. 333–347. MTP Press, Lancaster, United Kingdom.

Thomas, P., 1980, Studies on the mechanisms of biliary excretion of circulating glycoproteins, *Biochem. J.* **192:**837–843.

Traub, L. M., Evans, W. H., and Sagi-Eisenberg, R., 1990, A novel 100dKa protein, located to receptor enriched endosomes, is immunologically related to the signal transducing guanine-nucleotide-binding proteins G_t and G_i, *Biochem J.* **272:**453–458.

Traub, L. M., Shai, E., and Sagi-Eisenberg, R., 1991, Characterization of the interaction between p100, a novel G-protein-related protein, and rat liver endosomes. *Biochem. J.* **280:**171–178.

Tuma, D. J. and Sorrell, M. F., 1988, Effects of ethanol on protein trafficking in the liver, *Sem. Liver Disease* **8:**69–80.

Warren, G., 1985, Membrane traffic and organelle division, *Trends in Biochem. Sci.* **11:**439–443.

Warren, G., Davust, J., and Cockroft, A., 1984, Recycling of transferrin receptors in A431 cells is inhibited during mitosis, *EMBO Journal* **3:**2217–2225.

Whetton, A. D., Hauslay, M. D., Dodd, N. J. F., and Evans. W. H., 1983, The lipid fluidity of plasma membrane, endocytic, and Golgi apparatus subfractions isolated from rat liver, *Biochem. J.* **214:**851–854.

Willingham, M. C., and Pastan, I., 1985, An Atlas of Immunofluorescence in Cultured Cells, Academic Press, New York.

Wisse, E., 1977, Ultrastructure and function of Kupffer cells and other sinusoidal cells in the liver, in *Kupffer Cells and other Liver Sinusoidal Cells* (E. Wisse and D. L. Knook, eds.) pp. 33–60, Elsevier/North-Holland Biomedical Press, Amsterdam.

Wolkoff, A. W. and Chung, C. T., 1980, Identification, purification and partial characterization of an organic anion binding protein from rat liver cell plasma membrane, *J. Clin. Invest.* **65:**1152–1161.

Chapter 8

Potential Role of Endosomes in Transmembrane Signaling

M. N. Khan, W. H. Lai, J. W. Burgess, B. I. Posner, and J. J. M. Bergeron

1. INTRODUCTION

After ligands bind to their surface receptors on target cells, the ligand-receptor complexes undergo aggregation and internalization. This process of receptor-mediated endocytosis leads to the concentration of ligand-receptor complexes in a heterogeneous population of tubulovesicular structures that has been distinguished from plasma membrane, lysosomes, and Golgi elements. These structures, now called endosomes, are not clearly characterized by a specific enzyme or protein marker but rather by their unique property to concentrate internalized ligand-receptor complexes. It is widely appreciated that, within endosomes, ligand-receptor complexes are directed toward dissociation and receptor recycling or toward transcytosis, with ligand undergoing degradation in the former situation or transcellular passage in the latter. It has been suggested that, in addition to this sorting function, endosomes play a role in signal transduction.

M. N. Khan, W. H. Lai, J. W. Burgess, and J. J. M. Bergeron Department of Anatomy, McGill University, Montreal, Canada H3A 2B2. **B. I. Posner** Department of Medicine, McGill University, Montreal, Canada H3A 2B2. *Current address for M. N. K.:* Phoenix International Life Sciences, Montreal, Canada H4R 2N6.

Subcellular Biochemistry, Volume 19: Endocytic Components: Identification and Characterization, edited by J. J. M. Bergeron and J. R. Harris. Plenum Press, New York, 1993.

Thus, internalized hormone-receptor complexes may, within endosomes, be brought into contact with an intracellular effector system whose activities are subsequently modulated. Furthermore, it is possible that a biological property of the receptor (e.g., kinase activity) is altered in endosomes, permitting an important aspect of cellular signaling to transpire. In this article, we review our knowledge of endosomes with an emphasis on their potential roles in transmembrane signaling.

2. ENDOCYTOSIS

Many physiologically important molecules bind to their receptors on target cells and are internalized into these cells by receptor-mediated endocytosis. Biologically important molecules, taken up by this process, include plasma (nutrient) transport proteins [such as low-density lipoprotein (LDL), transferrin, and transcobalamin II], polypeptide hormones (viz. insulin and prolactin), peptide growth factors [viz. epidermal growth factor (EGF)], asialoglycoproteins, and lysosomal enzyme precursors. Many viruses and bacterial toxins also enter cells following binding to specific cellular receptors (reviewed in Posner *et al.*, 1982a; Helenius *et al.*, 1983; Bergeron *et al.*, 1985; Wileman *et al.*, 1985). Several years ago, endocytosis was virtually synonymous with the incorporation of newly internalized material into lysosomes (DeDuve and Wattiaux, 1966). One of the first observations that proteins entering the cell by the endocytic process pass through prelysosomal vesicles before reaching secondary lysosomes was made in the early 1960s (Straus, 1964). In studies of the endocytic uptake of intravenously administered horseradish peroxidase by rat-kidney epithelium, internalized peroxidase was stained with a blue benzidine stain, and acid phosphatase–positive granules (lysosomes) were stained with a red naphthol dye. At early time points the two enzymes were in separate vesicles; the peroxidase-containing vesicles (which Straus termed *phagosomes*) in the peripheral cytoplasm, and the lysosomes in the perinuclear area. Later, phagosomes and lysosomes combined, as judged by the purple color reaction for the two enzymes in the same granules. Straus (1964) proposed that the function of these prelysosomal vesicles was to deliver newly internalized material to the secondary lysosomes. Modern techniques involving the use of ^{125}I-labeled ligands (Bergerson *et al.*, 1979, 1983; Josefsberg *et al.*, 1979; Khan *et al.*, 1981, 1982; Posner *et al.*, 1980; Lai *et al.*, 1989a), ligand molecules coupled to horseradish peroxidase (Wall *et al.*, 1980; Willingham *et al.*, 1981; Kay *et al.*, 1984), ferritin (Wall *et al.*, 1980), or colloidal gold (Maxfield *et al.*, 1978; Handley *et al.*, 1981; Geuze *et al.*, 1983b) have greatly enhanced our understanding of the pathways taken by ligand as it passes through the cell. These labeled tracers have been used in conjunction with electron microscopy and improved cell-fractionation techniques to localize the ligand in various cellular organelles (cf. Bergeron *et al.*, 1985).

3. ORGANELLES INVOLVED IN RECEPTOR-MEDIATED ENDOCYTOSIS

3.1. Coated Pits

The distribution of receptors on the cell surface has been extensively studied by electron microscopy. Generally, plasma-membrane receptors, including those for insulin and other peptide hormones, tend to be diffusely spread over the cell surface (Bergeron *et al.*, 1980, 1983, 1985; Pastan and Willingham, 1981; Lai *et al.*, 1989a). In contrast, the LDL receptor tends to precluster in coated pits of fibroblasts (Brown and Goldstein, 1986). Roth and Porter (1964) observed that specialized regions of the plasma membrane consisting of pits coated with clathrin bristles were involved in endocytosis. There is evidence that some receptors, for example, the transferrin receptor, may form large aggregates prior to their concentration in coated pits (Cheng, 1986). Coated pits function as selective membrane filters, concentrating certain membrane proteins within their interiors while excluding others (Bretscher *et al.*, 1980). Morphological evidence suggests that a number of ligands can enter the cell via the same coated pit (Maxfield *et al.*, 1978; Carpentier *et al.*, 1982; Via *et al.*, 1982; Geuze *et al.*, 1982). The studies of Roth and Porter (1964) have shown that coated pits pinch off from the surface membrane to form coated vesicles. Soon after entering the cell, clathrin coats are shed, forming uncoated vesicles that carry extracellular fluid and some surface membrane components into the cell (Anderson *et al.*, 1977; Wall *et al.*, 1980; Wall and Hubbard, 1981; Willingham *et al.*, 1981; Carpentier *et al.*, 1982; Geuze *et al.*, 1983b).

The role of coated pits in the receptor-mediated internalization of LDL has been well documented (cf. Brown and Goldstein, 1986). Thus, an LDL receptor mutated at tyrosine 807 in the cytoplasmic tail cannot migrate to coated pits and cannot internalize (see below). However, attempts to document the participation of coated pits in the *in vivo* internalization of insulin, prolactin, and other peptide hormones have been disappointing perhaps because the early clustering event in coated pits occurred too quickly and transiently to be observed in these studies (Bergeron *et al.*, 1979, 1980, 1983; Smith and Jarett, 1983). Subcellular fractionation studies have suggested a small but significant accumulation of insulin in a coated vesicle fraction prepared from liver homogenates (Pilch *et al.*, 1983). Electron-microscopic radioautography of [^{125}I]insulin binding to 3T3-L1 adipocytes has also suggested that ligand localizes preferentially to microvilli and coated pits prior to internalization (Fan *et al.*, 1982).

3.2. Intracellular Vesicles

Bergeron *et al.* (1973) demonstrated the binding of insulin to Golgi fractions isolated from rat liver. Subsequently, Pricer and Ashwell (1976) demonstrated the binding of asialo-orosomucoid to Golgi and lysosomal elements from the same tissue source. It is now apparent that a variety of cells have intracellular pools of ligand-specific binding sites (Bergeron *et al.*, 1985; Posner *et al.*, 1985). Early studies by the *in vivo* radioautographic technique defined the internalization of [^{125}I]insulin into a heterogenous series of vesicular and tubular components often filled with lipoprotein particles and associated with stacked Golgi saccules near centrioles (Bergeron *et al.*, 1979). These structures were not lysosomal as defined by acid phosphatase cytochemistry but their intracellular location and lipoprotein content at first indicated that they might be Golgi vacuoles (Bergeron *et al.*, 1979). Our early studies, using cell-fractionation techniques, suggested that insulin and prolactin receptors are concentrated in lipoprotein-containing vesicles in rat-liver Golgi fractions (cf. Posner *et al.*, 1981). In attempts to distinguish between internalization into these structures and true lysosomes we showed that purified lysosomes did not accumulate internalized insulin even when degradation was blocked by chloroquine. Furthermore, ligand uptake into Tritosome preparations (Triton WR 1339–filled lysosomes) was accounted for completely by ligand internalization into contaminating lipoprotein-filled structures (Posner *et al.*, 1982b).

Since luminal lipoprotein particles were considered morphological markers for the Golgi compartment (Ehrenreich *et al.*, 1973) we suggested at that time that internalization was into Golgi components (see also Bergeron *et al.*, 1985; Posner *et al.*, 1982a, 1985). Subsequently, we demonstrated that Golgi fractions of rat liver are heterogenous and resolvable into Golgi and non-Golgi components (Khan *et al.*, 1982; Kay *et al.*, 1984; R. J. Khan *et al.*, 1985). These studies suggested that peptide hormones are largely internalized into unique non-Golgi, nonlysosomal endocytic structures. On the basis of such studies and detailed electron-microscopic observations (Bergeron *et al.*, 1983) we were able to rule out any appreciable internalization of ligand (or receptor) into components of the Golgi apparatus. Concomitant studies following the uptake of α_2-macroglobulin by fibroblasts (Pastan and Willingham, 1981) and asialoglycoproteins by hepatocytes (Wall *et al.*, 1980) have provided evidence that incoming vesicles harboring ligand-receptor complexes fuse with a complicated arrangement of large smooth-surfaced vesicles and tubular structures in the Golgi region of the cell. However, the fusion between endocytic vesicles and lysosomes is inhibited below 20°C without inhibiting the endocytosis of [^{125}I]asialofetuin although endocytosis was inhibited when the temperature was below 10°C (Dunn *et al.*, 1980). The lipoprotein content of these vesicles is now considered characteristic of liver endocytic components and it has been thought to be largely derived from endocytosed lipoproteins (Bergeron *et al.*, 1985).

3.3. Endosomes: Terminology and Definition

The discovery of new intracellular structures involved in the internalization, sorting, and transportation of receptors and ligands by a number of investigators has posed problems of terminology and definition. Table I illustrates the plethora of terms coined for these structures by various investigators. The term *receptosome* has been used to distinguish the organelle from vesicles formed during fluid-phase pinocytosis and phagocytosis (Willingham and Pastan, 1980; Pastan and Willingham, 1981). This terminology implies the presence of receptors within the organelle. However, it is possible that some vesicles in the endocytic pathway do not contain receptors. In our own studies we described *unique* vesicles (Khan *et al.*, 1981). These are intracellular elements intermediate in density between Golgi and lysosomes that are greatly enriched in receptors for insulin and prolactin. Geuze *et al.* (1983a) used the terminology *compartment for the uncoupling of receptor and ligand* (CURL) to describe endosomal structures. This terminology may exclude some endocytic elements, for example, those

Table I
Terminology for Structures Involved in Internalization

Terms	References
Compartment for uncoupling receptor and ligand (CURL)	Geuze *et al.*, 1983b
Diacytosomes	Debanne *et al.*, 1982
Endosomal reticulum	Hopkins *et al.*, 1990
Endosomes	Helenius *et al.*, 1983
Golgi fractions (elements)	Posner *et al.*, 1981
Golgi vacuoles	Bergeron *et al.*, 1979
Golgi vesicles	Bergeron *et al.*, 1979; Josefsberg *et al.*, 1979
Ligandosomes	Smith and Peters, 1982
Lipoprotein-containing vesicles	Posner *et al.*, 1981
Low density vesicles	Courtoy *et al.*, 1984
Lysosomes	Gorden *et al.*, 1980
Multivesicular bodies	McKanna *et al.*, 1979; Walsh *et al.*, 1984
Receptosomes	Willingham and Pastan, 1980
Tubular vesicles	Wall *et al.*, 1980
"Unique" vesicles	Khan *et al.*, 1981

involved in the transcytotic movement of ligand wherein dissociation of ligand from its receptor does not occur. As first defined by Stochem and Wohlfarth-Botterman (1969) and more recently by Helenius *et al.* (1983), *endosome* is a convenient term to cover all of the intracellular nonlysosomal components involved in the uptake of exogenous substances into cells. There is now a growing catalog of endosomes with names assigned on the basis of position of the organelle within the cell, for example, peripheral versus juxtanuclear endosomes (Hopkins, 1983, 1985). Alternatively, endosomes have been named on the basis of their morphology, such as multivesicular or tubulovesicular endosomes (Harding *et al.*, 1984). Another popular terminology is based on the chronological appearance of ligand in endosome populations. Early endosomes, which are probably analogous to peripheral endosomes, may mediate such events as receptor recycling (Schmid *et al.*, 1988), which occur shortly after internalization. In contrast, the delivery of dissociated ligand to lysosomes is slow relative to the rate of receptor recycling and is likely to require the participation of a late endosome population (Wolkoff *et al.*, 1984; Wall and Hubbard, 1984; Dunn *et al.*, 1986).

Indeed, fractionation studies were the first to indicate distinct nonlysosomal compartments participating sequentially in endocytosis. Thus, using Percoll gradient centrifugation, we were able to resolve Golgi-enriched fractions into low density (d = 1.040–1.050) vesicles that cosedimented with galactosyltransferase, the marker enzyme for Golgi elements, and high density (d = 1.053–1.064) vesicles devoid of this enzyme (Khan *et al.*, 1982). Kinetics of [^{125}I]insulin uptake into these vesicles revealed that the early uptake (5 min postinjection) was largely into low-density vesicles followed by uptake (10 min postinjection) predominantly into high-density vesicles. The integrity of internalized ligand was significantly retained in these vesicles (Khan *et al.*, 1982). Later, we observed that chloroquine effected significant and prolonged accumulation of radiolabeled insulin into high-density vesicles (Khan *et al.*, 1985). In contrast, the distribution of internalized prolactin that also accumulated sequentially in low- and then high-density vesicles of Golgi-enriched fractions was minimally influenced by chloroquine treatment (R. J. Khan *et al.*, 1985). In subsequent studies differential and analytical centrifugation was employed to demonstrate that early and late endosomal components could be found in the rat-liver microsomes and large granule (ML) fractions respectively (Bergeron *et al.*, 1986). Thus, late endosomal structures of relatively small (Golgi-enriched fractions) and relatively larger (ML fraction) size could be defined. Of interest was our observation that chloroquine treatment augmented the accumulation of prolactin (as well as insulin) in the larger late endosomes of the ML fraction (Bergeron *et al.*, 1986; Khan *et al.*, 1986a). As noted above this is in contrast to the failure of chloroquine to augment prolactin uptake into the late endosomes of Golgi-enriched fractions (R. J. Khan *et al.*, 1985). Thus, a compartment within the late endosome population (viz. multivesicular bodies) can be distinguished where neutralization of the vesicular acidic milieu with the acidotropic agent chloroquine leads to prolactin accumulation. This suggests that prolactin degradation occurs within or proximate to these elements.

3.4. Endosomes: Distinct or Temporally Continuous Component

Most groups studying endocytosis have identified intracellular components consisting of small coated and uncoated vesicles, tubulovesicular components as well as multivesicular bodies that accumulate ligand and/or receptor and/or ligand-receptor complexes in a temporally defined order. The clearly different morphological features of the components as well as their operational subcompartmentation into, for example, *in vitro* fusion-competent endosomes (Gruenberg and Howell, 1987; Braell, 1987; Diaz *et al.*, 1988, 1989) microtubule disruption-sensitive endosomes (Gruenberg and Howell, 1987; Hopkins *et al.*, 1990) has implied distinct compartments where communication among the barriers separating the components could be via vesicular carriers or a poorly defined transformation of one component into another (Helenius *et al.*, 1983). Very recent observations suggest, however, that both views may be incorrect. By adroit use of immunofluorescence confocal microscopy in living cells, Hopkins and colleagues (1990) have documented (in Hep-2 cells) a transient and continuous membranous network encompassing all the subcompartments of the endosomal apparatus. Although these studies do not vitiate the essence of any of the biochemical observations, that is, endosomal functions, they do change profoundly our view of endocytic elements and limit considerably hypotheses concerning their biogenesis.

4. ISOLATION AND PURIFICATION OF ENDOSOMES

At present there is no distinct enzymic activity recognized as a marker for endosomal membranes. Purity is usually assessed by resolving vesicles containing a labeled ligand from marker-enzyme activities associated with Golgi, plasma membrane, and lysosomes. In some instances organelle purity can be assessed by electron microscopy if endosomes contain an identifiable endocytosed ligand. Table II summarizes some principal studies on the isolation and identification of endosomes from animal tissues and cultured cells.

Traditional fractionation procedures, based on differences in the size and density of organelles, have been applied successfully to the purification of endosomes (cf. Mullock *et al.*, 1987). Most existing methods are based on centrifugation primarily using sucrose (Dickson *et al.*, 1983; Quintart *et al.*, 1983; Courtoy *et al.*, 1984; Evans and Flint, 1985; Khan *et al.*, 1986b) or Percoll gradients

Table II
Isolation and Identification of Endosomal Vesicles in Animal Tissues and Cultured Cells

Component	Source	Parent fraction	Gradient technique	Endocytosed tracer	Uptake time (min)	RSA (fold)	Reference
Small vesicle	Rat liver	Microsome	Sucrose	$[^{125}I]$insulin	1	1.42	Khan *et al.*, 1982
Early	Rat liver	Golgi	Percoll	$[^{125}I]$insulin	5	94.6	Khan *et al.*, 1982
Late	Rat liver	Golgi	Percoll	$[^{125}I]$insulin	10	182.3	Khan *et al.*, 1982
Late	Rat liver	Microsome (L)	Sucrose	$[^{125}I]$prolactin	10	>100	Khan *et al.*, 1986b
Combined endosomes	Rat liver	Light mitochondria (L)	Sucrose	$[^{125}I]$prolactin	10	>30	Khan *et al.*, 1986b
Combined endosomes	Rat liver	L + P	Sucrose	$[^{125}I]$insulin			Khan *et al.*, 1989
Early	CHO	Microsomes	FFE	HRP	4	11	Schmid *et al.*, 1988
Late	CHO	Microsomes	FFE	HRP	10	13	Schmid *et al.*, 1988, 1989
Late	CHO	Golgi	FFE	HRP FITC-dextran	10	22	Marsh *et al.*, 1987
Late	Rat liver	Golgi	FFE	FITC-Asialo oro-mucoid	10	22	Fuchs *et al.*, 1989a
Early	BHK	Post-nuclear superna-tant	Immunoaffinity	HRP $[^{35}S]$-G protein	5	–	Gruenberg and Howell, 1987

Early	Rat liver	Homogenate	DAB	[^{3}H]-gal-BSA-HRP	10	200	Kay *et al.*, 1984
Early	HeLa	Post-nuclear supernatant	DAB	[^{125}I]-Tfn-HRP	30 (27°C)	–	Ajioka and Kaplan, 1986
Early	Hep G2	Post-nuclear supernatant	DAB/Percoll	[^{125}I]-ASOR/HRP	2	–	Stoorvogel *et al.*, 1987
Early	Hep G2	Post-nuclear supernatant	DAB/Percoll	[^{125}I]-Tfn	3.5	–	Stoorvogel *et al.*, 1989
Early	Macrophage	Microsomes	Percoll	[^{125}I]-PTH	5	–	Diment *et al.*, 1989
Early	Macrophage	Microsomes	Percoll	[^{125}I]-man BSA	6	–	Diment and Stahl, 1985
Early	Macrophage	Microsomes	Percoll	Lactoperoxidase	5	–	Diment *et al.*, 1988
Early	Macrophage	Microsomes	Percoll	man-IgG	5	–	Diaz *et al.*, 1988
Early	Macrophage	Microsomes	Percoll	man-IgG	5	–	Diaz *et al.*, 1988
Early	Macrophage	Post-nuclear supernatant	Percoll Nycodenz	DNP-B-gluconidase [^{125}I]-man BSA [^{125}I]macroglobulin trypsin complex [^{125}I]-man BSA	4	–	Ward *et al.*, 1989
Late	Rat liver	Post-mitochondrial supernatants	Nycodenz	[^{125}I]asialofetulin [^{125}I]polymeric IgA	10	–	Mullock *et al.*, 1989

(Khan *et al.*, 1982; Dickson *et al.*, 1983; Merion and Sly, 1983), although Nycodenz (Evans and Flint, 1985; Kindberg *et al.*, 1984), and metrizamide (Khan *et al.*, 1981; Mullock *et al.*, 1983) and Ficoll (Mullock *et al.*, 1983) have also been used (Table II). Our own studies have used discontinuous sucrose-density gradients to prepare endosomes from light mitochondria (L) and microsomal fractions (Khan *et al.*, 1986b). These endosomal fractions were dramatically enriched (30–175-fold) in endocytosed insulin and prolactin. The bulk of marker enzymes assayed did not copurify with the endosomal components, and our combined acid-phosphatase cytochemistry and electron-microscope radioautography studies established that about 80% of endosomes purified from parent L fractions contained no acid phosphatase.

Many of the smooth-membrane vesicles involved in membrane traffic share common densities and are difficult to isolate by traditional procedures. Density-modification techniques, first pioneered by DeDuve and colleagues who used Triton WR 1339 to isolate lysosomes from rat liver by floatation (Leighton *et al.*, 1968) have proven useful because of the ease with which density perturbants can be loaded into the endosomal compartment from the extracellular space. Cells can take up a ligand already bound to a dense molecule such as colloidal gold (Beardmore *et al.*, 1987). In the work of Courtoy *et al.* (1985) ligand bound to horseradish peroxidase is internalized and the density of the endosomal compartment is increased by incubation with 3,3-diaminobenzidine and H_2O_2. The 3,3-diaminobenzidine, polymerized in vesicles containing horseradish peroxidase, effects a major shift in vesicle density. These denser vehicles can now be separated from nonendosomal elements by conventional density centrifugation techniques. Our group has successfully used this procedure to resolve endocytic vesicles containing newly internalized [^{125}I]insulin (i.e. early endosomes) from Golgi components (Kay *et al.*, 1984). This technique has been applied to HeLa cells to demonstrate the constitutive internalization of unoccupied transferrin receptors (Ajioka and Kaplan, 1986) and to HepG2 cells to study the cointernalization of two receptors into an early endosomal compartment (Stoorvogel *et al.*, 1987) as well as fusion and recycling events among different (early and late) endosomal populations (Ajioka and Kaplan, 1987). A recent modification of density-perturbation techniques to isolate endosomes from cultured cells has been the use of gold-labeled ligand to alter the density of endocytic components (Beardmore *et al.*, 1987).

Free flow electrophoresis, which separates subcellular components on the basis of charge, has proven useful for separating acidic compartments such as endosomes and lysosomes from the bulk of cellular constituent. A two-step procedure in which lysosomes and endosomes were first separated from other cell organelles by free flow electrophoresis and then from each other by Percoll gradient centrifugation has been described in CHO and BHK-21 cells (Marsh *et al.*, 1987). Endosomes prepared by this procedure were enriched up to 70-fold relative to the initial homogenate and were capable of ATP-dependent acidification. This work has been extended to the separation of two functionally distinct endosomal populations, corresponding to early and late endosomes (Schmid *et al.*, 1988). The procedure is rapid but requires expensive apparatus and trypsin pretreatment of samples to maximize resolution. Such treatment may alter the properties of the endosome preparations.

Immunoisolation procedures have been applied successfully to the preparation of endosomes (Mullock *et al.*, 1987). Immunoisolation relies on the interaction of an antibody bound to a solid support and an antigen exposed on the outer endosomal surface. Thus, endocytic compartments can be labeled with a foreign (e.g. viral) antigen internalized by endocytosis. For example, immunoisolated endosome fractions, labeled with the transmembrane glycoprotein G of vesicular stomatitus virus, have been used in a cell-free system to demonstrate that early endosome fractions (isolated 5 min after internalization) have maximal fusion activity with other endosomal vesicles (Gruenberg and Howell, 1987, Gruenberg *et al.*, 1989). This methodology could eventually allow fractionation methods of great specificity to be developed but require the preparation of suitable specific antibodies and immunoadsorbents.

4.1. Biochemical Properties of Endosomes

Early studies, designed to compare the polypeptide composition of endosomes with other intracellular organelles, were performed using pinocytosed lactoperoxidase (LPO) and LPO-conjugates to catalyse the radioiodination of intraendosomal proteins. These studies suggested that endosomes, from a variety of cell types, contain a subset of proteins similar to those on the plasma membrane (Mellman *et al.*, 1980; Storrie *et al.*, 1981; Watts, 1984; Baenziger and Fiete, 1986). In contrast, our studies suggested that the electrophoretic pattern of plasmalemmal proteins is markedly different from that of endosomes, and few, if any, of the bands are shared with those in the endosomal preparations (Khan *et al.*, 1986b). Recent studies have suggested that early and late endosomes, isolated by free flow electrophoresis and labeled with lactoperoxidase *in vitro*, have distinct protein compositions (Schmid *et al.*, 1988). Only a small subset of polypeptides were common to the plasmalemma and early endosomes. The iodinated polypeptides common to the plasma membrane and early endosomes were not detected in late endosomes. This suggests that most internalized cell-surface components are recycled or sorted prior to the formation of late endosomes. Examination of the patterns of ^{35}S-methionine-labeled proteins in purified lysosomal fractions have strongly suggested that lysosomes are again quite distinct from both endosome subpopulations and the plasma membrane (Marsh *et al.*, 1987).

An important characteristic of endosomes is their low internal pH. Tycko and Maxfield (1982) used fluorescein-labeled α_2-macroglobulin to measure the pH of the microenvironment of ligands undergoing receptor-mediated endocytosis. The excitation spectrum of fluorescein is strongly pH dependent between pH 4 and pH 7.4 and this allowed these investigators to conclude that ligand reached an acidic endocytic compartment prior to fusion of these vesicles with lysosomes. Similar results were obtained using fluorescein-labeled transferrin (Van Renswoude *et al.*, 1982). The suggestion that endosomes are acidic received direct support as well from *in vitro* studies (Galloway *et al.*, 1983). These workers used isolated endosomes labeled with the pH-sensitive fluorochrome fluorescein-dextran to demonstrate that endosomes can lower their internal pH in an ATP-dependent manner. It is now apparent that endosomes, like lysosomes and many secretory organelles, lower their internal pH via an NEM-sensitive ATPase (Mellman *et al.*, 1986; Al-Awqati, 1986). The ATP-dependent acidification of liver endosomes is electrogenic, with proton pumping being accompanied by the generation of an interior-positive membrane potential opposing further acidification (Fuchs *et al.*, 1989a,b). It is well documented that endocytic tracers encounter compartments of gradually decreasing pH en route to the lysosomes (Merion *et al.*, 1983; Murphy *et al.*, 1984; Kielian *et al.*, 1986). In addition, recent studies have demonstrated that the initial rate and extent of proton pumping are lowest in early endosomes and gradually increased in late endosomes and lysosomes (Schmid *et al.*, 1988, 1989). A recent and interesting finding his been the description of a second electrogenic ion-transport system, the Na^+/K^+ ATPase, in early endosomes isolated from CHO cells (Fuchs *et al.*, 1989a; Cain *et al.*, 1989). The activity of the Na^+/K^+ ATPase would also result in the generation of an interior-positive potential and may explain the limited capacity of this subpopulation for acidification *in vitro* and *in vivo*.

4.2. Functions of Endosomes

Three functions can be attributed to endosomes: (1) processing of internalized ligand, (2) sorting and recycling of receptors, and (3) facilitation and transmembrane signaling.

4.2.1. Processing of Internalized Ligands

Degradation of internalized ligands has been thought to occur solely within secondary lysosomes. Proteolysis of mannosylated BSA, however, occurs very rapidly in macrophages with kinetics that are not compatible with ligand degradation in secondary lysosomes (Stahl *et al.*, 1980). It is now apparent that endosomes contain proteolytic enzymes. Diment and Stahl (1985) first demonstrated prelysosomal degradation of mannose-BSA by a pepstatin A–sensitive protease in macrophage endosomes. This intraendosomal protease has recently been identified as cathepsin D (Diment *et al.*, 1988). Additional protease activities that have been reported in endosomal fractions include cathepsin B (Wiley *et al.*, 1985; Pease *et al.*, 1985; Roederer *et al.*, 1987), carboxypeptidases (Schaudies *et al.*, 1987), and leucyl β-naphthylamidase (Ajioka and Kaplan, 1987). Thus, proteolytic processing within the endosomal apparatus may be quite complex. During receptor-mediated internalization of EGF in mouse-embryo 3T3 fibroblasts, the ligand undergoes sequential proteolytic processing resulting in the removal of six amino-acid residues from its carboxy terminus (Schaudies *et al.*, 1987). Addition of methylamine in the extracellular medium results in the retention of ligand in an early endosomal compartment and processing is limited to the removal of the carboxyterminal arginine residues. Comparable studies have examined the internalization and processing of parathyroid hormone in macrophage endosomes (Diment *et al.*, 1989). Parathyroid hormone, normally 84 amino-acid residues in length, is cleaved into fragments that include a bioactive peptide (residues 1–34). Degraded fragments are then returned to the extracellular medium without delivery to lysosomes. Other studies, involving the internalization of cholera toxin in rat liver, have demonstrated that endosomes represent the major site for the proteolytic generation of the active A1 peptide from the A subunit (Janicot *et al.*, 1988). Our own studies have demonstrated that insulin is degraded in both early and late endosomes (Hamel *et al.*, 1988; Doherty *et al.*, 1990) whereas no degradation of internalized EGF or prolactin was observed (Doherty *et al.*, 1990). This may reflect the existence of a specific protease within hepatic endosomes with a relatively high preference for insulin as a substrate.

4.2.2. Sorting and Recycling of Receptors

Signals for Endocytosis. Various studies have indicated that cytoplasmic domains of receptors contain signals required for endocytosis. This was originally recognized in studies on the molecular basis of familial hypercholesterolemia in which mutations in the gene for the LDL receptor (839 amino acids) impair LDL uptake into cells and hence clearance from the circulation. Several families were observed in which the defect was caused by a receptor that could bind LDL normally but failed to mediate internalization of bound LDL adequately (Goldstein *et al.*, 1985). In several subjects the internalization-defective mutations resulted in marked truncation of the COOH-terminal cytoplasmic domain, that is, residues 790 to 839 (Lehrman *et al.*, 1985). In one young subject (JD), a single base substitution (tyrosine to cysteine) at residue 807 markedly impaired internalization of the LDL receptor (Davis *et al.*, 1986). Subsequent studies by Mostov *et al.* (1986) on the polymeric immunoglobulin (poly-Ig) receptor (755 amino acid residues; 653–755 cytoplasmic) showed that a mutated receptor with a markedly truncated cytoplasmic domain (only residues 653 and 654), expressed in Madin-Darby canine kidney cells, could not be endocytosed, unlike the wild-type receptor in these cells. Comparable observations were made on the EGF receptor (Prywes *et al.*, 1986). Similarly, a deletion of residues six to 41 of the 65 N-terminal amino acids constituting the cytoplasmic domain of the transferrin receptor resulted in an internalization-defective phenotype (Rothenberger *et al.*, 1987). The studies of Lobel *et al.* (1989) further extended these observations to the mannose-6-phosphate (M-6-P)/IGF-II receptor. These workers showed that a deletion of ≥143 amino acids in the 163 amino acid cytoplasmic domain resulted in an internalization-defective mutant. They also made the interesting observation that a double mutant in which tyrosine residues 24 and 26 of the cytoplasmic domain were converted to alanine were equally impaired in endocytosis.

The nature of the signals in the cytoplasmic domain that determine receptor endocytosis are not clear. Receptors such as those for insulin, transferrin, LDL, poly-Ig, and M-6-P/IGF-II show virtually no homology in the primary sequence of their cytoplasmic domains. Nevertheless, the above studies are consistent with a special role for tyrosine, perhaps within a noncontiguous epitope formed by amino acids of the cytoplasmic domain, in providing a signal for endocytosis. This view is supported by the study of Lazarovits and Roth (1988) who showed that substituting tyrosine for cysteine at residue 543 (residue 5 of the cytoplasmic domain) changed the influenza virus hemagglutinin (HA) from a slowly to a rapidly internalized protein. Recent studies on the transferrin receptor have identified a critical 10 amino-acid residue (amino acids 19–28) of the cytoplasmic domain, whose deletion greatly impaired receptor endocytosis. The sole tyrosine residue within the cytoplasmic domain of transferrin receptor is located at position 20. The replacement of this tyrosine by glycine yielded a mutant that was about 20% as active as the wild-type receptor in endocytosis. Similarly, Breitfeld *et al.* (1990) have shown that efficient internalization of the poly-Ig receptor is impaired by deletion of the COOH-terminal 30 amino acids (residues 726–755). A comparable phenotype was generated by substitution of serine for tyrosine at position 734. Thus, a number of observations indicate that a tyrosine residue in the cytoplasmic domain is part of signal-directing receptor endocytosis. However, the studies of Davis *et al.* (1989) suggest that the requirement for a tyrosine in the cytoplasmic domain may not be highly stringent or universally applicable. In this study it was found that substituting tyrosine 807 in the LDL receptor with phenylalanine did not impair endocytosis whereas tryptophan substitution led to modest endocytotic efficiency (about 60% of wild-type). By contrast, substitution with ten other different amino acids impaired endocytosis by about 80%.

Recent studies on the LDL receptor identified a tetramer at positions 804 to 807 whose sequence was conserved as NPXY (where X stands for any amino acid) in receptors from six species (Chen *et al.*, 1990). Replacement of Asn 804, Pro 805 or Tyr 807 with alanine markedly impaired internalization of the LDL receptor. Chen *et al.* (1990) also noted that 11 other intrinsic membrane proteins including the insulin, IGF-I, and EGF receptors contained the NPXY sequence in their cytoplasmic domains. In the case of the insulin receptor, the NPXY 960 sequence is located in a juxtamembrane cytoplasmic domain coded by the 16th exon of the insulin-receptor gene (Ullrich *et al.*, 1985). Backer *et al.* (1990) have shown that replacing tyrosine 960 with phenylalanine had a minimal effect on endocytosis but deletion of the 12 amino acids encompassing residues 954 to 965 (including the NPXY 960 sequence) generated a receptor significantly impaired in endocytosis. Thies *et al.* (1990) deleted all 22 amino acids coded by Exon 16 (residues 944 to 965) and generated a mutant receptor [ΔEX16] of comparable efficiency in endocytosis. Previous studies with a kinase-deficient mutant receptor [alanine for lysine 1018] demonstrated a marked decrease in endocytic capacity (Russell *et al.*, 1987). Furthermore, internalization was shown to be deficient in a receptor where the tyrosine autophosphorylation sites were substituted (White *et al.*, 1988a). Thus, kinase-dependent autophosphorylation appears to be required for endocytosis. The ΔEX16 receptor, however, showed near-normal autophosphorylation activity (Thies *et al.*, 1990). Thus, two ingredients, namely the juxtamembrane NPXY sequence and receptor autophosphorylating capacity, appear to be necessary but individually not sufficient to promote insulin-receptor endocytosis. The molecular events underlying these requirements remain to be defined. Perhaps autophosphorylation determines an appropriate orientation for the NPXY sequence. That a stringent positioning of the NPXY sequence may be critical for internalization is indicated by the observation of Chen *et al.* (1990) that deletion of residues 797 to 800 in the LDL receptor markedly impairs internalization (9% of control) even though the NPXY sequence is unaltered.

In contrast to the insulin receptor, deletion experiments with the EGF receptor rule out a requirement for the NPXY sequence in receptor endocytosis (Prywes *et al.*, 1986). In many of the receptors, such as LDL, LDL receptor-related protein (LRP), insulin, and IGF-I, NPXY sequence is located within 50 amino-acid residues of the plasma membrane, even though insulin and IGF-I receptors have very long cytoplasmic tails with tyrosine kinase-domains (Figure 1). This is very different from the LDL receptor, where the NPXY sequence mediates ligand-independent internalization. On the contrary, in the EGF receptor, the three NPXY sequences are located on the carboxyl-terminal side of the tyrosine-kinase domain, more than 500 amino acids from the plasma membrane. Nevertheless, the activity of this sequence in EGF receptor may become important for internalization under certain controlled conditions.

Downregulation and Recycling. The LDL and M-6-P receptors recycle constitutively but target their ligands to lysosomes (Goldstein *et al.*, 1985; Lobel *et al.*, 1989). In contrast, the transferrin receptor recycles together with its ligand, which remains attached during the entire transit through the cell with only the iron being released (Klausner *et al.*, 1983). The EGF receptor, which recycles and is downregulated (net receptor loss) in a ligand-dependent fashion,

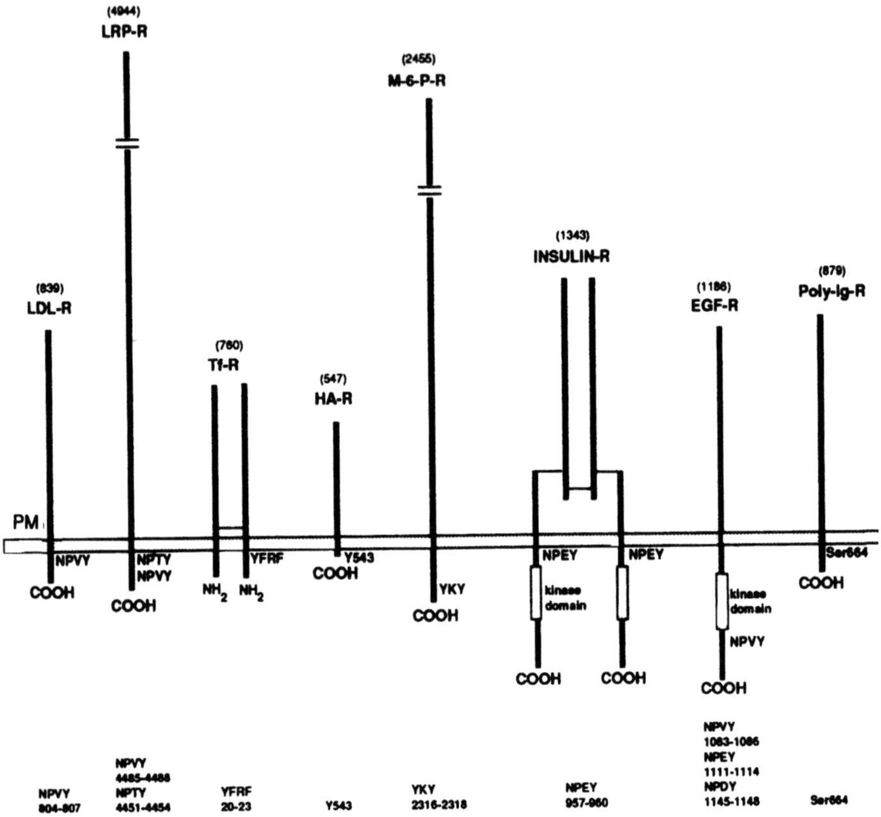

FIGURE 1. Structure comparison of cell surface receptors involved in receptor-mediated endocytosis. The total number of amino acids for each receptor and the number of amino acids involved in sorting are indicated accordingly. The cytoplasmic domain of the EGF and insulin receptors expresss tyrosine kinase activity.

targets itself as well as its ligands to lysosomes at saturating doses of ligand (Lai *et al.*, 1989a). The poly-Ig receptor represents a unique mechanism in which constitutive transcytosis of its ligand results in cleavage and ultimate loss of the receptor (Mostov *et al.*, 1986).

The insulin and EGF receptors possess intrinsic tyrosine kinase activity that appears to be active after internalization into endosomes (Khan *et al.*, 1986b; Kay *et al.*, 1986; Lai *et al.*, 1989b). Subsequently, receptors such as those for insulin and EGF enter distinct cellular sorting pathways leading either to recycling to the cell surface or degradation. The kinase activity of these receptors appears to be necessary for their targeting to the degradation pathway and hence receptor downregulation. EGF-receptor tyrosine kinase activity has been postulated to control downregulation (Honegger *et al.*, 1987; Lai *et al.*, 1989a; Felder *et al.*, 1990). Thus, in NIH3T3 cells transfected with the wild-type EGF receptor, receptor downregulation (degradation) was demonstrated. In contrast, a kinase-negative mutant expressed in the same cells showed no downregulation but instead demonstrated constitutive receptor recycling (Honegger *et al.*, 1987). The role of the tyrosine-kinase domain of the EGF receptor in regulating sorting of internalized receptor for degradation or recycling has been proposed to occur via tyrosine-kinase-mediated spatial segregation within multivesicular bodies (Felder *et al.*, 1990). *In vivo* studies (Lai *et al.*, 1989b) have demonstrated the inhibition of cell-surface phosphorylation activity and internalization coincident with downregulation. This further suggested that EGF-receptor sorting is governed by ligand-mediated receptor-phosphorylation activity. In addition, Beguinot *et al.* (1985) have shown that when EGF receptor was phosphorylated by protein kinase C on threonine (residue 654) close to the cytosolic face of the plasma membrane, the EGF receptor was internalized but not degraded. These studies indicate that kinase activities control receptor sorting even though the chain of events that is initiated by phosphorylation is still poorly understood. The sorting signals that have thus far been located on receptors are summarized in Figure 1.

4.2.3. Facilitation of Transmembrane Signaling

In addition to the above functions it has been suggested that the endosomal system plays a role in transmitting hormonal signals within the cell. Based on studies of insulin internalization, we suggested the possibility that an activated form of receptor may extend the signaling process initiated at the cell surface to a substantially larger cytosolic volume and/or carry out specific functions intracellularly (Posner *et al.*, 1980, 1987).

As alluded to above, biochemical and genetic studies have led to elucidation of the structure of the insulin receptor and the demonstration that the β-subunit contains a tyrosine kinase activity in its cytoplasmic domain (Ebina *et al.*, 1985; Ullrich *et al.*, 1985). Rosen and her collaborators showed that insulin, on binding to the α-subunit of the receptor, augmented the kinase activity of the β-subunit (Rosen *et al.*, 1983). Furthermore, it was demonstrated that this augmented activity persisted after the removal of insulin, reflecting an activated state of the receptor kinase (Khan *et al.*, 1986b; Klein *et al.*, 1986; Yu and Czech, 1986). The activated state has shown to be dependent on autophosphorylation of the receptor on its β-subunit (Khan *et al.*, 1986b).

Studies involving the transfection of cells with mutagenized receptor have proven useful for defining the mechanisms of insulin action. Mutant receptors, in which tyrosine residues 1150 and 1151 of the putative regulatory domain were

Table III
Role of the Insulin Receptor Kinase in Mediating the Insulin Response

A. IRK is necessary:
1. Mutated receptors with decreased or absent kinase activity cannot mediate the insulin response [e.g., lys 1018 → ala; tyr 1150 and 1151 → phe] (Ellis *et al.*, 1986; Chou *et al.*, 1987; McClain, 1987; Hari and Roth, 1987).
2. An antibody which inhibits IRK blocks biological response to insulin in CHO, hepatoma cells and Xenopus oocytes (Morgan *et al.*, 1986; Morgan and Roth, 1987).
3. Microinjection of activated IRK into Xenopus oocytes mimics insulin like activity (Maller *et al.*, 1986).
4. Microinjection of a protein-tyrosin phosphatase inhibits insulin action in Xenopus oocytes (Circirelli *et al.*, 1990).
5. Agents which activate IRK [e.g., vanadate, pervanadate, and trypsin] mimic insulin action (Tamura *et al.*, 1984; Kadota *et al.*, 1987a, 1987b; Shoelson *et al.*, 1988).

B. IRK is not sufficient:
1. Insulin receptors mutated at tyr 960 have normal kinase activity but appear to be defective in mediating the insulin response (White *et al.*, 1988b).
2. Insulin receptors with a C-terminal 43-amino acid deletion have normal kinase activity but are defective in realizing metabolic effects of insulin (Maegawa *et al.*, 1988).

replaced with phenylalanine, bound insulin normally but could not be significantly activated (Ellis *et al.*, 1986). Cells transfected with these receptors would not support insulin-stimulated 2-deoxyglucose uptake or glycogen synthase activity (Ellis *et al.*, 1986) whereas normal thymidine incorporation into DNA was observed (Debant *et al.*, 1988). Substitution of tyrosine 1146 of the regulatory region for phenylalanine resulted in receptors with greatly reduced tyrosine-kinase activity (Wilden *et al.*, 1990). These receptors did not internalize insulin normally or stimulate DNA synthesis, although insulin-dependent glycogen synthase activity was unaffected. Replacement of lysine 1018 of the ATP binding site of the β-subunit inactivated the receptor kinase and interfered with receptor-ligand internalization and transmembrane signaling (Russell *et al.*, 1987). Evidence supporting a critical role for the receptor kinase for realizing the biological effects of insulin are summarized in Table III. Of particular interest are studies involving alterations of the β-subunit submembrane region that includes the NPXY internalization sequence. Substitution of tyrosine 960, the terminal residue of this sequence, resulted in mutant receptors that exhibit normal insulin-stimulated tyrosine-kinase activity. However, other insulin-dependent processes, including the transport of glucose and amino acids as well as DNA synthesis, were severely reduced in these cells (White *et al.*, 1988a,b). These studies imply that insulin-receptor kinase activation is necessary but not sufficient for insulin action; receptor internalization is also required for insulin-dependent processes.

There appears to be a relationship between kinase activation and internalization into endosomes of the insulin receptor kinase (IRK). Thus, *in vivo* insulin administration augmented the IRK in rat-liver endosomes with respect to both autophosphorylation (Khan *et al.*, 1986b, 1989) and exogenous-substrate phosphorylation (Posner *et al.*, 1987). Further, as described above, alterations of the insulin-receptor β-subunit that impair tyrosine kinase activity also dramatically reduce ligand-induced internalization (Russell *et al.*, 1987; Ebina *et al.*, 1987; Chou *et al.*, 1987; McClain *et al.*, 1987).

Given the above considerations, we have examined the impact of *in vivo* insulin on the temporal changes of the IRK activity during ligand-induced internalization into a combined endosomal (C-EN) fraction derived from hepatic light-mitochonrial and microsomal fractions. We demonstrated *in vitro* conditions that promote transmembrane signaling in intact membranes and showed that the insulin receptor β-subunit is phosphorylated at both serine and tyrosine residues in intact plasmalemma (PM), but is phosphorylated exclusively at tyrosine residues in an intact endosomal fraction (Khan *et al.*, 1989). We observed that insulin injection caused a time- and dose-dependent augmentation of the IRK with maximal activity of IRK in PM at 30 sec and in ENs at 2 to 5 min (Figure 2a). The demonstration that a dose of insulin as low as 15 ng/100 g bodyweight augmented IRK in ENs attests to the physiologic relevance of these observations (Figure 2b). The elevated IRK persisted after lectin purification, which removes any receptor-associated insulin, indicating that the IRK had been truly activated. Of particular importance was the observation that the maximum specific activity of endosomal IRK was three- to fourfold that of PM (Khan *et al.*, 1989). This could arise from selected concentration within ENs of activated IRK or from augmented activation associated with the internalization process. In addition, we showed that the activated state of endosomal IRK was a result of phosphorylation of the receptor kinase; alkaline phosphatase treatment *in vitro* abolished all receptor kinase activity. These data provide further support for our long-standing hypothesis (Posner *et al.*, 1980) that internalization plays a role in the realization of insulin action. As depicted in Figure 3, the internalization process is associated with a continuing cytoplasmic orientation of the kinase domain of the insulin receptor. The key role for receptor internalization may be to permit interaction of the activated IRK with topographically distinct substrates whose phosphorylation is crucial for signal transduction.

Our most recent studies have used immunoblotting techniques to evaluate β-subunit tyrosine phosphorylation at the cell surface and within the cell as a possible mechanism of augmentation of kinase activity in endosomes. Surprisingly, β-subunit phosphotyrosine levels were two- to three-fold higher in plasmamembrane fractions than in corresponding endosomal fractions, suggesting that internalization is associated with rapid dephosphorylation of β-subunit tyrosine residues. Thus, receptors derived from the two membrane systems exhibit an

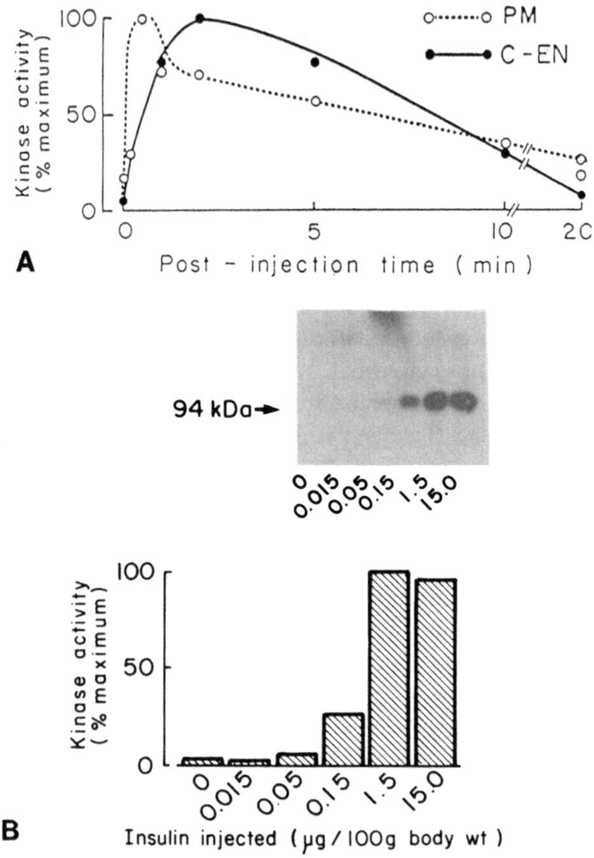

FIGURE 2. (A) Time course of insulin receptor activation in plasmalemma (PM) and combined endosomes (C-EN). Autophosphorylation was carried out in intact membrane fractions after *in vivo* insulin (1.5 μg/100 g body weight). Subsequently, membrane receptors were solubilized, immunoprecipitated and subjected to SDS-PAGE. Phosphorylation of the β-subunit (94 kDa) was quantitated by densitometry and plotted as percent maximum autophosphorylation. (B) Insulin dose dependence of β-subunit (94 kDa) autophosphorylation in endosomes. Rats were killed 5 min after insulin injection. The protocol noted in Figure 1 was followed. Autophosphorylation (upper panel) is plotted as percent maximum (lower panel).

inverse relationship between β-subunit phosphotyrosine content and tyrosine-kinase activity; endosomal receptors displayed significantly greater autophosphorylation activity despite decreased amounts of β-subunit phosphotyrosine (Burgess *et al.*, 1992).

In this regard, the insulin receptor β-subunit may function in a manner

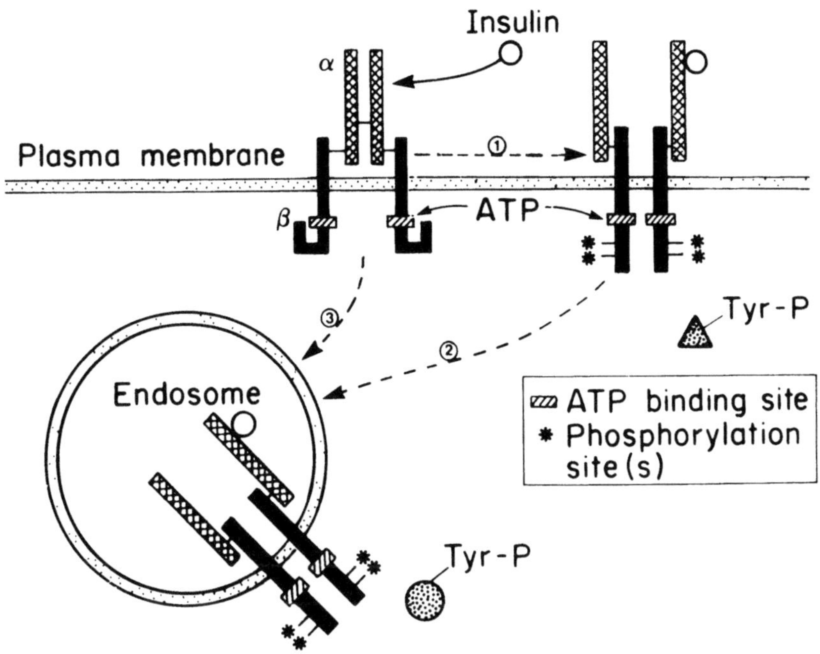

FIGURE 3. Activation and internalization of the insulin receptor kinase following the binding of insulin to the α-subunit. The activated β-subunit is depicted as having a changed conformation. Note: following internalization, the activated kinase remains cytosolically oriented and may interact with novel substrates intracellularly.

comparable to the src family of tyrosine kinases. These soluble enzymes contain tyrosine residues, notably those associated with the carboxyterminal tail, that inhibit the enzymatic activity when phosphorylated (Mustelin *et al.*, 1989; Okada and Nakagawa, 1989; Okamoto *et al.*, 1991; White, 1991). Dephosphorylation of these residues would affect activation. Similar events may affect the activation state of the insulin-receptor tyrosine kinase. In Figure 4, we present a model of phosphorylation- and dephosphorylation-dependent activation of the receptor kinase. Insulin binding initiates autophosphorylation and preliminary activation of the receptor kinase at the cell surface. Phosphorylation of tyrosines, which are inhibitory to kinase activity either by autophosphorylation or by a separate cellular-tyrosine kinase, may also occur at this time. Internalization to the endosomal apparatus is associated with selective dephosphorylation of the inhibitory tyrosine residues and complete activation of the intrinsic-receptor kinase activity. Further, dephosphorylation, mediated by either endosomal or soluble P-Tyr phosphatases, would lead to inactivation of the kinase activity prior

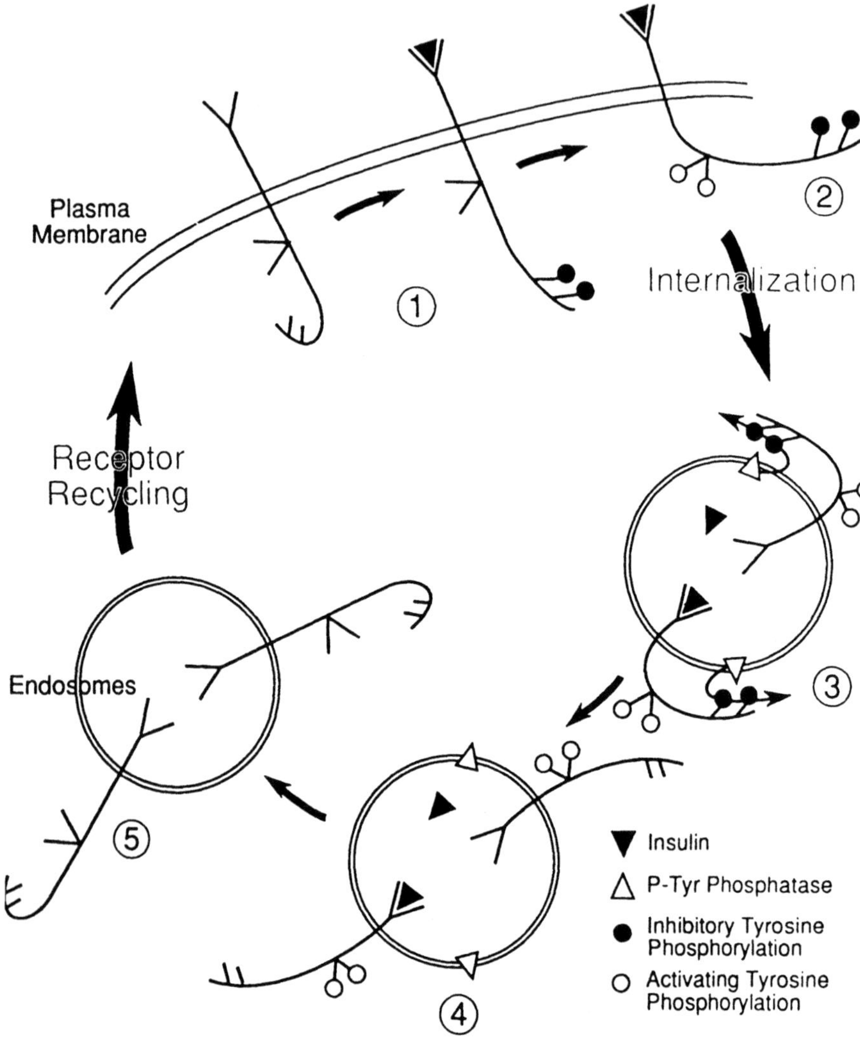

FIGURE 4. Dephosphorylation-mediated augmentation of the activation state of the insulin receptor tyrosine kinase in endosomes. Following the binding of insulin (step 1), autophosphorylation at the cell surface activates the IR-TK activity (step 2). Phosphotyrosine residues inhibitory to IR-kinase activity may be phosphorylated at this time, or prior to the binding of insulin by a protein TK other than the IR-TK (step 1). Internalization to the endosomal system results in the removal of inhibitory P-Tyr by an associated P-Tyr phosphatase (step 3) and augmentation of IR-TK activity (step 4). Phosphorylation of endogenous substrates or association with other mediators of insulin action now occurs prior to full dephosphorylation and inactivation of TK activity (step 5). The inactive receptor is then recycled to the cell surface.

to recycling of the receptor to the cell surface. Thus, full activation of the receptor kinase may require the participation of P-Tyr phosphatases. We have recently noted P-Tyr phosphatase activity in our endosome fractions that can dephosphorylate the insulin-receptor β-subunit. Thus, the potential for a regulatory role of dephosphorylation must be seriously considered and explored in future studies (Fause *et al.*, 1992).

In the last few years much data have accumulated that strongly suggest that, besides processing and/or sorting of ligand-receptor complexes, endosomes play a crucial role in facilitating the transmembrane signaling for hormones and/or growth factors whose receptors possesses intrinsic tyrosine kinase activity. In particular, the early events following insulin binding to its receptor involve activation, internalization, and concentration of the IRK into ENs (Khan *et al.*, 1989). Similarly, following the binding to target cells, EGF is internalized along with its receptors into the endosomal compartment (Lai *et al.*, 1989a). Cohen and Fava (1985) and Kay *et al.* (1986) have demonstrated the active EGF receptor kinase in the endocytic components of cells. In our laboratory, we have further extended these findings to elements containing activity of EGF receptor kinase during internalization (Lai *et al.*, 1989b; Wada *et al.*, 1992). These observations are consistent with an important role for the endosomal apparatus in facilitating transmembrane signaling and in mediating the biological effect of insulin and growth factors.

5. REFERENCES

Ajioka, R. S. and Kaplan, J., 1986, Intracellular pools of transferrin receptors result from constitutive internalization of unoccupied receptors, *Proc. Natl. Acad. Sci. USA* **83:**6445–6449.

Ajioka, R. S. and Kaplan, J., 1987, Characterization of endocytic compartments using the horseradish peroxidase—diaminobenzidine density shift technique, *J. Cell Biol.* **104:**77–85.

Al-Awqati, Q., 1986, Proton-translocating ATPases, *Ann. Rev. Cell Biol.* **2:**179–199.

Anderson, R. G. W., Brown, M. S., and Goldstein, J. L., 1977, Role of the coated endocytic vesicle in the uptake of receptor-bound low density lipoprotein in human fibroblasts, *Cell* **10:**351–364.

Backer, J. M., Kahn, C. R., Cahill, D. A., Ullrich, A., and White, M. F., 1990, Receptor-mediated internalization of insulin requires a 12-amino-acid sequence in the juxtamembrane region of the insulin receptor β-subunit, *J. Biol. Chem.* **265:**16450–16454.

Baenziger, J. U. and Fiete, D., 1986, Separation of two populations of endocytic vesicles involved in receptor-ligand sorting in rat hepatocytes, *J. Biol. Chem.* **261:**7445–7454.

Beardmore, J., Howell, K. E., Miller, K., and Hopkins, C. R., 1987, Isolation of an endocytic compartment from A431 cells using density modification procedure employing a receptor-specific monoclonal and antibody complexed with colloidal gold, *J. Cell Sci.* **87:**495–506.

Beguinot, L., Hanover, J. A., Ito, S., Richert, N. D., Willingham, M. C., and Pastan, I., 1985, Phorbol esters induce transit internalization without degradation of unoccupied epidermal growth factor receptors, *Proc. Natl. Acad. Sci USA* **82:**2774–2778.

Bergeron, J. J. M., Evans, W. H., and Geschwind, I. I., 1973, Insulin binding to rat liver Golgi fractions, *J. Cell Biol.* **59:**771–776.

Bergeron, J. J. M., Sikstrom, R., Hand, A. R., and Posner, B. I., 1979, Binding and uptake of ^{125}I-insulin into rat liver hepatocytes and endothelium: an *in vivo* radioautographic study, *J. Cell Biol.* **80:**427–443.

Bergeron, J. J. M., Rachubinski, R., Searle, N., Sikstrom, R., Borts, D., Bastian, P., and Posner, B. I., 1980, Radioautographic visualization of *in vivo* insulin binding to the exocrine pancreas, *Endocrinology* **107:**1069–1080.

Bergeron, J. J. M., Resch, L., Rachubinski, R., Patel, B. A., and Posner, B. I., 1983, Effect of colchicine on internalization of prolactin in female rat liver: an *in vivo* radioautographic study, *J. Cell Biol.* **96:**875–886.

Bergeron, J. J. M., Cruz, J., Khan, M. N., and Posner, B. I., 1985, Uptake of insulin and other ligands into receptor-rich endocytic components of target cells: the endosomal apparatus, *Ann. Rev. Physiol.* **47:**383–403.

Bergeron, J. J. M., Searle, N., Khan, M. N., and Posner, B. I., 1986, Differential and analytical subfractionation of rat liver components internalizing insulin and prolactin, *Biochemistry* **25:**1756–1764.

Braell, W. A., 1987, Fusion between endocytic vesicles in a cell-free system, *Proc. Natl. Acad. Sci. USA* **84:**1137–1141.

Breitfeld, P. P., Casanova, J. E., McKinnon, W. C., and Mostov, K. E., 1990, Deletions in the cytoplasmic domain of the polymeric immunoglobulin receptor differentially affect endocytotic rate and postendocytotic traffic, *J. Biol. Chem.* **265:**13750–13757.

Bretscher, M. S., Thomson, J. N., and Pearse, B. M. F., 1980, Coated pits act as molecular filters, *Proc. Natl. Acad. Sci. USA* **77:**4156–4159.

Brown, M. S., and Goldstein, J. L., 1986, A receptor-mediated pathway for cholesterol homeostasis, *Science* **232:**34–47.

Burgess, J. W., Wada, I., Ling, N., Khan, M. N., Bergeron, J. J. M., and Posner, B. I., 1992, Decrease in β-Subunit phosphotyrosine correlates with internalization and activation of the insulin receptor kinase, *J. Biol. Chem.* **267:**10077–10086.

Cain, C. C., Sipe, D. M., and Murphy, R. F., 1989, Regulation of endocytic pH by the Na^+, K^+, -ATPase in living cells, *Proc. Natl. Acad. Sci. USA* **86:**544–548.

Carpentier, J-L., Gorden, P., Anderson, R. G. W., Goldstein, J. L., Brown, M. S., Cohen, S., and Orci, L., 1982, Co-localization of ^{125}I-epidermal growth factor and ferritin-low density lipoprotein in coated pits: a quantitative electron microscopic study in normal and mutant human fibroblasts. *J. Cell Biol.* **95:**73–77.

Chen, W-J., Goldstein, J. L., and Brown, M. S., 1990, NPXY, a sequence often found in cytoplasmic tails, is required for coated pit-mediated internalization of the low density lipoprotein receptor. *J. Biol. Chem.* **265:**3116–3123.

Cheng, T. P., 1986, Redistribution of cell surface transferrin receptors prior to their concentration in coated pits as revealed by immunoferritin labels, *Cell Tissue Res.* **244:**613–619.

Chou, C. K., Dull, T. J., Russell, D. S., Gherzi, R., Lebwohl, D., Ullrich, A., and Rosen, O. M., 1987, Human insulin receptors mutated at the ATP-binding site lack protein tyrosine kinase activity and fail to mediate postreceptor effects of insulin, *J. Biol. Chem.* **262:**1842–1847.

Circirelli, M. F., Tonks, N. K., Diltz, C. D., Weiel, J. E., Fischer, E. H., and Krebs, E. G., 1990, Microinjection of a protein-tyrosine-phosphatase inhibits insulin action in *Xenopus* oocytes, *Proc. Natl. Acad. Sci. USA* **87:**5514–5518.

Cohen, S. and Fava, R. A., 1985, Internalization of functional epidermal growth factor: receptor/kinase complexes in A431 cells, *J. Biol. Chem.* **260:**12351–12358.

Courtoy, P. J., Quintart, J., and Baudhuin, P., 1984, Shift of equilibrium density induced by 3,3′-diaminobenzidine cytochemistry: a new procedure for the analysis and purification of peroxidase-containing organelles, *J. Cell Biol.* **98:**870–876.

Courtoy, P. J., Quintart, J., Limet, J. N., De Roe, C., and Baudhuin, P., 1985, Polymeric IgA and galactose-specific pathways in rat hepatocytes: evidence for intracellular ligand sorting, in *Endocytosis* (I. Pastan and M. C. Willingham, eds.), pp. 163–194, Plenum Press, New York.

Davis, C. G., Lehrman, M. A., Russell, D. W., Anderson, R. G. W., Brown, M. S., and Goldstein, J. L. 1986, The J. D. mutation in familial hypercholesterolemia: Amino acid substitution in cytoplasmic domain impedes internalization of LDL receptors, *Cell* **45:**15–24.

Davis, C. G., van Driel, I. R., Russell, D. W., Brown, M. S., and Goldstein, J. L., 1989, The low density lipoprotein receptor: identification of amino acids in cytoplasmic domain required for rapid endocytosis, *J. Biol. Chem.* **262:**4075–4082.

Debanne, M. T., Evans, W. H., Flint, N., and Regoeczi, E., 1982, Receptor-rich intracellular membrane vesicles transporting asialotransferrin and insulin in liver, *Nature* **298:**398–400.

Debant, A., Clauser, E., Ponzio, G., Filloux, C., Auzan, C., Contreres, J. O., and Rossi, B., 1988, Replacement of insulin receptor tyrosine residues 1162 and 1163 does not alter the mitogenic effect of the hormone, *Proc. Natl. Acad. Sci. USA* **85:**8032–8036.

DeDuve, C. and Wattiaux, R., 1966, Functions of lysosomes, *Ann. Rev. Physiol.* **28:**435–492.

Diaz, R., Mayorga, L., and Stahl, P., 1988, *In vitro* fusion of endosomes following receptor-mediated endocytosis, *J. Biol. Chem.* **263:**6093–6100.

Diaz, R., Mayorga, L. S., Mayorga, L. E., and Stahl, P., 1989, *In vitro* clustering and multiple fusion among macrophage endosomes, *J. Biol. Chem.* **264:**13171–13180.

Dickson, R. B., Beguinot, L., Hanover, J. A., Richert, N. D., Willingham, M. C., and Pastan, I., 1983, Isolation and characterization of a highly enriched preparation of receptosomes (endosomes) from a human cell line, *Proc. Natl. Acad. Sci. USA* **80:**5335–5339.

Diment, S. and Stahl, P., 1985, Macrophage endosomes contain proteases which degrade endocytosed protein ligands, *J. Biol. Chem.* **260:**15311–15317.

Diment, S., Leech, M. S., and Stahl, P., 1988, Cathepsin D is membrane-associated in macrophage endosomes, *J. Biol. Chem.* **263:**6901–6907.

Diment, S., Martin, K. J., and Stahl, P. D., 1989, Cleavage of parathyroid hormone in macrophage endosomes illustrates a novel pathway for intracellular processing of proteins, *J. Biol. Chem.* **264:**13403–13406.

Doherty, J-J. II, Kay, D. G., Lai, W. H., Posner, B. I., and Bergeron, J. J. M., 1990, Selective degradation of insulin within rat liver endosomes, *J. Cell Biol.* **110:**35–42.

Dunn, W. A., Hubbard, A. L., and Aronson, N. N., 1980, Low temperature selectively inhibits fusion between pinocytic vesicles and lysosomes during heterophage of ^{125}I-asialofetuin by the perfused rat liver, *J. Biol. Chem.* **255:**5971–5978.

Dunn, W. A., Connolly, T. P., and Hubbard, A. L., 1986, Receptor-mediated endocytosis of epidermal growth factor by rat hepatocytes: receptor pathway, *J. Cell Biol.* **102:**24–36.

Ebina, Y., Ellis, L., Jarnagin, K., Edery, M., Graf, L., Clauser, E., Ou, J-H., Masiarz, F., Kan, Y. W., Goldfine, I. D., Roth, R. A., and Rutter, W. J., 1985, The human insulin receptor cDNA: the structural basis for hormone-activated transmembrane signaling, *Cell* **40:**747–758.

Ebina, Y., Araki, E., Taira, M., Shimada, F., Mori, M., Craik, C. S., Siddle, K., Pierce, S. A., and Roth, R. A., 1987, Replacement of lysine residue 1030 in the putative ATP-binding region of the insulin receptor abolishes insulin- and antibody-stimulated glucose uptake and receptor kinase activity, *Proc. Natl. Acad. Sci. USA.* **84:**704–708.

Ehrenreich, J. H., Bergeron, J. J. M., Siekevitz, P., and Palade, G. E., 1973, Golgi fractions prepared from rat liver homogenates: isolation procedure and morphological characterization, *J. Cell Biol.* **59:**45–72.

Ellis, L., Clauser, E., Morgan, D., Edery, M. Roth, R. A., and Rutter, W. J., 1986, Replacement of insulin receptor tyrosine residues 1162 and 1163 compromises insulin-stimulated kinase activity and uptake of 2-deoxyglucose, *Cell* **45:**721–732.

Evans, W. H. and Flint, N., 1985, Subfractionation of hepatic endosomes in Nycodenz gradients and by free flow electrophoresis: separation of ligand-transporting and receptor-enriched membranes, *Biochem. J.* **232:**25–32.

Fan, J. Y., Carpentier, J-L., Gorden, P., Van Obberghen, E., Blackett, N. M., Grunfeld, C., and Orci, L., 1982, Receptor-mediated endocytosis of insulin: role of microvilli, coated pits, and coated vesicles, *Proc. Natl. Acad. Sci. USA* **79:**7788–7791.

Faure, R., Baquiran, G., Bergeron, J. J. M., and Posner, B. I., 1992, The dephosphorylation of insulin and epidermal growth factor receptors: Role of endosome-associated phosphotyrosin phosphatase(s), *J. Biol. Chem.* **267:**11215–11221.

Felder, S., Miller, K., Moehren, G., Ullrich, A., Schlessinger, J., and Hopkins, C. R., 1990, Kinase activity controls the sorting of the epidermal growth factor receptor within the multivesicular body, *Cell* **61:**623–634.

Fuchs, R., Schmid, S., and Mellman, I., 1989a, A possible role for Na^+, K^+-ATPase in regulating ATP-dependent endosome acidification, *Proc. Natl. Acad. Sci. USA* **86:**539–543.

Fuchs, R., Mâle, P., and Mellman, I., 1989b, Acidification and ion permeabilities of highly purified rat liver endosomes, *J. Biol. Chem.* **264:**2212–2220.

Galloway, C. J., Dean, G. E., Marsh, M., Rudnick, G., and Mellman, I., 1983, Acidification of macrophage and fibroblast endocytic vesicles *in vitro, Proc. Natl. Acad. Sci. USA* **80:**3334–3338.

Geuze, H., Slot, J. W., Strous, G. J. A. M., Lodish, H. F., and Scwartz, A. L., 1982, Immunocytochemical localization of the receptor for asialoglycoprotein in rat liver cells, *J. Cell Biol.* **92:**865–870.

Geuze, H. J., Slot, J. W., Strous, G. J. A. M., Lodish, H. F., and Schwartz, A. L., 1983a, Intracellular site of asialoglycoprotein receptor-ligand uncoupling: double-label immunoelectron microscopy during receptor-mediated endocytosis, *Cell* **32:**277–287.

Geuze, H. J., Slot, J. W., Strous, G. J. A. M., and Schwartz, A. L., 1983b, The pathway of the asialoglycoprotein-ligand during receptor-mediated endocytosis: a morphological study with colloidal gold/ligand in the human hepatoma cell line, Hep G2, *Eur. J. Cell Biol.* **32:**38–44.

Goldstein, J. L., Brown, M. S., Anderson, R. G., Russell, D. W., and Schneider, W. J., 1985, Receptor-mediated endocytosis: concepts emerging from the LDL receptor system, *Annu. Rev. Cell Biol.* **1:**1–39.

Gorden, P., Carpenter, J-L., Freychette, P., and Orci, L., 1980, Internalization of polypeptide hormone, *Diabatologia* **18:**263–274.

Gruenberg, J. and Howell, K. E., 1987, An internalized transmembrane protein resides in a fusion-competent endosome for less than 5 minutes, *Proc. Natl. Acad. Sci. USA* **84:**5758–5762.

Gruenberg, J., Griffiths, G., and Howell, K. E., 1989, Characterization of the early endosome and putative endocytic carrier vesicles in vivo and with an assay of vesicle fusion in vitro, *J. Cell Biol.* **108:**1301–1316.

Hamel, F. G., Posner, B. I., Bergeron, J. J. M., Frank, B. H., and Duckworth, W. C., 1988, Isolation of insulin degradation products from endosomes derived from intact rat liver, *J. Biol. Chem.* **263:**6703–6708.

Handley, D. A., Arbeeney, C. M., Witte, L. D., and Chien, S., 1981, Colloidal gold–low density lipoprotein conjugates as membrane receptor probes, *Proc. Natl. Acad. Sci. USA* **78:**368–371.

Harding, C., Heuser, J., and Stahl, P., 1984, Endocytosis and intracellular processing of transferrin and colloidal gold–transferrin in rat reticulocytes: demonstration of a pathway for receptor shedding, *Eur. J. Cell Biol.* **35:**256–263.

Hari, J. and Roth, R. A., 1987, Defective internalization of insulin and its receptor in cells expressing mutated insulin receptors lacking kinase activity, *J. Biol. Chem.* **262:**15341–15344.

Helenius, A., Mellman, I., Wall, D., and Hubbard, A., 1983, Endosomes, *Trends Biochem. Sci.* **8:**245–250.

Honegger, A. M., Dull, T. J., Felder, S., Van Obberghen, E., Bellot, F., Szapary, D., Schmidt, A., Ullrich, A., and Schlessinger, J., 1987, Point mutation at the ATP binding site of EGF receptor abolishes protein-tyrosine kinase activity and alters cellular routing, *Cell* **51**:199–209.

Hopkins, C. R., 1983, Intracellular routing of transferrin and transferrin receptors in epidermoid carcinoma A431 cells, *Cell* **35**:321–330.

Hopkins, C. R., 1985, The appearance and internalization of transferrin receptors at the margins of spreading human tumor cells, *Cell* **40**:199–208.

Hopkins, C. R., Gibson, A., Shipman, M., and Miller, K., 1990, Movement of internalized ligand-receptor complexes along a continuous endosomal reticulum, *Nature* **346**:335–339.

Janicot, M., Clot, J. P., and Desbuquois, B., 1988, Interactions of cholera toxin with isolated hepatocytes: effects of low pH, chloroquine and monensin on toxin internalization, processing and action, *Biochem. J.* **253**:735–743.

Josefsberg, Z., Posner, B. I., Patel, B., and Bergeron, J. J. M., 1979, The uptake or prolactin into female rat liver: concentration of intact hormone in the Golgi apparatus, *J. Biol. Chem.* **254**:209–214.

Kadota, S., Fantus, I. G., Deragon, G., Guyda, H., Hersh, B., and Posner, B. I., 1987a, Peroxide(s) of vanadium is a novel and potent insulin-mimetic agent which activates the insulin receptor kinase, *Biochem. Biophys. Res. Commun.* **147**:259–266.

Kadota, S., Fantus, I. G., Deragon, G., Guyda, H., and Posner, B. I., 1987b, Stimulation of insulin-like growth factor II receptor binding and insulin receptor kinase activity in rat adipocytes: effects of vanadate and H_2O_2, *J. Biol. Chem.* **262**:8252–8256.

Kay, D. G., Khan, M. N., Posner, B. I., and Bergeron, J. J. M., 1984, *In vivo* uptake of insulin into hepatic Golgi fractions: application of the diaminobenzidine-shift protocol, *Biochem. Biophys. Res. Commun.* **123**:1144–1148.

Kay, D. G., Lai, W. H., Uchihashi, M., Khan, M. N., Posner, B. I., and Bergeron, J. J. M., 1986, Epidermal growth factor receptor kinase translocation and activation *in vivo*, *J. Biol. Chem.* **261**:8473–8480.

Khan, M. N., Posner, B. I., Verma, A. K., Khan, R. J., and Bergeron, J. J. M., 1981, Intracellular hormone receptors: evidence for insulin and lactogen receptors in a unique vesicle sedimenting in lysosome fractions of rat liver, *Proc. Natl. Acad. Sci. USA.* **78**:4980–4984.

Khan, M. N., Posner, B. I., Khan, R. J., and Bergeron, J. J. M., 1982, Internalization of insulin into rat liver Golgi elements: evidence for vesicle heterogeneity and the path of intracellular processing, *J. Biol. Chem.* **257**:5969–5976.

Khan, M. N., Savoie, S., Bergeron, J. J. M., and Posner, B. I., 1985, Insulin and insulin receptor uptake into rat liver: evidence for site of chloroquine action on receptor recycling, *Diabetes.* **34**:1025–1030.

Khan, M. N., Savoie, S., Bergeron, J. J. M., and Posner, B. I., 1986a, Differential kinetics and sensitivity to chloroquine of receptor-mediated insulin and prolactin endocytosis in liver parenchymal cells, *Biochim. Biophys. Acta.* **888**:100–106.

Khan, M. N., Savoie, S., Bergeron, J. J. M., and Posner, B. I., 1986b, Characterization of rat liver endosomal fractions: *in vivo* activation of insulin stimulable receptor kinase in these structures, *J. Biol. Chem.* **261**:8462–8472.

Khan, M. N., Baquiran, G., Brule, C., Burgess, J., Foster, B., Bergeron, J. J. M., and Posner, B. I., 1989, Internalization and activation of the rat liver insulin receptor kinase *in vivo*, *J. Biol. Chem.* **264**:12931–12940.

Khan, R. J., Khan, M. N., Bergeron, J. J. M., and Posner, B. I., 1985, Prolactin uptake into liver endocytic components: reduced sensitivity to chloroquine, *Biochim. Biophys. Acta.* **838**:77–83.

Kielian, M. C., Marsh, M., and Helenius, A., 1986, Kinetics of endosome acidification detected by mutant and wild type Semliki forest virus, *EMBO J.* **5**:3103–3109.

Kindberg, G. M., Ford, T., Blomhoff, R., Rickwood, D., and Berg, T., 1984, Separation of endocytic vesicles in nycodenz gradients, *Anal. Biochem.* **142**:455–462.

Klausner, R. D., Ashwell, G., van Renswoude, J., Harford, J. B., and Bridges, K. R., 1983, Binding of apotransferrin to K562 cells: explanation of the transferrin cycle, *Proc. Natl. Acad. Sci. USA* **80**:2263–2266.

Klein, H. H., Friedenberg, G. R., Kladde, M., and Olefsky, J. M., 1986, Insulin activation of insulin receptor tyrosine kinase in intact rat adipocytes: an *in vitro* system to measure histone kinase activity of insulin receptors activated *in vivo, J. Biol. Chem.* **261**:4691–4697.

Lai, W. H., Cameron, P. H., Wada, I., Doherty, J-J. II, Kay, D. G., Posner, B. I., and Bergeron, J. J. M., 1989a, Ligand mediated internalization, recycling, and down regulation of the epidermal growth factor receptor *in vivo, J. Cell Biol.* **109**:2741–2749.

Lai, W. H., Cameron, P. H., Doherty, J-J. II, Posner, B. I., and Bergeron, J. J. M., 1989b, Ligand mediated autophosphorylation activity of the EGF receptor during internalization, *J. Cell Biol.* **109**:2751–2760.

Lazarovits, J. and Roth, M., 1988, A single amino acid change in the cytoplasmic domain allows the influenza virus hemagglutinin to be endocytosed through coated pits, *Cell* **53**:743–752.

Lehrman, M. A., Goldstein, J. L., Brown, M. S., Russell, D. W., and Schneider, W. J., 1985, Internalization-defective LDL receptors produced by genes with nonsense and frameshift mutations that truncate the cytoplasmic domain, *Cell* **41**:735–743.

Leighton, F., Poole, B., Beaufay, H., Baudhuin, P., Coffey, J. W., Folwer, S., and DeDuve, C., 1968, The large-scale separation of peroxisomes, mitochondria, and lysosomes from the livers of rats injected with Triton WR-1339, *J. Cell Biol.* **37**:482–513.

Lobel, P., Fujimoto, K., Ye, R. D., Griffiths, G., and Kornfeld, S., 1989, Mutations in the cytoplasmic domain of the 275 Kd mannose-6-phosphate receptor differentially alter lysosomal enzyme sorting and endocytosis, *Cell* **57**:787–796.

Maegawa, H., McClain, D. A., Freidenberg, G., Olefsky, J. M., Napier, M., Lipari, T., Dull, T. J., Lee, J., and Ullrich, A., 1988, Properties of a human insulin receptor with a COOH-terminal truncation, *J. Biol. Chem.* **263**:8912–8917.

Maller, J. L., Pike, L. J., Freidenberg, G. R., Cordera, R., Stith, B. J., Olefsky, J. M., and Krebs, E. G., 1986, Increased phosphorylation of ribosomal protein S6 following microinjection of insulin receptor-kinase into *Xenopus* oocytes, *Nature (London)* **320**:459–461.

Marsh, M., Schmid, S. L., Kern, H., Harms, E., Male, P., Mellman, I., and Helenius, A., 1987, Rapid analytical and preparative isolation of functional endosomes by free flow electrophoresis, *J. Cell Biol.* **104**:875–886.

Maxfield, F. R., Schlessinger, J., Schechter, Y., Pastan, I., and Willingham, M. C., 1978, Collection of insulin, EGF, and α_2-macroglobulin in the same patches on the surface of cultured fibroblasts and common internalization, *J. Cell Biol.* **14**:805–810.

McClain, D. A., Maegawa, H., Lee, J., Dull, T. J., Ullrich, A., and Olefsky, J. M., 1987, A mutant insulin receptor with defective tyrosine kinase displays no biologic activity and does not undergo endocytosis, *J. Biol. Chem.* **262**:14663–14671.

McKanna, J. A., Haigler, H. T., and Cohen, S., 1979, Hormone receptor topology and dynamics: morphological analysis using ferritin-labeled epidermal growth factor, *Proc. Natl. Acad. Sci. USA* **76**:5689–5693.

Mellman, I. S., Steinman, R. M., Unkeless, J. C., and Cohn, Z. A., 1980, Selective iodination and polypeptide composition of pinocytic vesicles, *J. Cell Biol.* **86**:712–722.

Mellman, I., Fuchs, R., and Helenius, A., 1986, Acidification of the endocytic and exocytic pathways, *Ann. Rev. Biochem.* **55**:663–700.

Merion, M., Schlesinger, P., Brooks, R. M., Moering, J. M., Moering, T. J., and Sly, W. S., 1983, Defective acidification of endosomes in Chinese hamster ovary cell mutants "cross-resistant" to toxins and viruses, *Proc. Natl. Acad, Sci. USA* **80**:5315–5319.

Merion, M. and Sly, W. S., 1983, The role of intermediate vesicles in the adsorptive endocytosis and transport of ligand to lysosomes by human fibroblasts, *J. Cell Biol.* **96**:644–650.

Morgan, D. A., Ho, L., Korn, L. J., and Roth, R. A., 1986, Insulin action is blocked by a monoclonal antibody that inhibits the insulin receptor kinase, *Proc. Natl. Acad. Sci. USA* **83**:328–332.

Morgan, D. A. and Roth, R. A., 1987, Acute insulin action requires insulin receptor kinase activity: introduction of an inhibitory monoclonal antibody into mammalian cells blocks the rapid effects of insulin, *Proc. Natl. Acad. Sci. USA* **84**:41–45.

Mostov, K. E., de Bruyn Kops, A., and Deitcher, D. L., 1986, Deletion of the cytoplasmic domain of the polymeric immunoglobulin receptor prevents basolateral localizationl and endocytosis, *Cell* **47**:359–364.

Mullock, B. M., Luzio, J. P., and Hinton, R. H., 1983, Preparation of a low-density species of endocytic vesicle containing immunoglobulin A, *Biochem. J.* **214**:823–827.

Bullock, B. M., Hinton, R. H., Peppard, J. V., Slots, J. W., and Luzio, J. P., 1987, The preparative isolation of endosome fractions: a review, *Cell Biochem. Function* **5**:235–243.

Mullock, B. M., Branch, W. J., van Schaik, M., Gilbert, L. K., and Luzio, J. P., 1989, Reconstitution of an endosome-lysosome interaction in a cell-free system. *J. Cell Biol.* **108**:2093–2099.

Murphy, R. F., Powers, S., and Cantor, C. R., 1984, Endosome pH measured in single cells by dual fluorescence flow cytometry: rapid acidification of insulin to pH 6, *J. Cell Biol.* **98**:1757–1762.

Mustelin, T., Coggeshall, K. M., and Altman, A., 1989, Rapid activation of the T-cell tyrosine protein kinase $pp56^{lck}$ by the CD45 phosphotyrosine phosphatase, *Proc. Natl. Acad. Sci. USA* **86**:6302–6306.

Okada, M. and Nakagawa, H., 1989, A protein tyrosine kinase involved in regulation of pp60*c-src* function, *J. Biol. Chem.* **264**:20886–20893.

Okamoto, M., Karasik, A., White, M. F., and Kahn, C. R., 1991, Coordinate phosphorylation of insulin-receptor kinase and its 175,000-Mr endogenous substrate in rat hepatocytes. *Diabetes* **40**:66–72.

Pastan, I. H., and Willingham, M. C., 1981, Journey to the center of the cell: role of the receptosome, *Science* **214**:504–509.

Pease, R. J., Smith, G. D., and Peters, T. J., 1985, Degradation of endocytosed insulin in rat liver is mediated by low-density vesicles, *Biochem. J.* **228**:137–146.

Pilch, P. F., Shia, M. A., Benson, R. J. J., and Fine, R. E., 1983, Coated vesicles participate in the receptor-mediated endocytosis of insulin, *J. Cell Biol.* **96**:133–138.

Posner, B. I., Patel, B., Verma, A. K., and Bergeron, J. J. M., 1980, Uptake of insulin by plasmalemma and Golgi subcellular fractions of rat liver, *J. Biol. Chem.* **255**:735–741.

Posner, B. I., Bergeron, J. J. M., Josefsberg, Z., Khan, M. N., Khan, R. J., Patel, B. A., Sikstrom, R. A., and Verma, A. K., 1981, Intracellular receptors and internalization, *Rec. Prog. Horm. Res.* **37**:539–579.

Posner, B. I., Khan, M. N., and Bergeron, J. J. M., 1982a, Endocytosis of peptide hormones and other ligands, *Endocrine Reviews* **3**:280–298.

Posner, B. I., Patel, B. A., Khan, M. N., and Bergeron, J. J. M., 1982b, Effect of chloroquine on the internalization of [^{125}I]insulin into subcellular fractions of rat liver: evidence of an effect of chloroquine on Golgi elements, *J. Biol. Chem.* **257**:5789–5799.

Posner, B. I., Khan, M. N., and Bergeron, J. J. M., 1985, Receptor-mediated uptake of peptide hormones and other ligands, in: *Polypeptide Hormone Receptors*, (Posner, B. I., ed.), pp. 61–90, Marcel Dekker, New York.

Posner, B. I., Khan, M. N., and Bergeron, J. J. M., 1987, Internalization of insulin and its receptor: role in signaling, in *Insulin, Insulin-like Growth Factors and Their Receptors in the Central Nervous System* (M. K. Raizada, M. I. Philips, and D. Leroith, eds.), pp. 1–10, Plenum Press, New York.

Pricer, W. E. and Ashwell, G., 1976, Subcellular distribution of a mammalian hepatic binding protein specific for asialoglycoproteins, *J. Biol. Chem.* **251**:7539–7544.

Prywes, R., Livnch, E., Ullrich, A., and Schlessinger, J., 1986, Mutations in the cytoplasmic domain of EGFR affect EGF binding and receptor internalization, *EMBO J.* **5**:2179–2190.

Quintart, J., Courtoy, P., Limet, J., and Baudhuin, P., 1983, Galactose-specific endocytosis in rat liver: biochemical and morphological characterization of a low-density compartment isolated from hepatocytes, *Eur. J. Biochem.* **131**:105–112.

Roederer, M., Bowser, R., and Murphy, R. F., 1987, Kinetics and temperature dependence of exposure of endocytosed material to proteolytic enzymes and low pH: evidence for a maturation model for the formation of lysosomes, *J. Cell Physiol.* **131**:200–209.

Rosen, O. M., Herrera, R., Olowe, Y., Petruzzelli, L. M., Cobb, M. H., 1983, Phosphorylation activates the insulin receptor tyrosine protein kinase, *Proc. Natl. Acad. Sci. USA* **80**:3237–3240.

Roth, T. F. and Porter, K. R., 1964, Yolk protein uptake in the oocyte of the mosquito *aedes aegypti.L.*, *J. Cell Biol.*, **64**:313–332.

Rothenberger, S., Iacopetta, B. J., and Kühn, L. C., 1987, Endocytosis of the transferrin receptor requires the cytoplasmic domain but not its phosphorylation site. *Cell* **49**:423–431.

Russell, D. S., Gherzi, R., Johnson, E. L., Chou, C-K., and Rosen, O. M., 1987, The protein-tyrosine kinase activity of the insulin receptor is necessary for insulin-mediated receptor down-regulation, *J. Biol. Chem.* **262**:11833–11840.

Schaudies, R. P., Gorman, R. M., Savage, C. R., and Portez, R. D., 1987, Proteolytic processing of epidermal growth factor within endosomes, *Biochem. Biophys. Res. Commun.* **143**:710–715.

Schmid, S. L., Fuchs, R., Male, P., and Mellman, I., 1988, Two distinct subpopulations of endosomes involved in membrane recycling and transport to lysosomes, *Cell* **52**:73–83.

Schmid, S., Fuchs, R., Kielian, M., Helenius, A., and Mellman, I., 1989, Acidification of endosome subpopulations in wild-type Chinese hamster ovary cells and temperature-sensitive acidification-defective mutants, *J. Cell Biol.* **108**:1291–1300.

Shoelson, S. E., White, M. F., and Kahn, C. R., 1988, Tryptic activation of the insulin receptor, *J. Biol. Chem.* **263**:4852–4860.

Smith, C. D. and Peters, T. J., 1982, Localization in rat liver of alkaline phosphodiesterase to a discrete organelle implicated in ligand internalization, *Biochim. Biophys. Acta* **716**:24–30.

Smith, R. M. and Jarett, L., 1983, Quantitative ultrastructural analysis of receptor-mediated insulin uptake into adipocytes, *J. Cell Physiol.* **115**:199–207.

Stahl, P., Schlessinger, P., Sigardson, L., Rodman, J. S., and Lee, Y. C., 1980, Receptor-mediated pinocytosis of mannose glycoconjugates by macrophages: characterization and evidence for receptor recycling, *Cell* **19**:207–215.

Stochem, W. and Wohlfarth-Botterman, K. E., 1969, Tinocytosis (Endocytosis), in *Handbook of Molecular Cytology* (A. Lima-de-Faria, ed.), pp. 1373–1400, Elsevier, North Holland, Amsterdam.

Stoorvogel, W., Geuze, H. J., and Strous, G. J., 1987, Sorting of endocytosed transferrin and asialoglycoprotein occurs immediately after internalization in Hep G2 cells, *J. Cell Biol.* **104**:1261–1268.

Stoorvogel, W., Geuze, H. J., Griffith, J. M., Schwartz, A. L., and Strous, G. J., 1989, Relations between the intracellular pathways of the receptors for transferrin, asialoglycoprotein, and mannose-6-phosphate in human hepatoma cells, *J. Cell Biol.* **108**:2137–2148.

Storrie, B., Deeson, T. D., and Maurey, K., 1981, Rapid cell surface appearance of endocytic membrane proteins in Chinese hamster ovary cells, *Mol. Cell. Biol.* **1**:261–268.

Straus, W., 1964, Occurrence of phagosomes and phago-lysosomes in different segments of the nephron in relation to the reabsorption, transport, digestion, and extrusion of intravenously injected horseradish peroxidase, *J. Cell Biol.* **21**:295–308.

Tamura, S., Brown, T. A., Whipple, J. H., Yamaguchi, Y. F., Dubler, R. E., Cheng, K., and Larner, J., 1984, A novel mechanism for the insulin-like effect of vanadate on glycogen synthase in rat adipocytes, *J. Biol. Chem.* **259:**6650–6658.

Thies, R. S., Webster, N. J., and McClain, D. A., 1990, A domain of the insulin receptor required for endocytosis in rat fibroblasts, *J. Biol. Chem.* **265:**10132–10137.

Tycko, B. and Maxfield, F. R., 1982, Rapid acidification of endocytic vesicles containing α_2-macroglobulin, *Cell* **28:**643–651.

Ullrich, B., Bell, J. R., Chen, E. Y., Herrera, R., Petruzzelli, L. M., Dull, T. J., Gray, A., Coussens, L., Liao, Y-C., Tsubokawa, M., Mason, A., Seeburg, P. H., Grunfeld, C., Rosen, O. M., and Ramachandran, J., 1985, Human insulin receptor and its relationship to the tyrosine kinase family of oncogenes, *Nature (Lond.),* **313:**756–761.

Van Renswoude, J., Bridges, K. R., Harford, J. B., and Klausner, R. D., 1982, Receptor-mediated endocytosis of transferrin and the uptake of Fe in K562 cells: identification of a nonlysosomal acidic compartment, *Proc. Natl. Acad. Sci. USA* **79:**6186–6190.

Via, D. P., Willingham, M. C., Pastan, I. H., Gotto, A. M., Smith, L. C., 1982, Co-clustering and internalization of low-density lipoproteins and α_2-macroglobulin in human skin fibroblasts, *Exp. Cell Res.* **141:**15–22.

Wada, I., Lai, W. H., Posner, B. I., and Bergeron, J. J. M., 1992, Association of the tyrosine phosphorylated epidermal growth tactor receptor with a 55-kDa tyrosine phosphorylated protein at the cell surface and in endosomes, *J. Cell Biol.* **116**:321–330.

Wall, D. A., Wilson, G., and Hubbard, A. L., 1980, The galactose-specific recognition system of mammalian liver: the route of ligand internalization in rat hepatocytes, *Cell* **21:**79–93.

Wall, D. A. and Hubbard, A. L., 1981, Galactose-specific recognition system of mammalian liver: receptor distribution on the hepatocyte cell-surface, *J. Cell Biol.* **90:**687–696.

Wall, D. A. and Hubbard, A. L., 1984, Receptor-mediated endocytosis of asialoglycoproteins by rat liver hepatocytes: biochemical characterization of the endosomal compartments, *J. Cell Biol.* **101:**2104–2112.

Walsh, R. J., Posner, B. I., and Patel, B., 1984, Binding and uptake of [^{125}I]Iodoprolactin by epithelial cells of the rat choroid plexus: an *in vivo* autoradiographic analysis, *Endocrinology* **114:**1496–1505.

Ward, D. M., Ajioka, R., and Kaplan, J., 1989, Cohort movement of different ligands and receptors in the intracellular endocytic pathway of alveolar macrophages, *J. Biol. Chem.* **264:**8164–8170.

Watts, C., 1984, *In situ* ^{125}I-labeling of endosome proteins with lactoperoxidase conjugates, *EMBO J.* **3:**1965–1970.

White, M. F., Shoelson, S. E., Keutmann, H., and Kahn, C. R., 1988a, A cascade of tyrosine autophosphorylation in the β-subunit activates the phosphotransferase of the insulin receptor, *J. Biol. Chem.* **263:**2969–2980.

White, M. F., Livingston, J. N., Backer, J. M., Lauris, V., Dull, T. J., Ullrich, A., and Kahn, C. R., 1988b, Mutation of the insulin receptor at tyrosine 960 inhibits signal transmission but does not affect its tyrosine kinase activity, *Cell* **54:**641–649.

White, M. F., 1991, Structure and function of tyrosine kinase receptors, *J. Bioenerg. Biomembr.* **23:**63–82.

Wilden, P. A., Backer, J. M., Kahn, C. R., Cahill, D. A., Schroeder, G. J., and White, M. F., 1990, The insulin receptor with phenylalanine replacing tyrosine-1146 provides evidence for separate signals regulating cellular metabolism and growth, *Proc. Natl. Acad. Sci. USA* **87:**3358–3362.

Wileman, T., Harding, C., and Stahl, P., 1985, Receptor-mediated endocytosis, *Biochem. J.* **232:**1–14.

Wiley, H. S., VanNostrand, W., McKinley, D. N., and Cummingham, D. D., 1985, Intracellular processing of epidermal growth factor and its effect on ligand-receptor interactions, *J. Biol. Chem.* **260:**5290–5295.

Willingham, M. C. and Pastan, I., 1980, The receptosomes: an intermediate organelle of receptor-mediated endocytosis in cultured fibroblasts, *Cell* **21**:67–77.

Willingham, M. C., Pastan, I. H., Sahagian, G. G., Jourdian, G. W., and Neufeld, E. F., 1981, Morphologic study of the internalization of a lysosomal enzyme by the mannose-6-phosphate receptor in cultured Chinese hamster ovary cells, *Proc. Natl. Acad. Sci. USA* **78**:6967–6971.

Wolkoff, A. W., Klausner, R. D., Ashwell, G., and Harford, J., 1984, Intracellular segregation of asialoglycoproteins and their receptor: a prelysosomal event subsequent to dissociation of the ligand receptor complex, *J. Cell Biol.* **98**:375–381.

Yu, K-T. and Czech, M. P., 1986, Tyrosine phosphorylation of insulin receptor β-subunit activates the receptor-tyrosine kinase in intact H-35 hepatoma cells, *J. Biol. Chem.* **261**:4715–4722.

Index

GPSR Compliance

The European Union's (EU) General Product Safety Regulation (GPSR) is a set of rules that requires consumer products to be safe and our obligations to ensure this.

If you have any concerns about our products, you can contact us on ProductSafety@springernature.com

In case Publisher is established outside the EU, the EU authorized representative is:

Springer Nature Customer Service Center GmbH
Europaplatz 3
69115 Heidelberg, Germany

Batch number: 10620163

Printed by Printforce, the Netherlands